ADULT NEUROGENESIS IN THE HIPPOCAMPUS

Dedication

To Ana and Oliver

ADULT NEUROGENESIS IN THE HIPPOCAMPUS

HEALTH, PSYCHOPATHOLOGY, AND BRAIN DISEASE

Edited by

Juan J. Canales

University of Leicester, Leicester, United Kingdom

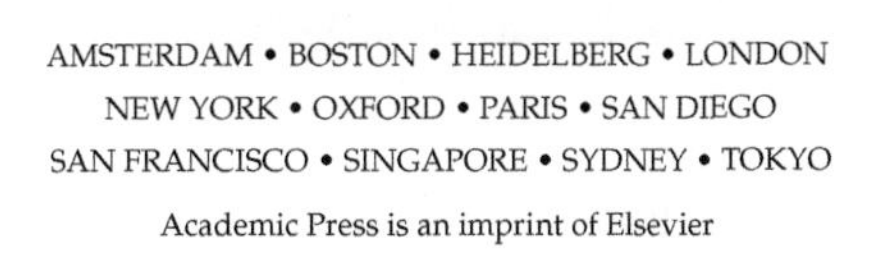
AMSTERDAM • BOSTON • HEIDELBERG • LONDON
NEW YORK • OXFORD • PARIS • SAN DIEGO
SAN FRANCISCO • SINGAPORE • SYDNEY • TOKYO
Academic Press is an imprint of Elsevier

Academic Press is an imprint of Elsevier
125 London Wall, London EC2Y 5AS, UK
525 B Street, Suite 1800, San Diego, CA 92101-4495, USA
50 Hampshire Street, 5th Floor, Cambridge, MA 02139, USA
The Boulevard, Langford Lane, Kidlington, Oxford OX5 1GB, UK

Cover image: An electron microscope photomicrograph showing an aspect of the dentate gyrus of the hippocampus from an adult mouse. Note the columnar organization of the granular hippocampal neurons and precursor cell types evident in the subgranular zone: a radial astrocyte (or type 1 cell) in blue, and three D cells (or type 2 cells) in red. (Photo courtesy of Prof. José M. García-Verdugo.)

British Library Cataloguing-in-Publication Data
A catalogue record for this book is available from the British Library

Library of Congress Cataloging-in-Publication Data
A catalog record for this book is available from the Library of Congress

ISBN: 978-0-12-801977-1

For information on all Academic Press publications
visit our website at https://www.elsevier.com/

Publisher: Mara Conner
Acquisition Editor: April Farr
Editorial Project Manager: Timothy Bennett
Production Project Manager: Edward Taylor
Designer: Mark Rogers

Typeset by TNQ Books and Journals
www.tnq.co.in

Contents

II

NEUROGENESIS IN HEALTH AND WELL-BEING

7. Aging

R. KÖNIG, P. ROTHENEICHNER, J. MARSCHALLINGER, L. AIGNER AND S. COUILLARD-DESPRES

III
NEUROGENESIS IN PSYCHOPATHOLOGY AND DISEASE

8. Adult Neurogenesis, Chronic Stress and Depression

P.J. LUCASSEN, C.A. OOMEN, M. SCHOUTEN, J.M. ENCINAS AND C.P. FITZSIMONS

9. Acute Stress and Anxiety

L. CULIG AND C. BELZUNG

10. Addiction

J.J. CANALES

11. Neurological Disorders

B.W. MAN LAU, S.-Y. YAU, K.-T. PO AND K.-F. SO

List of Contributors

L. Aigner Paracelsus Medical University, Salzburg, Austria

C. Belzung Francois Rabelais University, Tours, France

J.J. Canales University of Leicester, Leicester, United Kingdom

B.R. Christie University of Victoria, Victoria, BC, Canada

S. Couillard-Despres Paracelsus Medical University, Salzburg, Austria

L. Culig Francois Rabelais University, Tours, France

A. Di Antonio Stony Brook University, NY, United States

J.M. Encinas Achucarro Basque Center for Neuroscience, Bizkaia, Spain

C.P. Fitzsimons University of Amsterdam, Amsterdam, The Netherlands

J.M. García-Verdugo University of Valencia, Valencia, Spain

S. Ge Stony Brook University, NY, United States

S. Jessberger University of Zurich, Zurich, Switzerland

G.W. Kirschen Stony Brook University, NY, United States

R. König Paracelsus Medical University, Salzburg, Austria

D.C. Lie Friedrich-Alexander University Erlangen–Nuremberg, Erlangen, Germany

P.J. Lucassen University of Amsterdam, Amsterdam, The Netherlands

B.W. Man Lau The Hong Kong Polytechnic University, Hong Kong, China

J. Marschallinger Paracelsus Medical University, Salzburg, Austria

C.M. Merkley University of Toronto, Toronto, ON, Canada

M.M. Molina-Navarro University of Valencia, Valencia, Spain

T. Murphy King's College London, London, United Kingdom

C.A. Oomen Radboud University Medical Centre, Nijmegen, The Netherlands

A. Patten University of Victoria, Victoria, BC, Canada

K.-T. Po The Hong Kong Polytechnic University, Hong Kong, China

P. Rotheneichner Paracelsus Medical University, Salzburg, Austria

M. Schouten University of Amsterdam, Amsterdam, The Netherlands

Z. Sharp University of Victoria, Victoria, BC, Canada

K.-F. So The Jinan University, Guangzhou, China; The University of Hong Kong, Hong Kong, China

S. Thuret King's College London, London, United Kingdom

J.M. Wojtowicz University of Toronto, Toronto, ON, Canada

S.-Y. Yau University of Victoria, Victoria, BC, Canada

Preface

At the turn of the 20th century, Santiago Ramon y Cajal, one of the fathers of modern neuroscience and preeminent neuroanatomist, established that "once the development was ended, the founts of growth and regeneration of the axons and dendrites dried up irrevocably." Ramon y Cajal concluded that "in the adult centers, the nerve paths are something fixed, ended, and immutable. Everything may die, nothing may be regenerated." Acutely aware of the technical limitations of his time, Ramon y Cajal anticipated that "it is for the science of the future to change, if possible, this harsh decree." For many decades, numerous generations of neurologists and neuroscientists were taught that the key embryonic staging processes of cell proliferation, migration, and differentiation were not reenacted in adult life. As Ramon y Cajal intuited, science finally overturned this old dogma, albeit skepticism and stern resistance prevailed for many years, even after Joseph Altman's pioneering [^{3}H]thymidine autoradiography studies in the 1960s convincingly showed that newborn neurons continue to be formed postnatally in the rodent hippocampus. Currently, research into adult neurogenesis has become one of the most prolific and blossoming areas of neuroscience, with discoveries in this field now being recognized to have far-reaching ramifications for understanding brain function, both in the normal and in the diseased state.

Adult neurogenesis occurs both in the olfactory bulb and in the hippocampus. Owing to the prominent role of the hippocampus in the acquisition of new memories and the part it plays in the regulation of stress, it is currently believed that neurogenic processes in the adult hippocampus can not only help explain normal brain function but also shed light into the pathogenesis of certain psychopathologies and neurological conditions. This special volume is exclusively dedicated to describing the latest advances in the field of neurogenesis in the adult hippocampus and, therefore, should not be the only resource for neuroscientists with wider interests in the field of neurogenesis. The project for this compilation was born out of the need to create an up-to-date resource for advanced students and young scientists interested in health and psychopathology as they relate to adult hippocampal neurogenesis. The contributions to this volume have been thoughtfully prepared by a group of world leaders working in the area of neurogenesis and their reviews describe cutting-edge research which will also be of value to more senior investigators and academics, especially those involved in teaching undergraduate and graduate students.

Part I includes three chapters focused on the neurobiology, physiology, and plasticity of adult-generated hippocampal neurons, introducing the reader to the nature of the neurogenic process and the special morphological, molecular, and electrophysiological features of stems cells, neural progenitors, and young neurons generated in the adult hippocampus. Recent evidence indicates that neurogenesis is a tightly regulated process that can be positively influenced by the activity and behavior of the organism, undergoing marked changes throughout the life span. Part II is dedicated to such intriguing phenomena, with four chapters devoted to analyzing the involvement of adult neurogenesis in learning and memory, the beneficial effects of physical exercise, the importance of nutritional and dietary factors, and the age-related variations in the neurogenic capacity of the adult hippocampus. A burgeoning body of research has unveiled the significance of adult neurogenesis for understanding the wide array of abnormal processes associated with psychopathology and neurological disease. Part III of this volume consists of four chapters that describe the relationship between adult hippocampal neurogenesis and depression, chronic stress, anxiety, addiction, neurodegenerative diseases, neurodevelopmental disorders, and brain injury. This combined effort is aimed at shedding greater light and insight into the role of adult hippocampal neurogenesis in brain health and psychopathology from a multidisciplinary perspective.

I trust that this volume will encourage and enlighten all who have an interest in neurogenesis. The discovery of neurogenesis in the adult brain of rodents, monkeys, and humans has revolutionized modern neuroscience but many important questions remain unanswered. Future research should be oriented toward understanding neurogenesis as it relates to hippocampal circuits, cognition, and behavior and to learning to harness the extraordinary potential of neurogenesis to support adaptive behavior and treat human disease. It has been 100 years since Ramon y Cajal proclaimed that perhaps the science of the future could one day prove what had elusively remained invisible to his expert eye. Today, we live in exciting times for adult neurogenesis.

Juan J. Canales, DPhil.

CHAPTER

1

Neurobiology

M.M. Molina-Navarro, J.M. García-Verdugo
University of Valencia, Valencia, Spain

INTRODUCTION

Adult neurogenesis in mammals is the continuance of embryonic neurogenesis. The generation of new neurons only takes place in specific regions of the adult brain, including the ventricular–subventricular zone (V–SVZ) of the forebrain and the subgranular zone (SGZ) of the dentate gyrus (DG) of the hippocampus. Historically, cell proliferation/neurogenesis was first reported by Ezra Allen in 1912, showing cell divisions in the lateral ventricles of adult rodent brain. However, the hippocampus as a germinative zone was first discovered in the early 1960s by Joseph Altman, who applied a tritiated thymidine method to study cell divisions in the adult brain and showed a proliferative region of granule cells in the DG of the hippocampus (Altman, 1962, 1963). Although Altman's seminal discoveries in the field of adult neurogenesis were fundamental, they were initially met with skepticism by the wider scientific community and many years passed before they became accepted.

A key figure who contributed to refute the classic idea that no new nerve cells are born in the adult mammalian brain was Michael Kaplan who, in advance of concurrent commonly accepted ideas, conducted innovative electron microscopy analyses of neurogenic sites. Kaplan published detailed studies on adult neurogenesis in the hippocampus, the olfactory bulb, and the visual cortex (Kaplan, 1985; Kaplan & Hinds, 1977). However, Kaplan was forced to leave the field of neurogenesis, as did his colleague Altman, due to the general incredulity that reigned in this field. At this time, another prominent scientist, Fernando Nottebohm, pioneered the first functional studies on neurogenesis, showing the involvement of neurogenesis in learning in the avian brain. Working on songbirds, he used electron microscopy to identify synaptic terminals on newly born neurons

Adult Neurogenesis in the Hippocampus
http://dx.doi.org/10.1016/B978-0-12-801977-1.00001-5

in the forebrain (Burd & Nottebohm, 1985) and also introduced electrophysiology in this field (Paton & Nottebohm, 1984). Nottebohm was able to show that these new neurons are involved in learning and functionally integrate into existing circuitry. Nottebohm's studies on neurogenesis and neuronal replacement in birds paved the way for modern neurogenesis research in humans, and established a connection between neurogenesis and learning, which has become a major focus in current research.

After the experiments of Altman and Kaplan, adult neurogenesis in the hippocampus was rediscovered by Heather Cameron, Elizabeth Gould, and Bruce McEwen. These authors studied the regulation of adult hippocampal neurogenesis by hormonal stress and revealed that there was a relationship between the stress and the regulation of neurogenesis in the hippocampus (Cameron, Woolley, McEwen, & Gould, 1993; Gould, Cameron, Daniels, Woolley, & McEwen, 1992). It is likely that the coincidence of these studies with the discovery of neural stem/progenitor cells (NSPCs) in adult striatum (Reynolds & Weiss, 1992) led to the broader acceptance of adult neurogenesis by the research community, although it should be noted that Reynolds in fact isolated cells from the ventricles adjoining the striatum, and not from adult striatum as he initially thought. Consequently, at the end of the 20th century adult neurogenesis in the hippocampus became generally accepted. Recently neurogenesis in DG was reported in several animal species, including humans (Eriksson et al., 1998), and overwhelmingly accepted by scientists (Yu, Marchetto, & Gage, 2014).

In this chapter will first introduce the anatomy of the DG, the region of the hippocampus where adult neurogenesis occurs, and then discuss the morphological characteristics of the hippocampal neurogenic sites in rodents and primates.

NEUROGENESIS TAKES PLACE IN THE DENTATE GYRUS OF THE HIPPOCAMPUS

The DG is composed of three laminae or layers: the molecular layer, the granule cell layer, and the polymorphic cell layer (Fig. 1.1A). The molecular layer contains the dendrites of the dentate granule cells, fibers of the perforant path that originate in the entorhinal cortex, and a small number of interneurons and fibers from extrinsic inputs that terminate there. The granule cell layer is the principal cell layer. It is mainly formed by granule cells and there are some other neurons at the boundary of the granule and polymorphic layers. Finally, the polymorphic layer or hilus contains a number of cell types and the mossy cell is the most prominent.

This trilaminate region that forms the hippocampus has a characteristic V or U shape depending on the septotemporal position. The portion of the

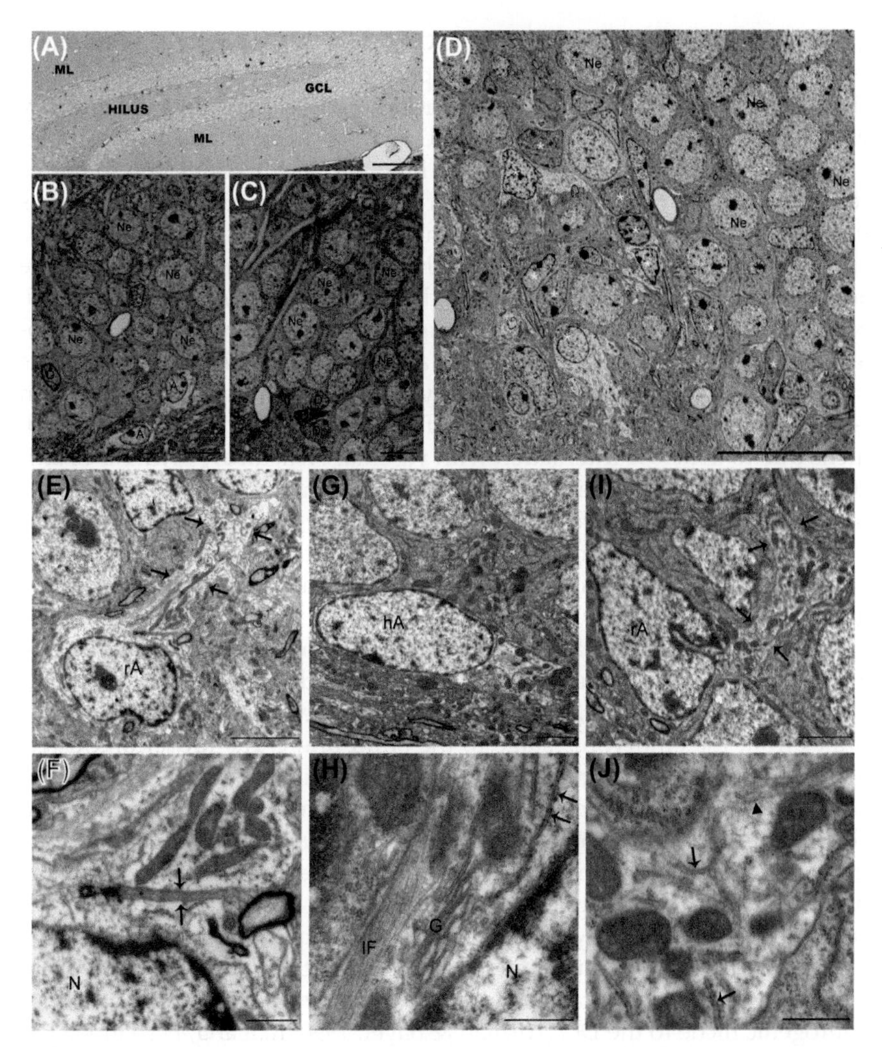

FIGURE 1.1 Semithin section of 1.5 μm from mouse DG. (A) In this figure is seen the granular cell layer (GCL), the molecular layer (ML), and the hilus that form the DG at a medium level. (B) An amplification of the DG at the level of the SGZ, where it is possible to observe both rA and hA, and a D cell type inserted into the GCL. (C) D cells of type 2 cells in the SGZ of the DG. (D) Electron microscopy image from a 1-month-old animal, where the granular neurons (Ne) and many cell groups with heterogeneous content and morphology (*) are shown. These last cells constitute the neurogenic niche. (E) rA with a long apical protrusion (*arrows*) that contains long mitochondria, and dictiosomes. (F) Detail of a characteristic primary cilium with its centriole in a rA (*arrows*). Mitochondria, intermediate filaments, and a Golgi apparatus are also shown. (G) hA from the SGZ of the DG. This type of astrocytes usually has the nucleus elongated and parallel to the granular neurons. (H) Detail of a hA where a bundle of intermediate filaments can be seen, as well as a Golgi apparatus and long cisterns of rough endoplasmic reticulum scarcely dilated (*arrows*). (I) Some rA can present a denser cytoplasmic matrix with many mitochondria, but with a lower number of intermediate filaments (*arrows*). (J) In this figure it is possible to observe a detail of these cells, which display a light cytoplasmic matrix, many mitochondria, and some endoplasmic reticulum cisterns without a clear organization (*arrows*). Occasionally some microtubules and intermediate filaments can be seen (*arrowhead*). Scale bars = 100 μm in A, 20 μm in D, 10 μm in B and C, 2 μm in E and I, 5 μm in G, 1 μm in H, and 0.5 μm in F and J. *Ne*, neuron; *rA*, radial astrocyte; *hA*, horizontal astrocyte; *D*, D cell or type 2 cell; *N*, nucleus.

granule cell layer located between the CA3 and the CA1 field is referred to as the suprapyramidal (or dorsal or upper) blade, and the portion opposite is the infrapyramidal (or ventral or lower) blade. The region bridging the two blades at the apex of the V or U shape is named the crest.

The subgranular zone is a narrow layer of cells between the granule cell layer and the polymorphic layer or hilus of the DG (Figs. 1.1B and C). In rodents, the transition from the granule cell layer and the hilus is sharp and in humans is a serrated border. The SGZ is a germinative matrix for adult neurogenesis characterized by different types of cells, notably the NSPCs, whose neuronal progeny migrate into the granular cell layer at varying distances, extending their axons and dendrites into the CA3 field and molecular layer, respectively. This intricate microenvironment is called the neurogenic niche, also referred to as vascular or angiogenic niche due to the close interaction with vascular structures (Tavazoie et al., 2008). Currently, the different cell types that form the SGZ and the stages of neuron differentiation have not yet been fully characterized. It is also a region with complex and diverse innervation that remains to be elucidated in terms of types of neuronal input and how information is integrated within it.

The DG is the hippocampal region where adult hippocampal neurogenesis occurs throughout the lifetime of an individual. To understand how this process unfolds, it is important to define how neurogenesis continues from the embryonic stage to the adult brain. The hippocampus is displaced by the growing cortical regions during embryonic development and is rolled into the shape of a seahorse, the name given to multiple species of small marine fishes in the genus *Hippocampus*. While in rodents the hippocampus remains dorsal, in humans there is a massive growth of the neocortex that displaces it to a ventral position. Barry et al. (2008) suggested that different types of radial glia participate in the development of the DG. Following embryonic development, the DG will become the new NSPC provider of granule cells instead of the V–SVZ, rearranging the neurogenic niche to the DG (Altman & Bayer, 1990; Li, Fang, Fernandez, & Pleasure, 2013). Thus, the origin of DG comes from an ectopic precursor cell pool.

Neurogenesis in the adult DG is spatially confined within a radius of about 100 μm from the precursor cell to the final location of the mature neuron (Kempermann, Gast, Kronenberg, Yamaguchi, & Gage, 2003). Furthermore, neurogenesis in the DG is cumulative and does not contribute to a turnover as in the olfactory bulb (Imayoshi et al., 2008; Ninkovic, Mori, & Gotz, 2007). This means that the DG has a local constraint and scarcity of new neurons. Despite these spatial limitations, the contribution of these new cells to the functions of the hippocampus remains important.

The SGZ has a special microenvironment, the so-called neurogenic niche, which has the capacity of both maintaining cells as stem cells and regulating their differentiation. As noted, the SGZ of the DG niche is composed of stem cells, intermediate progenitors, immature neurons, and blood vessels (Fig. 1.1D). The neurogenic niches form discontinuous cell groups in the SGZ, which is 20–25 μm wide. Recently, an effort has been made to elucidate the cell types that constituted the NSPCs of the SGZ. Applying the same experimental approach used to identify the neuronal stem cells in the V–SVZ it was also possible to identify the neuronal precursors in the SGZ. This approach consisted in distinguishing NSPCs by their proliferative activity as detected by the incorporation of tritiated thymidine or the thymidine analog bromodeoxyuridine (BrdU), in combination with neuron-specific markers.

NEUROGENESIS IN RODENTS

In addition to the standard strategy of detecting proliferating cells, Alvarez-Buylla's group developed an elegant method for surveying not only the cells that will undergo mitosis but also the progeny of these cells. This strategy consisted in using transgenic mice engineered to express the receptor for avian leukosis virus under the glial fibrillary acidic protein (GFAP) promoter (GFAP–Tva mice). These cells are susceptible to infection by the replication-competent avian leukosis retrovirus encoding the alkaline phosphatase (AP) protein. Once the retrovirus is injected into the brain it integrates into DNA, thereby allowing the permanent expression of AP and the follow-up of GFAP-expressing cells undergoing division. Using this methodology, it was shown that cells that express GFAP, a cell marker that labels astrocytes, could divide and create new cells in the DG that would later express neuronal markers (Seri, Garcia-Verdugo, McEwen, & Alvarez-Buylla, 2001). Furthermore these stem cells display a radial glia-like morphology and astrocytic properties (Filippov et al., 2003; Fukuda et al., 2003; Mignone, Kukekov, Chiang, Steindler, & Enikolopov, 2004; Seri et al., 2001). In addition, these cells also express the precursor cell-marker nestin (marker of neuroepithelial stem cells), which is a protein of an intermediate filament, indicating that nestin-expressing cells are the origin of adult neurons (Lagace et al., 2007; Ninkovic et al., 2007). However, its role has been taken over by the transcription factor involved in neurogenesis in precursor cells, Sox2 (SRY (Sex Determining Region Y)-Box 2) (Suh et al., 2007). Thus, these three markers (GFAP, nestin, and Sox2) give evidence of the existence of precursor cells in adult hippocampus. Interestingly, it has also been described that astrocytes secrete membrane-bound factors that promote neurogenesis, such as Notch, Sonic hedgehog, the bone morphogenetic proteins (BMPs), the Wnts (Wingless), and some growth factors and neurotrophic factors including FGF2 (fibroblast growth factor-2), VEGF (vascular endothelial growth factor), the VEGF receptor Flk1, insulin-like growth factor-binding proteins (IGFBPs), and the growth factor CNTF (ciliary neurotrophic factor), in addition to several cytokines (Lim & Alvarez-Buylla, 1999; Song, Stevens, & Gage, 2002; Taupin et al., 2000; Morrens, Van Den Broeck, & Kempermann, 2012).

Moreover, additional markers have been included for radial glia or neuronal precursor cells, such as BLBP (brain lipid-binding protein, stem cell marker), Id1 (inhibitor of DNA-binding 1, stem cell marker), and Hopx (Homeodomain-only protein, proliferation marker) (De Toni et al., 2008; Filippov et al., 2003; Nam & Benezra, 2009; Steiner et al., 2006). Finally, these cells are negative for S100β (calcium-binding protein), which is a reliable astrocytic marker (Savchenko, McKanna, Nikonenko, & Skibo, 2000; Seri, Garcia-Verdugo, Collado-Morente, McEwen, & Alvarez-Buylla, 2004; Steiner et al., 2004). This lack of immunoreactivity indicates that the population of progenitor cells is not completely identical to that of mature astrocytes.

Regardless, it is clear that NSPCs of the DG share certain astrocytic features. Indeed, Seri et al. (2001) reported that these cells present the following astrocytic features: light cytoplasm containing few ribosomes, intermediate filaments, and irregular contours with plasma membrane and processes that intercalate between adjoining cells.

As noted above, cells with characteristics of astrocytes can divide and create new neurons, consistent with the concept of astrocytes as early progenitors in the DG (Filippov et al., 2003; Fukuda et al., 2003), giving rise to transient progenitor cells that were originally described as D cells, characterized by smooth contours, dark scant cytoplasm with many ribosomes, and darker nuclei (Figs. 1.2A,B, and D) (Seri et al., 2001). Although these cells initially make clusters, they eventually line up along the SGZ separately and migrate radially to the adult DG granule cell layer (Dashtipour et al., 2002; Esposito et al., 2005; Jones, Rahimi, O'Boyle, Diaz, & Claiborne, 2003; Nacher, Crespo, & McEwen, 2001; Ribak, Korn, Shan, & Obenaus, 2004; Seki, 2002; Seki, Namba, Mochizuki, & Onodera, 2007; Seri et al., 2004; Shapiro & Ribak, 2005).

By using electron and confocal microscopy, the presence of subtypes of astrocytes and D cells as well as the three-dimensional organization of neuronal precursor cells in the SGZ was revealed (Seri et al., 2004). These authors characterized two types of astrocytes based on their orientation, morphology, and expression of molecular markers: radial astrocytes (rA) and horizontal astrocytes (hA). The ultrastructure and morphology of the astrocytes identified are as follows: rA have a large round, polygonal, or triangular cell body with a major radial process tangentially oriented along the SGZ that penetrates the granule cell layer intercalating extensively between granule neurons, as a way to protect them (Figs. 1.1E and F). In addition, the radial process branches profusely in the molecular layer, spreading out in numerous small branches, which gives the cells a

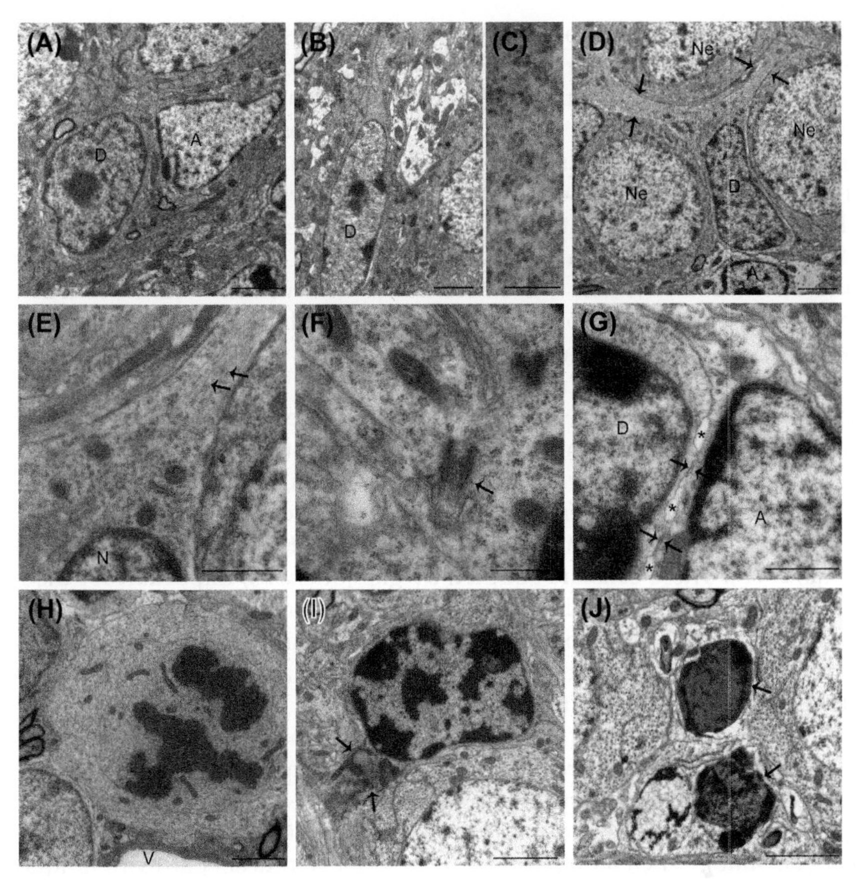

FIGURE 1.2 Electron microscopy of cells from the neurogenic niche of DG. (A) In this figure is seen an astrocyte (A) that can be clearly differentiated even at low magnification from type D cells (D). D cells have a very dense cytoplasmic matrix, especially rich in polyribosomes. Its nuclear matrix also presents greater electrodensity. (B) Elongated nucleus of a type D cell, where some nucleolus and a dense nuclear matrix are shown. (C) Appearance of the cytoplasm of the previous D cell with its characteristic abundance of free polyribosomes. (D) In this figure some nucleus of neurons (Ne), astrocytes with its clumps of heterochromatin, and a D cell are shown. This D cell presents two long protrusions (*arrows*). (E) Detail of figure D, where many polyribosomes and microtubules are shown (*arrows*) from one of the protrusions of the D cell. (F) In type D cells, in addition to the abundant polyribosomes and microtubules, a characteristic short cilium is also shown (*arrow*). (G) Another characteristic of the DG niches is the presence of free spaces between both D cells or D cells and astrocytes. In this figure it is possible to observe the contacts between a D cell and an astrocyte, where the free spaces (*) and the small adherens unions (*arrows*) are shown. (H) Sometimes mitotic figures can be seen clearly associated to the niches. Frequently these mitoses are close to the blood vessels (V). (I) In this figure a microglia cell is shown, with its nucleus characterized by abundant heterochromatin, and its cytoplasm with many heterogeneous bodies (*arrows*). (J) Occasionally pyknotic bodies (*arrows*) can be also present in niches, associated either to microglia cytoplasm or to astrocytes. Scale bars = 2 μm in A, B, D, H, I, and J; 1 μm in E and G; 0.5 μm in F; and 0.2 μm in C. *Ne*, neuron; *A*, astrocyte; *D*, D cell; *N*, nucleus.

treelike appearance. In contrast (Figs. 1.1G and H), hA are generally elongated, with no radial projection, but with extended branched processes parallel to the SGZ and thin short secondary branches into the granular cell layer and the hilus. Both types of astrocytes present the following ultrastructural characteristics under the electron microscope: they have a light cytoplasm, a dense network of intermediate filaments, irregularly shaped cell contours that intercalate between neighboring cells, heterochromatin clumps, a thin Golgi apparatus, small endoplasmic reticulum, and darker mitochondria than neurons and are in contact with other astrocytes through gap junctions (Seri et al., 2004). Although this is the characteristic ultrastructure of astrocytes (Peters, Palay, & Webster, 1991), there are also some ultrastructural differences between both types of astrocytes identified in the DG. rA present more organelles, elongated mitochondria, and more prominent bundles of intermediate filaments in the main process. rA are also characterized by a long cilium (Fig. 1.1F), which is necessary for the maintenance of the precursor cell pool in response to the neurogenic factor Sonic hedgehog signaling (Breunig et al., 2008; Han et al., 2008). These cells are coupled by gap junctions, important for their function as precursor cells (Kunze et al., 2009), and have a vascular connection in the SGZ (Filippov et al., 2003) and classical electrophysiological astrocytic properties such as passive membrane and potassium currents (Filippov et al., 2003; Fukuda et al., 2003).

Regarding the cellular markers, both types of astrocytes express GFAP, a marker for mature astrocytes, the transcription factor SOX2, vimentin (marker for immature astrocytes), musashi (RNA-binding protein specific to astrocytes), Mash1 (transcription factor thought to maintain the precursor cell state), 3-PGDH (enzyme in the serine synthesis pathway unique to neuroepithelial cells, radial glia, and astrocytes), Id1, Hopx, and BLBP (De Toni et al., 2008; Filippov et al., 2003; Nam & Benezra, 2009; Seri et al., 2004; Steiner et al., 2006). However, nestin was only present in rA, while conversely, hA were exclusively stained with S100β, and some of them were even stained with glutamine synthetase (enzyme that converts glutamate to glutamine) (Seri et al., 2004; Steiner et al., 2006). These results indicate that, on the one hand, the astrocytic nature of rA is convincingly demonstrated and, on the other, that these morphological differences and distinct set of regulatory signals imply a different function for both types of astrocytes. Thus, more studies are needed both to ascertain the molecular pathway by which they give rise to their offspring and to completely decipher the composition of the progeny of these precursor cells.

Indeed, it is possible to find rA in the SGZ with ultrastructural characteristics that differ slightly from the classical descriptions for rA (Figs. 1.1I and J). These cells have nuclear membranes that are much more irregular and with more heterochromatin clumps than rA and more nuclear invaginations. Their cytoplasm is more electrodense and has fewer intermediate filament bundles than rA, their mitochondria are bigger and less elongated, and they contain more polyribosomes and cisterns of rough endoplasmic reticulum. In addition, microtubules are easily visible in these cells. However, we do not know whether these cells present specific molecular markers and what role they play in the neurogenic niche.

It was suggested that rA are the NSPCs that divide both symmetrically to generate two astrocytes and asymmetrically to generate the D cells, which are highly proliferative (Kronenberg et al., 2003; Namba, Maekawa, Yuasa, Kohsaka, & Uchino, 2009; Seri et al., 2001; Seri et al., 2004). In addition, these precursor cells can originate in vitro in the three major neural lineages (neurons, astrocytes, and oligodendrocytes) (Suh et al., 2007).

D cells are considered the migratory neuroblasts and are characterized by a spherical nucleus that sometimes appears elongated with dense chromatin. Their cytoplasm is electrodense and contains abundant polyribosomes and their plasma membrane is smooth. They present extracellular spaces between them and also between astrocytes (Plumpe et al., 2006; Seri et al., 2004). D cells were classified in three subtypes (D1, D2, and D3) by using electron and confocal microscopy in combination with PSA–NCAM immunocytochemistry (Seri et al., 2004). D1 cells are small, with little cytoplasm, and with no processes or very thin protrusions usually in the SGZ. These cells are round, ovoid, or have the shape of an inverted drop. D2 cells have a short thick process that sometimes bifurcates, microtubules can be easily seen, and their cytoplasm is bigger. Finally, D3 cells have characteristics of immature granule neurons: a prominent, frequently branched, radial process that extended through the granule cell layer toward the molecular layer (emerging dendrite), and thin processes projecting into the hilus (the emerging axon). Their somata are round to polygonal, containing many polyribosomes (Fig. 1.2C), some dictiosomes, and short cilia close to the basal body and to the cell surface (Fig. 1.2F). These cells contain along their branches abundant microtubules (Fig. 1.2E). D3 cells were generally found in the interface of the granule cell layer and the SGZ, but occasionally the cell body could be found deeper in the granule cell layer (Seri et al., 2004). Alvarez-Buylla's group also provided evidence that D1 cells divide to give rise to D2 cells that then mature into D3 cells (Seri et al., 2004), although due to the fact that neuronal differentiation follows a gradual progression, a strict cellular classification is sometimes difficult.

Newborn granule neurons were identified by PSA–NCAM immunocytochemistry in proliferating cells combined with electron microscopy, since completely mature neurons were PSA–NCAM negative (as it was previously described in the formation of neurons (Seki & Arai, 1999)). Other cell markers helped to differentiate D cells from mature granule neurons, such as the transcription factors involved in differentiation of neurons NeuroD and Prox1, and the neuronal markers calbindin (calcium-binding protein) and NeuN. However, there was a gradual staining (Seri et al., 2004). Thus, D cells are transient cells from a potentially proliferative state to a postmitotic immature neuron.

A three-dimensional reconstruction by electron microscopy showed that the SGZ appears to be organized in clusters of cells that include astrocytes tightly associated to D cells and granule cells in the generation of new granule neurons in the DG (Seri et al., 2004; Shapiro, Korn, Shan, & Ribak, 2005). This cellular organization is very similar to the arrangement in the V–SVZ, where the astrocytes surround the chains of migrating cells in the rostral migratory stream (Lim & Alvarez-Buylla, 1999). The processes of these SGZ astrocytes seem to form a basket or nest where the D cells are insulated from the neuropil and have close membrane appositions with the somata of D cells, and sometimes with some protrusions of several D cells. Moreover, D cell processes made small electron-dense contacts with both astrocytes and mature granule neurons (Fig. 1.2G). In this sense, it seems that precursor cells are important niche factors, since they serve both as precursors and as guidance scaffolds.

In the neurogenic niche of the DG it is possible to observe mitosis. In young animals it is easier to observe mitosis in the SGZ than in old animals. Most of the cells undergoing mitosis contain a dense cytoplasmic matrix and are frequently close to blood vessels. Their cytoplasm is electrodense and more similar to D cells, in contrast to the astrocytic cytoplasm which is electrolucid (Seri et al., 2004) (Fig. 1.2H). This last type of mitosis can be easily detectable in the hilus. It has been described that the neurogenic niche of the hippocampus contains a vascular configuration that allows communication between blood vessels and DG cells (Licht & Keshet, 2015). This contact narrowly influences neurogenesis, although further studies are needed to understand the hippocampal vascular niche function. Finally, we can also observe microglia cells in the niches (Fig. 1.2I) that can occasionally be seen activated. Dense bodies/lysosomes are present not only in microglia cells but also in the cytoplasm of some astrocytes, probably pyknotic cell rests (Fig. 1.2J) (Amrein, Slomianka, & Lipp, 2004; Gould, Vail, Wagers, & Gross, 2001). To complete the description of the SGZ, few oligodendrocytes can be found in the SGZ (Seri et al., 2004) as well as some interneurons (Brandt et al., 2003).

According to new discoveries and knowledge, the terminology of the different cellular types of the DG has been changing. It is interesting to point out that some authors classify the SGZ cells in rA or B cells (Seri et al., 2001) or type 1 cells or radial glia-like precursor cells (Filippov et al., 2003; Fukuda et al., 2003), plus D cells (Seri et al., 2001) or type 2 cells or transiently amplifying progenitor cells (Filippov et al., 2003; Fukuda et al., 2003). Furthermore some authors subclassify type 2 in type 2a and type 2b according to some cell markers, although the sorting limits are not very clear. Type 2a express the transcription factors involved in neuronal differentiation neurogenin 2, and Tbr2 (T-Box Brain Protein 2), Ascl1/Mash1, and NeuroD1, while type 2b express DCX (doublecortin, a cytoskeletal-associated protein required for neuronal migration), PSA–NCAM (polysialylated–neural cell adhesion molecule, marker of developing and migrating neurons), NeuroD and NeuroD1 (transcription factors implicated in differentiation of neurons), β-III-tubulin (microtubule element from neurons), Prox1 (transcription factor involved in neurogenesis), and Tis21 (transcription coregulator of proneural genes) as markers of neuronal lineage determination as well as the precursor cells markers already described previously in this chapter (Attardo et al., 2009; Hodge et al., 2008; Kim, Leung, Reed, & Johnson, 2007; Kronenberg et al., 2003; Liu et al., 2010; Roybon, Deierborg, Brundin, & Li, 2009; Seki & Arai, 1993; Steiner, Zurborg, Horster, Fabel, & Kempermann, 2008). Finally type 3 (probably D3 cell subtype) cells or migratory neuroblasts of the adult SGZ are added to this classification (Steiner et al., 2006), since they are positive for DCX, PSA–NCAM, Tis21, or Prox1, which means that they are both initiating the neuronal differentiation and migration. But sometimes type 2b and type 3 collapse in only D cells, because there is an overlap of glia-like precursors and neural determination, as this is a gradual process.

NEUROGENESIS IN PRIMATES

All studies presented about adult neurogenesis in the DG have been performed using rodents as animal models. Although neurogenesis has also been shown in other mammals such as primates, including human, intriguingly it has not been found in most species of bat (Amrein, Dechmann, Winter, & Lipp, 2007).

Neurogenesis has been demonstrated in marmoset monkey (Gould, Tanapat, McEwen, Flugge, & Fuchs, 1998) and macaque monkey (Kornack & Rakic, 1999). Moreover, Ngwenya and colleagues described for the first time the ultrastructural features of newly generated cells in the DG of rhesus monkey. They suggest that the NSPCs show characteristics of immature astrocytes exhibiting irregular contour, indented nucleus with clumps of heterochromatin, voluminous cytoplasm with many organelles, and the presence of glial filaments. By contrast, the immature neurons had an irregularly shaped and small nucleus with slightly electrodense matrix, and scant cytoplasm with many polyribosomes. A relationship between the astrocytes and these immature neurons was also shown. These characteristics are very similar to those described in mouse (Ngwenya, Rosene, & Peters, 2008).

To describe adult neurogenesis in the DG, the common technique used currently to label progenitor cells is the thymidine analogue BrdU assay or the use of transgenic animals. In the BrdU strategy, this compound is incorporated into the DNA of an S-phase cell, indicating that these cells will subsequently divide and detection can be made in conjunction with a wide range of other antibodies raised against neuronal and nonneuronal cell types in the brain, a method that was instrumental in the discovery of neurogenesis. This technique can be used in animals, but this method is obviously not applicable to humans as a general experimental method. Nevertheless, Eriksson and colleagues took advantage of autopsy material from patients diagnosed with carcinomas who had been treated with BrdU to assess the proliferative activity of the tumor cells. The presence of progenitor cells in the human DG was shown, as well as cells morphologically similar to neurons that express neuronal markers such as NeuN (Eriksson et al., 1998). More recently, Knoth and colleagues demonstrated qualitative and semiquantitative information about neurogenesis in human using a combination of neurogenesis cell markers, which was consistent with the information in rodents previously described. Furthermore, like in rodents, their data indicate that human hippocampal neurogenesis diminishes with age (Knoth et al., 2010).

A novel approach for detecting neurogenesis is the retrospective ^{14}C birth dating. The advantage of this method is that it can provide information of neurogenesis in the adult human brain even for rare events. It consists in establishing the age of cells, since the whole individual incorporated the atmospheric ^{14}C after the testing of nuclear weapons between the 1940s and the mid-1960s. A study using this strategy showed that neurogenesis is produced in the DG of humans throughout life, decreasing modestly during aging (Spalding et al., 2013). However, some authors claim that there are some factors like DNA methylation or damage that could affect the results obtained by the retrospective ^{14}C birth dating. Therefore, this new approach is a cutting-edge method to date cells, but it should be properly validated.

Notwithstanding, neurogenesis studies in human brains are singular, and present some particularities as high sample variability and lack of standardized fixation and preservation methods. The observed structures at the electron microscope level and some immunohistochemistry experiments are not always reliable due to fixation artifacts. Thus, a thorough description of the cell types in the human neurogenic niche has not yet been provided, and the cellular type responsible for the human DG neurogenesis is still unknown.

CONCLUSIONS

Neurogenesis is a biological process that takes place in both the V–SVZ of the lateral ventricles and the DG of the hippocampus. Neurogenesis in DG was possible to prove thanks to different experimental approaches including experiments with the proliferation markers, tritiated thymidine and BrdU, the use of retrovirus for specific cell labeling, immunological staining with different cell markers, electrophysiological studies, and electron microscopy. The latter technique has been of great value to ultrastructurally characterize the cellular types present in the neurogenic niche, which are not possible to detect immunologically or with light microscopy. There are many morphological aspects, for example, the free spaces between D cells and between D cells and astrocytes or mature granule neurons, and the existence of cilia in rA, which undoubtedly could not have been discovered without electron microscopy.

The birth of new neurons has been principally shown in rodents, but also in higher animals such as primates. In all cases, the responsible cells for neurogenesis are the radial glia cells that express markers such as GFAP, highlighting their astrocytic nature. Their progeny traverse short itineraries to differentiate in granular neurons of the DG, showing an ultrastructure similar to the migratory neurons or undifferentiated neurons of the olfactory bulb, eventually differentiating to become functional mature neurons. In the years to come we can look forward to further advances in the characterization of the neurogenic niche in the hippocampus and the introduction of new methodologies that will allow a systematic and exhaustive description of neurogenesis in the adult human hippocampus.

Acknowledgments

We would like to acknowledge Arantxa Cebrián Silla for kindly providing the hippocampus material.

References

Allen, E. (1912). The cessation of mitosis in the central nervous system of the albino rat. *Journal of Comparative Neurology, 22*, 547–568.

Altman, J. (1962). Autoradiographic study of degenerative and regenerative proliferation of neuroglia cells with tritiated thymidine. *Experimental Neurology, 5*, 302–318.

Altman, J. (1963). Autoradiographic investigation of cell proliferation in the brains of rats and cats. *Anatomical Record, 145*, 573–591.

Altman, J., & Bayer, S. A. (1990). Migration and distribution of two populations of hippocampal granule cell precursors during the perinatal and postnatal periods. *Journal of Comparative Neurology, 301*, 365–381.

Amrein, I., Dechmann, D. K., Winter, Y., & Lipp, H. P. (2007). Absent or low rate of adult neurogenesis in the hippocampus of bats (Chiroptera). *PLoS One, 2*, e455.

Amrein, I., Slomianka, L., & Lipp, H. P. (2004). Granule cell number, cell death and cell proliferation in the dentate gyrus of wild-living rodents. *European Journal of Neuroscience, 20*, 3342–3350.

Attardo, A., Fabel, K., Krebs, J., Haubensak, W., Huttner, W. B., & Kempermann, G. (2009). Tis21 expression marks not only populations of neurogenic precursor cells but also new postmitotic neurons in adult hippocampal neurogenesis. *Cerebral Cortex, 20*, 304–314.

Barry, G., Piper, M., Lindwall, C., Moldrich, R., Mason, S., Little, E., et al. (2008). Specific glial populations regulate hippocampal morphogenesis. *Journal of Neuroscience, 28*, 12328–12340.

Brandt, M. D., Jessberger, S., Steiner, B., Kronenberg, G., Reuter, K., Bick-Sander, A., et al. (2003). Transient calretitin expression defines early postmitotic step of neuronal differentiation in adult hippocampal neurogenesis of mice. *Molecular and Cellular Neuroscience, 24*, 603–613.

Breunig, J. J., Sarkisian, M. R., Arellano, J. I., Morozov, Y. M., Ayoub, A. E., Sojitra, S., et al. (2008). Primary cilia regulate hippocampal neurogenesis by mediating sonic hedgehog signaling. *Proceedings of the National Academy of Sciences of the United States of America, 105*, 13127–13132.

Burd, G. D., & Nottebohm, F. (1985). Ultrastructural characterization of synaptic terminals formed on newly generated neurons in a song control nucleus of the adult canary forebrain. *Journal of Comparative Neurology, 240*, 143–152.

Cameron, H. A., Woolley, C. S., McEwen, B. S., & Gould, E. (1993). Differentiation of newly born neurons and glia in the dentate gyrus of the adult rat. *Neuroscience, 56*, 337–344.

Dashtipour, K., Yan, X. X., Dinh, T. T., Okazaki, M. M., Nadler, J. V., & Ribak, C. E. (2002). Quantitative and morphological analysis of dentate granule cells with recurrent basal dendrites from normal and epileptic rats. *Hippocampus, 12*, 235–244.

De Toni, A., Zbinden, M., Epstein, J. A., Ruiz i Altaba, A., Prochiantz, A., & Caille, I. (2008). Regulation of survival in adult hippocampal and glioblastoma stem cell lineages by the homeodomain-only protein HOP. *Neural Development, 3*, 13.

Eriksson, P. S., Perfilieva, E., Bjork-Eriksson, T., Alborn, A. M., Nordborg, C., Peterson, D. A., et al. (1998). Neurogenesis in the adult human hippocampus. *Nature Medicine, 4*, 1313–1317.

Esposito, M. S., Piatti, V. C., Laplagne, D. A., Morgenstern, N. A., Ferrari, C. C., Pitossi, F. J., et al. (2005). Neuronal differentiation in the adult hippocampus recapitulates embryonic development. *Journal of Neuroscience, 25*, 10074–10086.

Filippov, V., Kronenberg, G., Pivneva, T., Reuter, K., Steiner, B., Wang, L. P., et al. (2003). Subpopulation of nestin-expressing progenitor cells in the adult murine hippocampus shows electrophysiological and morphological characteristics of astrocytes. *Molecular and Cellular Neuroscience, 23*, 373–382.

Fukuda, S., Kato, F., Tozuka, Y., Yamaguchi, M., Miyamoto, Y., & Hisatsune, T. (2003). Two distinct subpopulations of nestin-positive cells in adult mouse dentate gyrus. *Journal of Neuroscience, 23*, 9357–9366.

Gould, E., Cameron, H. A., Daniels, D. C., Woolley, C. S., & McEwen, B. S. (1992). Adrenal hormones suppress cell division in the adult rat dentate gyrus. *Journal of Neuroscience, 12*, 3642–3650.

Gould, E., Tanapat, P., McEwen, B. S., Flugge, G., & Fuchs, E. (1998). Proliferation of granule cell precursors in the dentate gyrus of adult monkeys is diminished by stress. *Proceedings of the National Academy of Sciences of the United States of America, 95*, 3168–3171.

Gould, E., Vail, N., Wagers, M., & Gross, C. G. (2001). Adult-generated hippocampal and neocortical neurons in macaques have a transient existence. *Proceedings of the National Academy of Sciences of the United States of America, 98*, 10910–10917.

Han, Y. G., Spassky, N., Romaguera-Ros, M., Garcia-Verdugo, J. M., Aguilar, A., Schneider-Maunoury, S., et al. (2008). Hedgehog signaling and primary cilia are required for the formation of adult neural stem cells. *Nature Neuroscience, 11*, 277–284.

Hodge, R. D., Kowalczyk, T. D., Wolf, S. A., Encinas, J. M., Rippey, C., Enikolopov, G., et al. (2008). Intermediate progenitors in adult hippocampal neurogenesis: Tbr2 expression and coordinate regulation of neuronal output. *Journal of Neuroscience, 28*, 3707–3717.

Imayoshi, I., Sakamoto, M., Ohtsuka, T., Takao, K., Miyakawa, T., Yamaguchi, M., et al. (2008). Roles of continuous neurogenesis in the structural and functional integrity of the adult forebrain. *Nature Neuroscience, 11*, 1153–1161.

Jones, S. P., Rahimi, O., O'Boyle, M. P., Diaz, D. L., & Claiborne, B. J. (2003). Maturation of granule cell dendrites after mossy fiber arrival in hippocampal field CA3. *Hippocampus, 13*, 413–427.

Kaplan, M. S. (1985). Formation and turnover of neurons in young and senescent animals: an electronmicroscopic and morphometric analysis. *Annals of the New York Academy of Sciences, 457*, 173–192.

Kaplan, M. S., & Hinds, J. W. (1977). Neurogenesis in the adult rat: electron microscopic analysis of light radioautographs. *Science, 197*, 1092–1094.

Kempermann, G., Gast, D., Kronenberg, G., Yamaguchi, M., & Gage, F. H. (2003). Early determination and long-term persistence of adult-generated new neurons in the hippocampus of mice. *Development, 130*, 391–399.

Kim, E. J., Leung, C. T., Reed, R. R., & Johnson, J. E. (2007). In vivo analysis of Ascl1 defined progenitors reveals distinct developmental dynamics during adult neurogenesis and gliogenesis. *Journal of Neuroscience, 27*, 12764–12774.

Knoth, R., Singec, I., Ditter, M., Pantazis, G., Capetian, P., Meyer, R. P., et al. (2010). Murine features of neurogenesis in the human hippocampus across the lifespan from 0 to 100 years. *PLoS One, 5*, e8809.

Kornack, D. R., & Rakic, P. (1999). Continuation of neurogenesis in the hippocampus of the adult macaque monkey. *Proceedings of the National Academy of Sciences of the United States of America, 96*, 5768–5773.

Kronenberg, G., Reuter, K., Steiner, B., Brandt, M. D., Jessberger, S., Yamaguchi, M., et al. (2003). Subpopulations of proliferating cells of the adult hippocampus respond differently to physiologic neurogenic stimuli. *Journal of Comparative Neurology, 467*, 455–463.

Kunze, A., Congreso, M. R., Hartmann, C., Wallraff-Beck, A., Huttmann, K., Bedner, P., et al. (2009). Connexin expression by radial glia-like cells is required for neurogenesis in the adult dentate gyrus. *Proceedings of the National Academy of Sciences of the United States of America, 106*, 11336–11341.

Lagace, D. C., Whitman, M. C., Noonan, M. A., Ables, J. L., DeCarolis, N. A., Arguello, A. A., et al. (2007). Dynamic contribution of nestin-expressing stem cells to adult neurogenesis. *Journal of Neuroscience, 27*, 12623–12629.

Licht, T., & Keshet, E. (2015). The vascular niche in adult neurogenesis. *Mechanisms of Development, 138*, 56–62.

Li, G., Fang, L., Fernandez, G., & Pleasure, S. J. (2013). The ventral hippocampus is the embryonic origin for adult neural stem cells in the dentate gyrus. *Neuron, 78*, 658–672.

Lim, D. A., & Alvarez-Buylla, A. (1999). Interaction between astrocytes and adult subventricular zone precursors stimulates neurogenesis. *Proceedings of the National Academy of Sciences of the United States of America, 96*, 7526–7531.

Liu, Y., Namba, T., Liu, J., Suzuki, R., Shioda, S., & Seki, T. (2010). Glial fibrillary acidic protein-expressing neural progenitors give rise to immature neurons via early intermediate progenitors expressing both glial fibrillary acidic protein and neuronal markers in the adult hippocampus. *Neuroscience, 166*, 241–251.

Mignone, J. L., Kukekov, V., Chiang, A. S., Steindler, D., & Enikolopov, G. (2004). Neural stem and progenitor cells in nestin-GFP transgenic mice. *Journal of Comparative Neurology, 469*, 311–324.

Morrens, J., Van Den Broeck, W., & Kempermann, G. (2012). Glial cells in adult neurogenesis. *Glia, 60*, 159–174.

Nacher, J., Crespo, C., & McEwen, B. S. (2001). Doublecortin expression in the adult rat telencephalon. *European Journal of Neuroscience, 14*, 629–644.

Namba, T., Maekawa, M., Yuasa, S., Kohsaka, S., & Uchino, S. (2009). The Alzheimer's disease drug memantine increases the number of radial glia-like progenitor cells in adult hippocampus. *Glia, 57*, 1082–1090.

Nam, H. S., & Benezra, R. (2009). High levels of Id1 expression define B1 type adult neural stem cells. *Cell Stem Cell, 5*, 515–526.

Ngwenya, L. B., Rosene, D. L., & Peters, A. (2008). An ultrastructural characterization of the newly generated cells in the adult monkey dentate gyrus. *Hippocampus, 18*, 210–220.

Ninkovic, J., Mori, T., & Gotz, M. (2007). Distinct modes of neuron addition in adult mouse neurogenesis. *Journal of Neuroscience, 27*, 10906–10911.

Paton, J. A., & Nottebohm, F. N. (1984). Neurons generated in the adult brain are recruited into functional circuits. *Science, 225*, 1046–1048.

Peters, A., Palay, S. L., & Webster, H. D. F. (1991). *The fine structure of the nervous system: Neurons and their supporting cells.* New York: Oxford University Press.

Plumpe, T., Ehninger, D., Steiner, B., Klempin, F., Jessberger, S., Brandt, M., et al. (2006). Variability of doublecortin-associated dendrite maturation in adult hippocampal neurogenesis is independent of the regulation of precursor cell proliferation. *BMC Neuroscience, 7*, 77.

Reynolds, B. A., & Weiss, S. (1992). Generation of neurons and astrocytes from isolated cells of the adult mammalian central nervous system. *Science, 255*, 1707–1710.

Ribak, C. E., Korn, M. J., Shan, Z., & Obenaus, A. (2004). Dendritic growth cones and recurrent basal dendrites are typical features of newly generated dentate granule cells in the adult hippocampus. *Brain Research, 1000*, 195–199.

Roybon, L., Deierborg, T., Brundin, P., & Li, J. Y. (2009). Involvement of Ngn2, Tbr and NeuroD proteins during postnatal olfactory bulb neurogenesis. *European Journal of Neuroscience, 29*, 232–243.

Savchenko, V. L., McKanna, J. A., Nikonenko, I. R., & Skibo, G. G. (2000). Microglia and astrocytes in the adult rat brain: comparative immunocytochemical analysis demonstrates the efficacy of lipocortin 1 immunoreactivity. *Neuroscience, 96*, 195–203.

Seki, T. (2002). Hippocampal adult neurogenesis occurs in a microenvironment provided by PSA-NCAM-expressing immature neurons. *Journal of Neuroscience Research, 69*, 772–783.

Seki, T., & Arai, Y. (1993). Highly polysialylated neural cell adhesion molecule (NCAM-H) is expressed by newly generated granule cells in the dentate gyrus of the adult rat. *Journal of Neuroscience, 13*, 2351–2358.

Seki, T., & Arai, Y. (1999). Temporal and spacial relationships between PSA-NCAM-expressing, newly generated granule cells, and radial glia-like cells in the adult dentate gyrus. *Journal of Comparative Neurology, 410*, 503–513.

Seki, T., Namba, T., Mochizuki, H., & Onodera, M. (2007). Clustering, migration, and neurite formation of neural precursor cells in the adult rat hippocampus. *Journal of Comparative Neurology, 502*, 275–290.

Seri, B., Garcia-Verdugo, J. M., Collado-Morente, L., McEwen, B. S., & Alvarez-Buylla, A. (2004). Cell types, lineage, and architecture of the germinal zone in the adult dentate gyrus. *Journal of Comparative Neurology, 478*, 359–378.

Seri, B., Garcia-Verdugo, J. M., McEwen, B. S., & Alvarez-Buylla, A. (2001). Astrocytes give rise to new neurons in the adult mammalian hippocampus. *Journal of Neuroscience, 21*, 7153–7160.

Shapiro, L. A., Korn, M. J., Shan, Z., & Ribak, C. E. (2005). GFAP-expressing radial glia-like cell bodies are involved in a one-to-one relationship with doublecortin-immunolabeled newborn neurons in the adult dentate gyrus. *Brain Research, 1040*, 81–91.

Shapiro, L. A., & Ribak, C. E. (2005). Integration of newly born dentate granule cells into adult brains: hypotheses based on normal and epileptic rodents. *Brain Research Reviews, 48*, 43–56.

Song, H., Stevens, C. F., & Gage, F. H. (2002). Astroglia induce neurogenesis from adult neural stem cells. *Nature, 417*, 39–44.

Spalding, K. L., Bergmann, O., Alkass, K., Bernard, S., Salehpour, M., Huttner, H. B., et al. (2013). Dynamics of hippocampal neurogenesis in adult humans. *Cell, 153*, 1219–1227.

Steiner, B., Klempin, F., Wang, L., Kott, M., Kettenmann, H., & Kempermann, G. (2006). Type-2 cells as link between glial and neuronal lineage in adult hippocampal neurogenesis. *Glia, 54*, 805–814.

Steiner, B., Kronenberg, G., Jessberger, S., Brandt, M. D., Reuter, K., & Kempermann, G. (2004). Differential regulation of gliogenesis in the context of adult hippocampal neurogenesis in mice. *Glia, 46*, 41–52.

Steiner, B., Zurborg, S., Horster, H., Fabel, K., & Kempermann, G. (2008). Differential 24 h responsiveness of Prox1-expressing precursor cells in adult hippocampal neurogenesis to physical activity, environmental enrichment, and kainic acid-induced seizures. *Neuroscience, 154*, 521–529.

Suh, H., Consiglio, A., Ray, J., Sawai, T., D'Amour, K. A., & Gage, F. H. (2007). In vivo fate analysis reveals the multipotent and self-renewal capacities of $Sox2^+$ neural stem cells in the adult hippocampus. *Cell Stem Cell, 1*, 515–528.

Taupin, P., Ray, J., Fischer, W. H., Suhr, S. T., Hakansson, K., Grubb, A., et al. (2000). FGF-2-responsive neural stem cell proliferation requires CCg, a novel autocrine/paracrine cofactor. *Neuron, 28*, 385–397.

Tavazoie, M., Van der Veken, L., Silva-Vargas, V., Louissaint, M., Colonna, L., Zaidi, B., et al. (2008). A specialized vascular niche for adult neural stem cells. *Cell Stem Cell, 3*, 279–288.

Yu, D. X., Marchetto, M. C., & Gage, F. H. (2014). How to make a hippocampal dentate gyrus granule neuron. *Development, 141*, 2366–2375.

CHAPTER

2

Physiology and Plasticity

G.W. Kirschen, A. Di Antonio, S. Ge
Stony Brook University, NY, United States

INTRODUCTION

During embryogenesis, the maturing brain is inherently plastic, generating new neurons and synaptic connections that will serve to perform vegetative, perceptual, cognitive, motor, and other functions throughout life. The hippocampus faces the unique challenge of encoding and retrieving memories formed continuously from the postnatal period until senescence. Perhaps for this reason, it remains plastic past prenatal and early postnatal life and into adulthood. Unlike most brain structures, in which cells are terminally differentiated, the dentate gyrus (DG) of the hippocampus maintains a pool of multipotent progenitors that constantly enter the cell cycle, giving rise to new neurons and astrocytes throughout adulthood (Cameron, Woolley, McEwen, & Gould, 1993; Gage et al., 1998). In the prenatal and perinatal periods, new neurons destined to become dentate granule cells (DGCs) migrate from their birthplace in the subventricular zone to their final destination in the granule cell layer (GCL) of the DG (Altman & Bayer, 1990; Esposito et al., 2005; Gage, 2000). While the GCL remains the destination of newborn DGCs in the older organism, the stem cell niche relocates to the subgranular zone (SGZ) of the DG. In rodents, only about 20% of DGCs derive from the embryonic period, with the bulk generated in the first three postnatal months, and approximately 10% contributed during adulthood (Altman & Bayer, 1990; Altman & Das, 1965; Imayoshi et al., 2008). There is, of course, compensatory cell death; otherwise the DG would continue to swell to the point of causing brain herniation. In fact, the majority of newborn DGCs die by programmed cell death (apoptosis) within the first few weeks of their birth, with half dying by 4 days post-mitosis (Biebl, Cooper, Winkler, & Kuhn, 2000; Dayer, Ford, Cleaver, Yassaee, & Cameron, 2003; Sierra et al., 2010).

Adult Neurogenesis in the Hippocampus
http://dx.doi.org/10.1016/B978-0-12-801977-1.00002-7

However, various intrinsic or environmental conditions influence birth and death rates of DGCs (Dayer et al., 2003; Gould & Cameron, 1996). Moreover, synaptic connectivity within the DG is plastic, with continuous addition of new cells competing to integrate into the hippocampal circuit. Thus, the DG is plastic at both the cellular and the synaptic levels—with a constant turnover of neurons and the formation of synaptic input and output connections.

The function of the DG depends crucially on its adaptability in an unpredictable environment with different stimuli sharing similar characteristics (for example, the array of different faces that we encounter daily). As the principal afferent input to the hippocampus, the DG is unsurprisingly essential for hippocampus-dependent forms of memory, as demonstrated by DG lesion studies. Ablation of the DG results in impairments in spatial memory formation and pattern separation, the ability to separate distinct experiences (Czeh, Seress, Nadel, & Bures, 1998; Gilbert, Kesner, & Lee, 2001). Intriguingly, disabling adult-born DGCs while keeping other neurons intact impairs hippocampus-dependent memory, suggesting that adult-born neurons have an important role in hippocampal function. How the brain manages object/spatial memory appears to derive from neurogenesis and distinct activity patterns of newly generated DGCs, which may "sharpen the resolution" of our memory of past experiences (Aimone, Deng, & Gage, 2011). Thus, the pool of regenerating cells is able to hook into the hippocampal circuitry and contribute to memory encoding. Next, we will discuss advances in testing the roles of new DGCs in DG physiology and function.

THE GENERATION AND DEVELOPMENT OF ADULT-BORN DENTATE GRANULE CELLS

DGCs arise from radial glia-like (RGL) multipotent precursors located in the hilus of the DG, and are derived from neuroepithelial progenitors, the bona fide neural stem cells. RGL cells, so-called because of their morphological and functional resemblance to neocortical radial glia cells, are characterized by expression of intermediate filaments Nestin and GFAP, and the transcription factor Sox2, which promotes self-renewal (Kempermann, Jessberger, Steiner, & Kronenberg, 2004; Lagace et al., 2007; Rizzino, 2009). Radial glia cells express Pax6, a transcription factor involved in neuronal fate determination. RGL cells are Pax6 negative, however, highlighting a distinction between these two cell types (Gotz, Stoykova, & Gruss, 1998; Hack et al., 2005). In addition to being multipotent, RGL cells are believed to serve the crucial function of guiding newly born neurons from the SGZ to their final destination within the GCL. This migratory guidance occurs in a fashion similar to that of radial migration of embryonically born cortical neurons of the ventricular zone into the cortical plate, although the "schlep" is considerably shorter for DGCs—just a few micrometers—compared to several 100 μm for cortical neurons (Liebmann, Stahr, Guenther, Witte, & Frahm, 2013; Lois & Alvarez-Buylla, 1994; Zhu, Li, Zhou, Wu, & Rao, 1999). The radial process of radial glia cells serves as a guidepost around which newborn neurons wrap, climbing to get from birthplace to final destination (Shapiro & Ribak, 2005). In both cortex and DG, neuronal migration depends on gap junction coupling between radial glia (or RGL cell) and newborn neuron, as migration is disrupted following knockdown or deletion of gap junction-forming connexin proteins normally expressed by radial glia/RGL cells (Elias, Turmaine, Parnavelas, & Kriegstein, 2010; Kunze et al., 2009; Liebmann et al., 2013). In addition to using RGL cells as a guide, newborn neurons express polysialic acid neural cell adhesion molecule (PSA–NCAM), which aids in migration, neurite sprouting, and survival (Gascon, Vutskits, Jenny, Durbec, & Kiss, 2007; Gascon, Vutskits, & Kiss, 2007; Seki & Arai, 1993). Finally, Disrupted-in-Schizophrenia-1 (DISC1) and the DISC1-interacting protein girdin have been implicated in proper migration as well as axon targeting and dendrite development of adult-born DGCs (Enomoto et al., 2009; Faulkner et al., 2008).

It takes approximately 4 weeks for DGCs to reach functional maturity, with full maturity occurring by 8 weeks. RGL mother cells first divide asymmetrically (giving rise to daughter cells of two distinct types), producing another RGL cell and a transiently amplifying progenitor. Transiently amplifying cells can divide to produce more of themselves, or undergo another asymmetric division, producing the first post-mitotic immature neuron (Alvarez-Buylla, Garcia-Verdugo, & Tramontin, 2001; Kronenberg et al., 2006). While approximately 70% of daughter cells ultimately differentiate into neurons, the remaining 30% differentiate into glia (eg, astrocytes) or retain their RGL cell identity; the latter serves to replenish the regenerative pool of precursors (Namba et al., 2005). Astrocytes regulate proliferation and the developmental fate of newborn neurons through the secretion of neurotrophic factors and cytokines (Morrens, Van Den Broeck, & Kempermann, 2012; Seri, Garcia-Verdugo, Collado-Morente, McEwen, & Alvarez-Buylla, 2004; Song, Stevens, & Gage, 2002). From 1 to 2 weeks after birth, these immature neurons express doublecortin (DCX), a microtubule-associated protein important for migration (Brown et al., 2003; Filipovic, Santhosh Kumar, Fiondella, & Loturco, 2012; Gleeson, Lin, Flanagan, & Walsh, 1999).

As they grow, immature DGCs elaborate an apical dendrite that climbs through the GCL and into the molecular layer, where, by the second week after division, it forms synapses with layer II entorhinal cortex (EC) neurons to join the perforant pathway of afferent information into the DG. By the middle of the second week, newborn DGCs extend an axon into the hilus of the DG that synapses onto pyramidal neurons of area CA3 of the hippocampus, joining the mossy fiber pathway. Although dendritic spine density is sparse until weeks 3–4 of maturation, DGCs begin producing "immature" action potentials (APs) during the first week after birth, characterized by lower amplitude and longer duration as compared to those of mature neurons (Esposito et al., 2005; Schmidt-Hieber, Jonas, & Bischofberger, 2004; Zhao, Teng, Summers, Ming, & Gage, 2006). Just as babies babble, these APs are like the first "vocalizations" of newborn DGCs.

Early APs generated by newborn DGCs have a large calcium component attributable to the T-type calcium channel. Importantly, activity-dependent synaptic development and neuronal integration rely on calcium-mediated signaling and *N*-methyl D-aspartate receptor (NMDAR) activation (Gould & Tanapat, 1999; Schmidt-Hieber et al., 2004). Proper integration and survival depend on α-calcium/calmodulin-dependent protein kinase II (αCaMKII), a downstream target of intracellular calcium expressed by approximately 3 weeks after neuronal birth (Arruda-Carvalho et al., 2014).

At the subcellular level, developmental maturation is characterized by reorganization of organelles and cytoskeletal architecture. DCX plays an important role in neuronal migration by reducing microtubule bundling and inducing actin reorganization, necessary steps in forming a leading process (Toriyama et al., 2012). Newborn cells preparing for transit must become polarized. DCX, in conjunction with the microtubule- and dynein-associated protein Lis1, coordinates the movement of the centrosome and Golgi apparatus to the apical end and the nucleus to the basal end of the cell, such that the microtubule-bound centrosome "drags" the nucleus in the direction of migration (Koizumi et al., 2006; Tanaka et al., 2004; Tsai & Gleeson, 2005). As they approach their destination around the second week, DGCs elaborate an apical primary cilium, which mediates sonic hedgehog (Shh) and Wnt signaling to regulate glutamate-dependent synaptic maturation and integration (Breunig et al., 2008; Corbit et al., 2008; Kumamoto et al., 2012). Yang, Arnold, Habas, Hetman, and Hagg (2008) demonstrated that astrocytes promote neurogenesis by secreting ciliary neurotrophic factor (CNTF), which increases synaptic plasticity and enhances hippocampus-dependent spatial memory (Kazim et al., 2014; Muller, Chakrapani, Schwegler, Hofmann, & Kirsch, 2009).

Some intriguing questions remain unresolved. With which cells do primary cilia communicate, and how? How do maturing DGCs "know" where to send their dendrites and axons? And is the genesis of adult-born DGCs uniform throughout the DG? For at least this final question, evidence is beginning to accumulate.

Although the DG is often treated as a unitary, homogeneous structure, emerging evidence suggests physiological and functional divisibility. Anatomically, it consists of a dorsal and ventral layer of granule cells, named the supra- and infrapyramidal blades, respectively. Another distinction is made between septal DG (sDG) and temporal DG (tDG), the medial and lateral aspects of both blades, respectively. The sDG has been implicated in spatial learning while the tDG is involved in fear memory and anxiety-like behaviors (Bannerman et al., 2004; Moser, Moser, Forrest, Andersen, & Morris, 1995; Trivedi & Coover, 2004). DGCs of the infrapyramidal blade may exhibit increased proliferation while those of the suprapyramidal blade may exhibit increased survival; however these findings have been mixed (Bekiari et al., 2014; Snyder, Ferrante, & Cameron, 2012). Nevertheless, newborn neurons of the sDG mature slightly earlier than those of the tDG, with septal pole DGCs exhibiting earlier activity-dependent synaptic plasticity and expression of neuronal maturity marker NeuN, compared to their temporal counterparts (Piatti et al., 2011; Snyder, Ferrante, & Cameron, 2012), possibly relating to higher basal excitatory activity at the septal pole. Regardless of regional differences, the DG must accommodate new DGCs each day, raising the question of whether adult-born DGCs replace those of early life.

During the first 2 weeks after the birth of the organism, DGC proliferation continues, but with more cell death, decreasing the overall number of DGCs (Altman & Das, 1965; Gould, Woolley, & McEwen, 1991). An explosion of neurogenesis ensues over the next few months. During adulthood, however, a steady state of cell birth and death exists at baseline. Experiments using the thymidine analog bromodeoxyuridine (BrdU) to label dividing cells in the SGZ have shown that a majority of newborn cells are lost by 2 weeks postmitosis. However, those that survive and successfully integrate replace older cells, even those derived from embryonic life (Dayer et al., 2003; Gould et al., 1999). *Survival* of newborn cells and synaptic potentiation are affected by both intrinsic factors, such as adrenal steroids and NMDAR activation (Gould & Cameron, 1996), and extrinsic factors, including exercise, learning, and environmental stimulation (Kempermann, Kuhn, & Gage, 1998; Krugers et al., 2007; Leuner, Gould, & Shors, 2006; van Praag, Shubert, Zhao, & Gage, 2005). Survival is only one part of the equation. "Net" neurogenesis can also increase/decrease as a result of changes in precursor proliferation. Factors that decrease *proliferation* include ionizing radiation, which causes DNA damage, as well as chronic stress and sleep deprivation (Guzman-Marin et al., 2003; Heine, Zareno, Maslam, Joels, & Lucassen, 2005; Peissner, Kocher, Treuer, & Gillardon, 1999). Factors that positively regulate DGC proliferation include physical exercise, caloric restriction, and learning (Gould et al., 1999; Lee, Duan, Long, Ingram, & Mattson, 2000; van Praag, Christie, Sejnowski, & Gage, 1999; van Praag, Kempermann, & Gage, 1999). Proliferation is also hormonally modulated by ghrelin, oxytocin, estrogen, and glucocorticoids (Cameron & Gould, 1994; Gould, Tanapat, & McEwen, 1997; Johansson et al., 2008; Leuner, Caponiti, & Gould, 2012; McEwen et al., 1994; Tanapat, Hastings, Reeves, & Gould, 1999).

What causes newborn DGCs to die, as the majority do? The confluence of proapoptotic signals and lack of neurotrophic support contribute to their demise (Cameron & Gould, 1994; Gould et al., 1997; Linnarsson, Willson, & Ernfors, 2000). One proapoptotic signal is mediated by adrenal steroids, which circulate at low, diurnally fluctuating levels under physiological conditions. DGCs express glucocorticoid receptors (GR) and mineralocorticoid receptors (MR), whose ligands readily cross the blood–brain barrier (BBB). Corticosterone, the major rodent glucocorticoid, binds the GR and MR, and at high concentrations increases synaptic glutamate, activates NMDARs and causes calcium-mediated excitotoxicity and apoptosis (Bhatt, Feng, Wang, Famuyide, & Hersey, 2013; Jacobs, Trinh, Rootwelt, Lomo, & Paulsen, 2006; Reagan & McEwen, 1997). However, corticosterone can also stimulate neurogenesis, likely by inducing division and differentiation of RGL cells that express the GR and MR, thus maintaining a constant turnover of cells (Garcia, Steiner, Kronenberg, Bick-Sander, & Kempermann, 2004; Gould et al., 1997). So then what protects newborn DGCs destined to survive and thrive in the hostile and competitive environment of the DG? They express various receptors that bind growth factors including brain-derived neurotrophic factor, neurotrophin-3, CNTF, and basic fibroblast growth factor. Each plays a role in survival, maturation, and synaptic plasticity; thus only DGCs "lucky" enough to receive an adequate supply of trophic factors will survive (Lowenstein & Arsenault, 1996; Minichiello & Klein, 1996; Muller et al., 2009).

Population control in the DG is tightly maintained through different regulatory mechanisms, among them activity of the newborn DGCs themselves, signals from adjacent astrocytes, and hormonal influences. We will now detail the plastic properties of DGCs, emphasizing differences between mature and immature neurons.

ADULT-BORN DENTATE GRANULE CELLS REGULATE STRUCTURAL AND PHYSIOLOGICAL PLASTICITY OF THE DENTATE GYRUS

The DG is plastic at the cellular level: with constant birth, integration, and death of neurons. We turn our attention now to the synapse, where young adult-born DGCs exhibit enhanced plasticity compared to their elders. We will review electrophysiological findings in adult-born DGCs as they mature.

After the discovery of DGCs generated in adulthood, the next step was to characterize their electrophysiological and functional properties. By 1 week of age, newborn DGCs receive excitatory inputs from local interneurons in the GCL and begin to extend an axon through the hilus to reach region CA3, forming a fully mature excitatory output by 4 to 8 weeks (Faulkner et al., 2008; Hastings & Gould, 1999). Using

γ-aminobutyric acid (GABA) as their neurotransmitter, these interneurons excite newborn neurons, recapitulating the embryonic situation in which GABA acts to excite maturing neurons (Ge, Pradhan, Ming, & Song, 2007; Tozuka, Fukuda, Namba, Seki, & Hisatsune, 2005). In most brain regions, GABA switches to a purely inhibitory (hyperpolarizing) role after the embryonic period. However in the adult DG, GABA acts in excitatory *and* inhibitory fashions at low and high concentrations, respectively (Chiang et al., 2012; Ge et al., 2006; Sauer, Struber, & Bartos, 2012). GABA's excitatory effects on 0- to 2-week-old DGCs are due to the fact that these immature cells express the Na–K–Cl cotransporter-1 that pumps chloride, sodium, and potassium ions into the cell, maintaining a high intracellular chloride concentration. Chloride is tonically expelled through $GABA_A$ receptors ($GABA_ARs$), whose expression is regulated by adrenal corticosteroids, in response to low ambient GABA (Orchinik, Weiland, & McEwen, 1994). By contrast, mature DGCs have lower input resistance and contain less intracellular chloride, thus low GABA concentrations have smaller effects on these cells compared to newborn cells, while high GABA input from interneurons onto mature DGCs causes significant hyperpolarization (Ge et al., 2006; Mongiat, Esposito, Lombardi, & Schinder, 2009).

Newborn DGCs' early synaptic responses display a slow kinetic profile, with slow rise and decay phases. The slow speed is believed to be due to dendritic (rather than perisomatic) localization of $GABA_ARs$, although recent evidence suggests that it may also relate to low ambient GABA concentrations (Esposito et al., 2005; Markwardt, Wadiche, & Overstreet-Wadiche, 2009; Overstreet Wadiche, Bromberg, Bensen, & Westbrook, 2005; Soltesz, Smetters, & Mody, 1995). Functional GABAergic synapses that form during the second week exhibit faster synaptic response rates compared to those generated during the first week, possibly resulting from perisomatic localization of $GABA_ARs$ (Esposito et al., 2005; Ge, Sailor, Ming, & Song, 2008). The need for two distinct types of GABAergic input onto newborn neurons during their early development is yet to be fully understood, but given that dendritic synapses influence excitability while perisomatic synapses modulate output, the former may assist in activity-dependent integration, while the latter may amplify the efferent signal to CA3. Newborn neurons are critically dependent on GABA-induced excitation to integrate into the circuit, as silencing them with hyperpolarizing current impairs normal synapse formation and dendrite development (Ge et al., 2006). Early tonic depolarization of newborn DGCs in a background of mature DGC inhibition may facilitate newborn DGC integration. Immature DGCs' relative "loudness" in comparison to their mature counterparts stems from several factors: immature neurons' differing responses to GABA, their higher input resistance, and the presence of constitutively inward rectifying potassium (Kir) channels on older cells, making older cells less excitable

(Marin-Burgin, Mongiat, Pardi, & Schinder, 2012; Mongiat et al., 2009; Takigawa & Alzheimer, 2002). A report published just before the submission of this chapter showed that activity of mature DGCs triggers greater interneuron-mediated feedback inhibition onto mature DGCs versus immature DGCs (Temprana et al., 2015). Thus, several mechanisms converge to facilitate activity of newborn DGCs, allowing them to compete successfully for synaptic real estate. GABA-mediated depolarization during the first 2 weeks of DGC life paves the way for the next maturational step: glutamatergic synapse formation (Chancey et al., 2013).

Starting at 2 weeks and peaking between 3 and 4 weeks of age, those DGCs that have survived, outcompeting their sisters and forming functional GABAergic synapses, start to form functional excitatory glutamatergic synapses with EC afferents (Chancey, Poulsen, Wadiche, & Overstreet-Wadiche, 2014; Faulkner et al., 2008). Dendritic spines proliferate during the third week, corresponding to the initiation of glutamatergic innervation (Esposito et al., 2005; Ge et al., 2006; Zhao et al., 2006). Although DGCs express the prerequisites for glutamatergic synapses—α-amino-3-hydroxy-5-methyl-4-isoxazolepropionic acid receptors (AMPARs) and NMDARs— these synapses appear not to be fully functional during the "GABAergic period," as AMPA or NMDA puffing evokes smaller responses than would be expected from mature glutamateric neurons, likely due to low AMPAR expression (Overstreet Wadiche et al., 2005). By 4 weeks, the electrophysiological properties of newborn neurons, including spontaneous postsynaptic responses and intrinsic membrane properties, mirror those of embryonic-derived DGCs (Laplagne et al., 2006).

A molecular switch gets thrown between 2 and 3 weeks post-mitosis, as GABA-mediated excitation and NMDAR activation cause glutamatergic *"synapse unsilencing,"* a process of marked increase in AMPAR incorporation at the synapse that establishes functional glutamatergic synapses (Ben-Ari, Khazipov, Leinekugel, Caillard, & Gaiarsa, 1997; Chancey et al., 2013; Leinekugel, Medina, Khalilov, Ben-Ari, & Khazipov, 1997; Wang & Kriegstein, 2008). After synapse unsilencing, GABAR opening causes hyperpolarization as intracellular chloride falls relative to extracellular levels. There is a concomitant change in NMDAR subunits. The NMDAR is a diheteromer consisting of two GluN1 and two GluN2 subunits. GluN2 has two isoforms, with GluN2B predominating in the fetal brain, and a postnatal switch to the GluN2A subunit triggered transcriptionally in an activity-dependent fashion (Paoletti, Bellone, & Zhou, 2013; Rodenas-Ruano, Chavez, Cossio, Castillo, & Zukin, 2012). Compared to synapses containing NMDARs with the "mature" GluN2A subunit, those containing GluN2B exhibit smaller amplitudes of depolarization on activation and slower deactivation. Importantly, synapses containing receptors with GluN2B exhibit enhanced plasticity (measured by long-term potentiation (LTP) induction), compared to those with the GluN2A subunit, due to the relatively higher binding affinity of GluN2B for CaMKII, which facilitates plasticity subsequent to calcium entry (Barria & Malinow, 2005; Barth & Malenka, 2001). Recapitulating the embryonic maturational sequence, newborn DGCs first express NMDARs with the GluN2B subunit and switch to GluN2A between 4 and 8 weeks after division, consistent with the time at which glutamatergic synapses become functionally mature (Ge et al., 2007; Snyder et al., 2001; van Praag et al., 2002). Fig. 2.1 depicts this process.

Enhanced plasticity in the form of LTP was originally discovered at the perforant pathway–granule cell synapse in the rabbit hippocampus in the laboratory of Per Anderson (Bliss & Lomo, 1973; Lomo, 1971). Terje Lømo described how brief trains of excitatory current delivered at presynaptic cells led to enhanced transmission efficiency to postsynaptic cells. Downstream transcriptional changes can result in formation of new synapses that further facilitate transmission. Since these findings, there has been an explosion of studies on hippocampal synaptic plasticity, exploring the synapses at which it occurs, its molecular underpinnings, and its functional significance. Such plasticity also occurs between CA3 and CA1 pyramidal

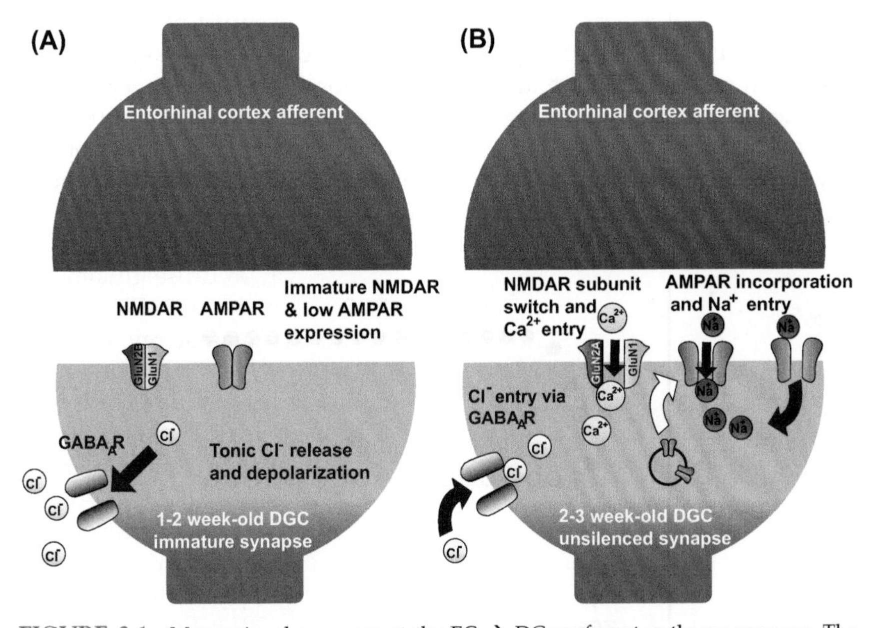

FIGURE 2.1 Maturational sequence at the EC → DG perforant pathway synapse. The first 2 weeks after DGC division are characterized by GABAergic depolarization and hypofunctional glutamatergic receptor signaling (A). Glutamatergic "synapse unsilencing" occurs between 2 and 3 weeks, and is characterized by the NR2B → NR2A NMDAR subunit switch, increased AMPAR expression and incorporation at the synapse, and Ca^{2+} and Na^{+} entry via these receptors, respectively (B). After the switch, $GABA_AR$ opening results in Cl^--mediated hyperpolarization. Black arrows represent direction of ion flow; white arrow represents direction of AMPAR trafficking.

neurons of the Schaffer collateral pathway, where it is critical for long-lasting spatial/object memory (Alger & Teyler, 1976; Baudry et al., 2014; Morris, Hagan, & Rawlins, 1986). At the perforant pathway, enhanced synaptic plasticity occurs more readily at synapses of newborn DGCs compared to those of mature DGCs, suggesting that newborn DGCs are more plastic than their elders (Schmidt-Hieber et al., 2004; Snyder et al., 2001; Wang, Scott, & Wojtowicz, 2000).

Compared to mature neurons, newborn DGCs are hyperexcitable and exhibit more robust synaptic plasticity and a lower threshold for its induction. For example, 2-week-old DGCs are characterized by greater mean depolarization amplitudes after repeated stimulation compared to 4-month-old DGCs (Ge et al., 2007). What makes newborn DGCs so plastic? As discussed above, adult-born neurons, like their embryonic predecessors, initially express GluN2B. This is vital for their plasticity, with LTP abolished in retrovirally labeled newborn DGCs in mouse hippocampal slice preparations treated with the GluNRB-specific antagonist ifenprodil. Additionally, expression of GluN2B is limited to the first 1–1.5 months after mitosis, with the subsequent switch to GluN2A. These findings have led to the idea of a "critical period for enhanced synaptic plasticity" (Ge et al., 2007). Further supporting this critical period hypothesis, synapses of adult-born neurons that are fully mature (2 months old) no longer exhibit enhanced plasticity. Even though we refer to 1-month-old adult-born neurons as "mature" due to their morphology and functional integration, they are not completely identical to embryonic-derived DGCs, as they still exhibit hyperexcitability and enhanced plasticity (see Fig. 2.2).

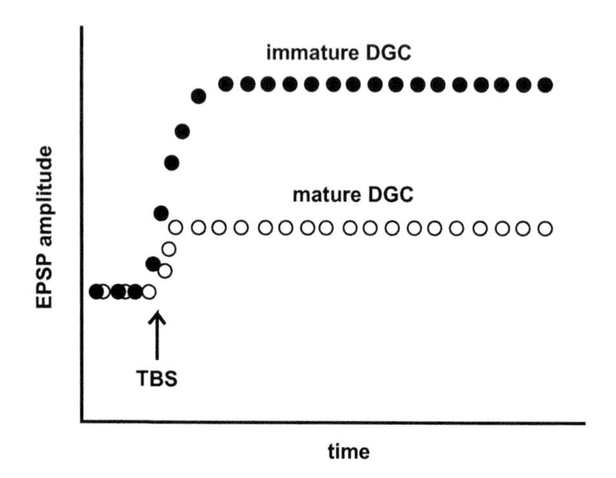

FIGURE 2.2 Example of hyperplasticity of newborn DGCs. Theta burst stimulation (TBS) results in more robust LTP at immature (2-week-old) compared to mature (4-month-old) DGCs, as measured by amplitude of excitatory postsynaptic potential (EPSP). *Adapted from Ge, S., et al. (2007). A Critical Period for Enhanced Synaptic Plasticity in Newly Generated Neurons of the Adult Brain. Neuron, 54(4), 559–566.*

Darcy, Trouche, Jin, and Feig (2014) recently elucidated another mediator of hyperplasticity in newborn DGCs. Using mice deficient in the calcium-activated guanine nucleotide exchange factor Ras-GRF2, the authors showed that GRF2 controls early survival as well as enhanced LTP of DGCs in the critical period via activation of the Erk MAP kinase pathway. Interestingly, they claim that this pathway is "parallel" and "distinct" from the NMDAR activation described above. While GluN2B inhibition blocked LTP (replicating previous findings), Erk activation was unaffected in this case. Further work will be necessary to identify downstream targets of GRF2 and the immature NMDAR in adult-born DGCs.

Long-term depression (LTD) also occurs at synapses within the hippocampus and plays an important role in the overall plasticity, physiology, and function of the DG. In contrast to the large depolarization required for LTP, LTD occurs in the context of small presynaptic currents that lead to only a modest amount of vesicular glutamate release onto the postsynaptic terminal. The outcome is dephosphorylation (deactivation) and internalization of AMPARs, such that a greater presynaptic signal is subsequently required to elicit a postsynaptic response of equal magnitude to that of the original, ie, a weakening of synaptic strength. The two basic types of LTD are heterosynaptic, in which the strength of connection is only depressed at inactive synapses, and homosynaptic, in which the strength of connection is only depressed at active synapses. Whereas heterosynaptic LTD is mediated by protein kinase C and the phosphatidylinositol-4,5-bisphosphate 3-kinase pathway, homosynaptic LTD involves protein kinase A and its interaction with the A kinase-anchoring protein AKAP79/150 (Daw et al., 2002; Snyder et al., 2005). Heterosynaptic LTD occurs in the DG between EC cells and DGCs of the perforant pathway and homosynaptic LTD occurs at the Schaffer collateral synapse between CA3 and CA1 pyramidal neurons (Bear & Abraham, 1996).

LTD appears to have an important effect on hippocampal neurogenesis under experimental conditions. Chun, Sun, and Jung (2009) reported that LTD "priming" via low-frequency stimulation at the perforant pathway–granule cell synapse in adult rats impaired LTP induction the following day, and blunted progenitor cell proliferation, which is normally significantly elevated following LTP induction. The researchers did not distinguish between newborn versus mature neurons with their LTD/LTP induction, however, leaving open the question of what combination of synapses (mature and immature) mediated this effect. Lisman (2011) proposed a possible role of LTD in DG physiology. As newborn neurons start to compete for connections within the hippocampal circuit, some will be more successful than others. LTP and LTD may serve to winnow the relatively more successful from the relatively less successful. To "weed out" neurons that have formed fewer and/or weaker connections, these may get even weaker over time via LTD, eventually leading to the isolation and apoptosis of these neurons. Conversely, neurons that have formed more and/or stronger connections may be selectively preserved and their connections enhanced via LTP. Thus, small initial differences in number/strength of synapses between newborn DGCs and input/output neurons become exaggerated over time, allowing a few "lucky" neurons to survive and "join the family," while the rest perish alone, rejected by their family members.

So far, we have focused on inputs onto newborn DGC dendrites. Equally important and equally plastic during the critical period are their axonal outputs. Using a retrovirus to birth date and deliver the stimulatory light-gated ion channel Channelrhodopsin-2 to dividing neural precursors, we found that newborn DGCs form glutamatergic synaptic output connections with CA3 pyramidal neurons by 2 weeks of age, which become stable by 4 weeks (Gu et al., 2012). Theta burst optical stimulation revealed enhanced plasticity (as measured by efficacy of LTP induction) of 4-week-old compared to 8-week-old neurons. Importantly, functionally silencing 4-week-old (but not 2- or 8-week-old) DGCs with the inhibitory light-driven proton pump archaerhodopsin-3 led to impairment of spatial learning, suggesting a critical age-specific role for these cells in hippocampus-dependent memory.

In sum, immature DGCs exhibit a critical period during the first few weeks of life, characterized by hyperexcitability and enhanced synaptic plasticity that appear to be due to GABAergic stimulation and NMDAR subtype specificity and stimulation. The switch to the more mature glutamatergic phenotype occurs in an activity-dependent fashion, reinforcing the idea that increased connectivity and excitation lead to enhanced maturation and survival. Demonstration of this hyperplasticity allows us to ask the next important question: what significance does it have on the circuit and the behavior of the organism?

THE FUNCTION OF ADULT-BORN DENTATE GRANULE CELLS IN CIRCUITS AND BEHAVIOR

The more we study adult neurogenesis, plasticity, and behavior, the more we have come to realize that these factors are inextricably linked. Just as enhanced plasticity can promote proliferation and survival of newborn DGCs, so too can neurogenesis affect plasticity. And not only do neurogenesis and plasticity affect behavior, but behavior can also feedback to modulate the other two. Proliferation and survival of adult-born DGCs are differentially regulated, with proliferation depending on NMDAR activation, and survival mediated by upregulation of the transcription factor Zif268. The fact that adult-born DGCs express Zif268, a marker of neuronal activity, during the learning process suggests that these cells may be particularly involved in processing of stimuli and memory encoding (Bruel-Jungerman, Davis, Rampon, & Laroche, 2006; Chun, Sun, Park, & Jung, 2006; Snyder, Clifford, Jeurling, & Cameron, 2012; Veyrac et al., 2013). Comparing rats that learned hippocampus-dependent memory tasks to those that did not receive training, Gould et al. (1999) found that learning increased neurogenesis, as measured with BrdU labeling of cells in the DG. That learning increases neurogenesis has been replicated in various studies (Dobrossy et al., 2003; Leuner et al., 2006, 2004). Experience thus regulates functional integration and recruitment of adult-born DGCs.

The logic may seem counterintuitive. Increased neurogenesis, we might reason, should increase learning, rather than the other way around, as more newborn neurons going "on line" could mean greater potential to encode memories. The reality is a reciprocal relationship: learning increases neurogenesis, which increases learning. A correlation between neurogenesis and learning has been established, as strains of mice that naturally exhibit lower levels of neurogenesis tend to perform worse on tasks of spatial learning compared to those with higher baseline neurogenesis (Kempermann & Gage, 2002). Interestingly, neurogenesis can increase learning, with factors that increase neurogenesis, such as exercise or ablation of the proapoptosis gene BAX, leading to improvements on hippocampus-dependent memory tasks (Sahay et al., 2011; van Praag, Christie, et al., 1999; van Praag et al., 2005). Additionally, ablation of neurogenesis with the antimitotic agent MAM impairs learning acquisition (Shors, Townsend, Zhao, Kozorovitskiy, & Gould, 2002). In sum, plasticity, neurogenesis, and learning appear to be mutually interdependent (Bruel-Jungerman et al., 2006; Sahay et al., 2011; Yau & So, 2014).

An additional mechanism influencing plasticity and potentially also neurogenesis is the contribution of the hippocampal theta rhythm (Vanderwolf, 1969). This slow (4–10 Hz) oscillation reflects global synaptic activity and contributions of individual cells, including newborn DGCs. Theta waves occur during locomotor activity and have been implicated in memory processing (Raghavachari et al., 2001). They are observed in several brain structures, but are strongest in the DG → CA3 → CA1 circuit (Bullock, Buzsaki, & McClune, 1990). Rhythmic stimulation at theta frequency has been shown to be optimal for the induction of LTP (Greenstein, Pavlides, & Winson, 1988). Why theta stimulation is so efficient at inducing LTP is unclear, but it has been proposed to synchronize depolarization of pyramidal cell dendrites with fast spikes, which is necessary for LTP (Harris, Hirase, Leinekugel, Henze, & Buzsaki, 2001; Paulsen & Sejnowski, 2000). Hippocampal theta therefore likely contributes to synaptic plasticity and possibly has a role in neurogenesis.

At this point, we might ask: how do newborn cells contribute to existing circuits to modify behaviors such as learning? The DG receives its primary input from the EC and then projects to CA3, which in turn projects to

CA1, which then projects back to cortex to form a loop (Amaral & Witter, 1989; Yeckel & Berger, 1990). Recent work has shed light on how newborn DGCs fit within this framework. Vivar, Traub, and Gutierrez (2012) combined retroviral labeling of newborn cells with retrograde synaptic tracing via the rabies virus to discover that input to newborn DGCs goes beyond EC, and evolves as these cells mature. The 2- to 3-week-old cells receive most of their excitatory input from mature DGCs, mossy cells of the hilus, cholinergic cells of the septum, and area CA3 (thus connections to CA3 are bidirectional, since DGCs also send their axons to CA3). By 4 weeks, newborn DGCs receive comparatively less input from mature DGCs while input from EC begins to strengthen, and by 2–3 months, input from mature DGCs appears to be absent. Cells at this age continue to receive input from hilar mossy cells and CA3, but their predominant input is EC. EC projections can be divided into the lateral (LPP) and medial (MPP) perforant pathways, arising from the lateral (LEC) and medial (MEC) EC, respectively. The distinction between LPP and MPP is important, since LEC is thought to integrate sensory information about the environment and provide information on familiarity, while MEC is thought to provide more direct spatial information (Knierim, Neunuebel, & Deshmukh, 2014). Newborn DGCs, in contrast to mature DGCs, respond preferentially to LEC projections (Shimazu et al., 2006; Vivar et al., 2012). This preferential input from LEC may implicate newborn DGCs as specialized, relative to mature DGCs, in behaviors such as novel object recognition and pattern separation (Guo et al., 2011; Sahay et al., 2011). The changing inputs to newborn DGCs as they mature may explain the findings of an age-dependent role of these cells in hippocampus-dependent behaviors (Gu et al., 2012).

CONCLUSIONS

Plasticity is a defining property of the hippocampus, an intrinsic part of its physiology, and the root of its function. A constant supply of new neurons dividing and hooking into the established circuit is necessary for normal memory formation. What makes newborn neurons of the DG uniquely suited to contribute to hippocampal function are their "eager" and "extraverted" personalities—their hyperexcitability and the ease with which they can form synapses and undergo hyperplasticity in an environment of older, more "brittle," and less "talkative" granule neurons. While much remains to be discovered about the roles and properties of hippocampal plasticity, perhaps the most pressing issues involve how this process becomes disrupted in pathological conditions of memory impairment such as Alzheimer disease or traumatic brain injury, and how insights into hippocampal physiology may guide the development of novel treatments that rescue plasticity and function.

Acknowledgment

We would like to acknowledge Yan Gu for his discussion and feedback on this work.

References

Aimone, J. B., Deng, W., & Gage, F. H. (2011). Resolving new memories: a critical look at the dentate gyrus, adult neurogenesis, and pattern separation. *Neuron, 70*, 589–596.

Alger, B. E., & Teyler, T. J. (1976). Long-term and short-term plasticity in the CA1, CA3, and dentate regions of the rat hippocampal slice. *Brain Research, 110*, 463–480.

Altman, J., & Bayer, S. A. (1990). Migration and distribution of two populations of hippocampal granule cell precursors during the perinatal and postnatal periods. *The Journal of Comparative Neurology, 301*, 365–381.

Altman, J., & Das, G. D. (1965). Autoradiographic and histological evidence of postnatal hippocampal neurogenesis in rats. *The Journal of Comparative Neurology, 124*, 319–335.

Alvarez-Buylla, A., Garcia-Verdugo, J. M., & Tramontin, A. D. (2001). A unified hypothesis on the lineage of neural stem cells. *Nature Reviews Neuroscience, 2*, 287–293.

Amaral, D. G., & Witter, M. P. (1989). The three-dimensional organization of the hippocampal formation: a review of anatomical data. *Neuroscience, 31*, 571–591.

Arruda-Carvalho, M., Restivo, L., Guskjolen, A., Epp, J. R., Elgersma, Y., Josselyn, S. A., et al. (2014). Conditional deletion of alpha-CaMKII impairs integration of adult-generated granule cells into dentate gyrus circuits and hippocampus-dependent learning. *The Journal of Neuroscience: The Official Journal of the Society for Neuroscience, 34*, 11919–11928.

Bannerman, D. M., Rawlins, J. N., McHugh, S. B., Deacon, R. M., Yee, B. K., Bast, T., et al. (2004). Regional dissociations within the hippocampus–memory and anxiety. *Neuroscience and Biobehavioral Reviews, 28*, 273–283.

Barria, A., & Malinow, R. (2005). NMDA receptor subunit composition controls synaptic plasticity by regulating binding to CaMKII. *Neuron, 48*, 289–301.

Barth, A. L., & Malenka, R. C. (2001). NMDAR EPSC kinetics do not regulate the critical period for LTP at thalamocortical synapses. *Nature Neuroscience, 4*, 235–236.

Baudry, M., Zhu, G., Liu, Y., Wang, Y., Briz, V., & Bi, X. (2014). Multiple cellular cascades participate in long-term potentiation and in hippocampus-dependent learning. *Brain Research, 1621*, 73–81.

Bear, M. F., & Abraham, W. C. (1996). Long-term depression in hippocampus. *Annual Review of Neuroscience, 19*, 437–462.

Bekiari, C., Giannakopoulou, A., Siskos, N., Grivas, I., Tsingotjidou, A., Michaloudi, H., et al. (2014). Neurogenesis in the septal and temporal part of the adult rat dentate gyrus. *Hippocampus, 25*(4), 511–523.

Ben-Ari, Y., Khazipov, R., Leinekugel, X., Caillard, O., & Gaiarsa, J. L. (1997). $GABA_A$, NMDA and AMPA receptors: a developmentally regulated 'menage a trois'. *Trends in Neurosciences, 20*, 523–529.

Bhatt, A. J., Feng, Y., Wang, J., Famuyide, M., & Hersey, K. (2013). Dexamethasone induces apoptosis of progenitor cells in the subventricular zone and dentate gyrus of developing rat brain. *Journal of Neuroscience Research, 91*, 1191–1202.

Biebl, M., Cooper, C. M., Winkler, J., & Kuhn, H. G. (2000). Analysis of neurogenesis and programmed cell death reveals a self-renewing capacity in the adult rat brain. *Neuroscience Letters, 291*, 17–20.

Bliss, T. V., & Lomo, T. (1973). Long-lasting potentiation of synaptic transmission in the dentate area of the anaesthetized rabbit following stimulation of the perforant path. *The Journal of Physiology, 232*, 331–356.

Breunig, J. J., Sarkisian, M. R., Arellano, J. I., Morozov, Y. M., Ayoub, A. E., Sojitra, S., et al. (2008). Primary cilia regulate hippocampal neurogenesis by mediating sonic hedgehog signaling. *Proceedings of the National Academy of Sciences of the United States of America, 105*, 13127–13132.

Brown, J. P., Couillard-Despres, S., Cooper-Kuhn, C. M., Winkler, J., Aigner, L., & Kuhn, H. G. (2003). Transient expression of doublecortin during adult neurogenesis. *The Journal of Comparative Neurology, 467*, 1–10.

Bruel-Jungerman, E., Davis, S., Rampon, C., & Laroche, S. (2006). Long-term potentiation enhances neurogenesis in the adult dentate gyrus. *The Journal of Neuroscience: The Official Journal of the Society for Neuroscience, 26*, 5888–5893.

Bullock, T. H., Buzsaki, G., & McClune, M. C. (1990). Coherence of compound field potentials reveals discontinuities in the CA1-subiculum of the hippocampus in freely-moving rats. *Neuroscience, 38*, 609–619.

Cameron, H. A., & Gould, E. (1994). Adult neurogenesis is regulated by adrenal steroids in the dentate gyrus. *Neuroscience, 61*, 203–209.

Cameron, H. A., Woolley, C. S., McEwen, B. S., & Gould, E. (1993). Differentiation of newly born neurons and glia in the dentate gyrus of the adult rat. *Neuroscience, 56*, 337–344.

Chancey, J. H., Adlaf, E. W., Sapp, M. C., Pugh, P. C., Wadiche, J. I., & Overstreet-Wadiche, L. S. (2013). GABA depolarization is required for experience-dependent synapse unsilencing in adult-born neurons. *The Journal of Neuroscience: The Official Journal of the Society for Neuroscience, 33*, 6614–6622.

Chancey, J. H., Poulsen, D. J., Wadiche, J. I., & Overstreet-Wadiche, L. (2014). Hilar mossy cells provide the first glutamatergic synapses to adult-born dentate granule cells. *The Journal of Neuroscience: The Official Journal of the Society for Neuroscience, 34*, 2349–2354.

Chiang, P. H., Wu, P. Y., Kuo, T. W., Liu, Y. C., Chan, C. F., Chien, T. C., et al. (2012). GABA is depolarizing in hippocampal dentate granule cells of the adolescent and adult rats. *The Journal of Neuroscience: The Official Journal of the Society for Neuroscience, 32*, 62–67.

Chun, S. K., Sun, W., Park, J. J., & Jung, M. W. (2006). Enhanced proliferation of progenitor cells following long-term potentiation induction in the rat dentate gyrus. *Neurobiology of Learning and Memory, 86*, 322–329.

Chun, S. K., Sun, W., & Jung, M. W. (2009). LTD induction suppresses LTP-induced hippocampal adult neurogenesis. *Neuroreport, 20*, 1279–1283.

Corbit, K. C., Shyer, A. E., Dowdle, W. E., Gaulden, J., Singla, V., Chen, M. H., et al. (2008). Kif3a constrains beta-catenin-dependent Wnt signalling through dual ciliary and non-ciliary mechanisms. *Nature Cell Biology, 10*, 70–76.

Czeh, B., Seress, L., Nadel, L., & Bures, J. (1998). Lateralized fascia dentata lesion and blockade of one hippocampus: effect on spatial memory in rats. *Hippocampus, 8*, 647–650.

Darcy, M. J., Trouche, S., Jin, S. X., & Feig, L. A. (2014). Ras-GRF2 mediates long-term potentiation, survival, and response to an enriched environment of newborn neurons in the hippocampus. *Hippocampus, 24*, 1317–1329.

Daw, M. I., Bortolotto, Z. A., Saulle, E., Zaman, S., Collingridge, G. L., & Isaac, J. T. (2002). Phosphatidylinositol 3 kinase regulates synapse specificity of hippocampal long-term depression. *Nature Neuroscience, 5*, 835–836.

Dayer, A. G., Ford, A. A., Cleaver, K. M., Yassaee, M., & Cameron, H. A. (2003). Short-term and long-term survival of new neurons in the rat dentate gyrus. *The Journal of Comparative Neurology, 460*, 563–572.

Dobrossy, M. D., Drapeau, E., Aurousseau, C., Le Moal, M., Piazza, P. V., & Abrous, D. N. (2003). Differential effects of learning on neurogenesis: learning increases or decreases the number of newly born cells depending on their birth date. *Molecular Psychiatry, 8*, 974–982.

Elias, L. A., Turmaine, M., Parnavelas, J. G., & Kriegstein, A. R. (2010). Connexin 43 mediates the tangential to radial migratory switch in ventrally derived cortical interneurons. *The Journal of Neuroscience: The Official Journal of the Society for Neuroscience, 30*, 7072–7077.

Enomoto, A., Asai, N., Namba, T., Wang, Y., Kato, T., Tanaka, M., et al. (2009). Roles of disrupted-in-schizophrenia 1-interacting protein girdin in postnatal development of the dentate gyrus. *Neuron, 63*, 774–787.

Esposito, M. S., Piatti, V. C., Laplagne, D. A., Morgenstern, N. A., Ferrari, C. C., Pitossi, F. J., et al. (2005). Neuronal differentiation in the adult hippocampus recapitulates embryonic development. *The Journal of Neuroscience: The Official Journal of the Society for Neuroscience, 25*, 10074–10086.

Faulkner, R. L., Jang, M. H., Liu, X. B., Duan, X., Sailor, K. A., Kim, J. Y., et al. (2008). Development of hippocampal mossy fiber synaptic outputs by new neurons in the adult brain. *Proceedings of the National Academy of Sciences of the United States of America, 105*, 14157–14162.

Filipovic, R., Santhosh Kumar, S., Fiondella, C., & Loturco, J. (2012). Increasing doublecortin expression promotes migration of human embryonic stem cell-derived neurons. *Stem Cells, 30*, 1852–1862.

Gage, F. H., Kempermann, G., Palmer, T. D., Peterson, D. A., & Ray, J. (1998). Multipotent progenitor cells in the adult dentate gyrus. *Journal of Neurobiology, 36*, 249–266.

Gage, F. H. (2000). Mammalian neural stem cells. *Science, 287*, 1433–1438.

Garcia, A., Steiner, B., Kronenberg, G., Bick-Sander, A., & Kempermann, G. (2004). Age-dependent expression of glucocorticoid- and mineralocorticoid receptors on neural precursor cell populations in the adult murine hippocampus. *Aging Cell, 3*, 363–371.

Gascon, E., Vutskits, L., Jenny, B., Durbec, P., & Kiss, J. Z. (2007). PSA-NCAM in postnatally generated immature neurons of the olfactory bulb: a crucial role in regulating p75 expression and cell survival. *Development, 134*, 1181–1190.

Gascon, E., Vutskits, L., & Kiss, J. Z. (2007). Polysialic acid-neural cell adhesion molecule in brain plasticity: from synapses to integration of new neurons. *Brain Research Reviews, 56*, 101–118.

Ge, S., Goh, E. L., Sailor, K. A., Kitabatake, Y., Ming, G. L., & Song, H. (2006). GABA regulates synaptic integration of newly generated neurons in the adult brain. *Nature, 439*, 589–593.

Ge, S., Pradhan, D. A., Ming, G. L., & Song, H. (2007). GABA sets the tempo for activity-dependent adult neurogenesis. *Trends in Neurosciences, 30*, 1–8.

Ge, S., Sailor, K. A., Ming, G. L., & Song, H. (2008). Synaptic integration and plasticity of new neurons in the adult hippocampus. *The Journal of Physiology, 586*, 3759–3765.

Gilbert, P. E., Kesner, R. P., & Lee, I. (2001). Dissociating hippocampal subregions: double dissociation between dentate gyrus and CA1. *Hippocampus, 11*, 626–636.

Gleeson, J. G., Lin, P. T., Flanagan, L. A., & Walsh, C. A. (1999). Doublecortin is a microtubule-associated protein and is expressed widely by migrating neurons. *Neuron, 23*, 257–271.

Gotz, M., Stoykova, A., & Gruss, P. (1998). Pax6 controls radial glia differentiation in the cerebral cortex. *Neuron, 21*, 1031–1044.

Gould, E., & Cameron, H. A. (1996). Regulation of neuronal birth, migration and death in the rat dentate gyrus. *Developmental Neuroscience, 18*, 22–35.

Gould, E., & Tanapat, P. (1999). Stress and hippocampal neurogenesis. *Biological Psychiatry, 46*, 1472–1479.

Gould, E., Woolley, C. S., & McEwen, B. S. (1991). Adrenal steroids regulate postnatal development of the rat dentate gyrus: I. Effects of glucocorticoids on cell death. *The Journal of Comparative Neurology, 313*, 479–485.

Gould, E., Tanapat, P., & McEwen, B. S. (1997). Activation of the type 2 adrenal steroid receptor can rescue granule cells from death during development. *Brain Research Developmental Brain Research, 101*, 265–268.

Gould, E., Reeves, A. J., Fallah, M., Tanapat, P., Gross, C. G., & Fuchs, E. (1999). Hippocampal neurogenesis in adult Old World primates. *Proceedings of the National Academy of Sciences of the United States of America, 96*, 5263–5267.

Greenstein, Y. J., Pavlides, C., & Winson, J. (1988). Long-term potentiation in the dentate gyrus is preferentially induced at theta rhythm periodicity. *Brain Research, 438*, 331–334.

Gu, Y., Arruda-Carvalho, M., Wang, J., Janoschka, S. R., Josselyn, S. A., Frankland, P. W., et al. (2012). Optical controlling reveals time-dependent roles for adult-born dentate granule cells. *Nature Neuroscience, 15*, 1700–1706.

Guo, W., Allan, A. M., Zong, R., Zhang, L., Johnson, E. B., Schaller, E. G., et al. (2011). Ablation of Fmrp in adult neural stem cells disrupts hippocampus-dependent learning. *Nature Medicine, 17*, 559–565.

Guzman-Marin, R., Suntsova, N., Stewart, D. R., Gong, H., Szymusiak, R., & McGinty, D. (2003). Sleep deprivation reduces proliferation of cells in the dentate gyrus of the hippocampus in rats. *The Journal of Physiology, 549*, 563–571.

Hack, M. A., Saghatelyan, A., de Chevigny, A., Pfeifer, A., Ashery-Padan, R., Lledo, P. M., et al. (2005). Neuronal fate determinants of adult olfactory bulb neurogenesis. *Nature Neuroscience, 8*, 865–872.

Harris, K. D., Hirase, H., Leinekugel, X., Henze, D. A., & Buzsaki, G. (2001). Temporal interaction between single spikes and complex spike bursts in hippocampal pyramidal cells. *Neuron, 32*, 141–149.

Hastings, N. B., & Gould, E. (1999). Rapid extension of axons into the CA3 region by adult-generated granule cells. *The Journal of Comparative Neurology, 413*, 146–154.

Heine, V. M., Zareno, J., Maslam, S., Joels, M., & Lucassen, P. J. (2005). Chronic stress in the adult dentate gyrus reduces cell proliferation near the vasculature and VEGF and Flk-1 protein expression. *The European Journal of Neuroscience, 21*, 1304–1314.

Imayoshi, I., Sakamoto, M., Ohtsuka, T., Takao, K., Miyakawa, T., Yamaguchi, M., et al. (2008). Roles of continuous neurogenesis in the structural and functional integrity of the adult forebrain. *Nature Neuroscience, 11*, 1153–1161.

Jacobs, C. M., Trinh, M. D., Rootwelt, T., Lomo, J., & Paulsen, R. E. (2006). Dexamethasone induces cell death which may be blocked by NMDA receptor antagonists but is insensitive to Mg^{2+} in cerebellar granule neurons. *Brain Research, 1070*, 116–123.

Johansson, I., Destefanis, S., Aberg, N. D., Aberg, M. A., Blomgren, K., Zhu, C., et al. (2008). Proliferative and protective effects of growth hormone secretagogues on adult rat hippocampal progenitor cells. *Endocrinology, 149*, 2191–2199.

Kazim, S. F., Blanchard, J., Dai, C. L., Tung, Y. C., LaFerla, F. M., Iqbal, I. G., et al. (2014). Disease modifying effect of chronic oral treatment with a neurotrophic peptidergic compound in a triple transgenic mouse model of Alzheimer's disease. *Neurobiology of Disease, 71*, 110–130.

Kempermann, G., & Gage, F. H. (2002). Genetic influence on phenotypic differentiation in adult hippocampal neurogenesis. *Brain Research Developmental Brain Research, 134*, 1–12.

Kempermann, G., Kuhn, H. G., & Gage, F. H. (1998). Experience-induced neurogenesis in the senescent dentate gyrus. *The Journal of Neuroscience: The Official Journal of the Society for Neuroscience, 18*, 3206–3212.

Kempermann, G., Jessberger, S., Steiner, B., & Kronenberg, G. (2004). Milestones of neuronal development in the adult hippocampus. *Trends in Neurosciences, 27*, 447–452.

Knierim, J. J., Neunuebel, J. P., & Deshmukh, S. S. (2014). Functional correlates of the lateral and medial entorhinal cortex: objects, path integration and local-global reference frames. *Philosophical Transactions of the Royal Society of London Series B, Biological Sciences, 369*, 20130369.

Koizumi, H., Higginbotham, H., Poon, T., Tanaka, T., Brinkman, B. C., & Gleeson, J. G. (2006). Doublecortin maintains bipolar shape and nuclear translocation during migration in the adult forebrain. *Nature Neuroscience, 9*, 779–786.

Kronenberg, G., Bick-Sander, A., Bunk, E., Wolf, C., Ehninger, D., & Kempermann, G. (2006). Physical exercise prevents age-related decline in precursor cell activity in the mouse dentate gyrus. *Neurobiology of Aging, 27*, 1505–1513.

Krugers, H. J., van der Linden, S., van Olst, E., Alfarez, D. N., Maslam, S., Lucassen, P. J., et al. (2007). Dissociation between apoptosis, neurogenesis, and synaptic potentiation in the dentate gyrus of adrenalectomized rats. *Synapse, 61*, 221–230.

Kumamoto, N., Gu, Y., Wang, J., Janoschka, S., Takemaru, K., Levine, J., et al. (2012). A role for primary cilia in glutamatergic synaptic integration of adult-born neurons. *Nature Neuroscience, 15*, 399–405, S391.

Kunze, A., Congreso, M. R., Hartmann, C., Wallraff-Beck, A., Huttmann, K., Bedner, P., et al. (2009). Connexin expression by radial glia-like cells is required for neurogenesis in the adult dentate gyrus. *Proceedings of the National Academy of Sciences of the United States of America, 106*, 11336–11341.

Lagace, D. C., Whitman, M. C., Noonan, M. A., Ables, J. L., DeCarolis, N. A., Arguello, A. A., et al. (2007). Dynamic contribution of nestin-expressing stem cells to adult neurogenesis. *The Journal of Neuroscience: The Official Journal of the Society for Neuroscience, 27*, 12623–12629.

Laplagne, D. A., Esposito, M. S., Piatti, V. C., Morgenstern, N. A., Zhao, C., van Praag, H., et al. (2006). Functional convergence of neurons generated in the developing and adult hippocampus. *PLoS Biology, 4*, e409.

Lee, J., Duan, W., Long, J. M., Ingram, D. K., & Mattson, M. P. (2000). Dietary restriction increases the number of newly generated neural cells, and induces BDNF expression, in the dentate gyrus of rats. *Journal of Molecular Neuroscience, 15*, 99–108.

Leinekugel, X., Medina, I., Khalilov, I., Ben-Ari, Y., & Khazipov, R. (1997). Ca^{2+} oscillations mediated by the synergistic excitatory actions of GABA(A) and NMDA receptors in the neonatal hippocampus. *Neuron, 18*, 243–255.

Leuner, B., Mendolia-Loffredo, S., Kozorovitskiy, Y., Samburg, D., Gould, E., & Shors, T. J. (2004). Learning enhances the survival of new neurons beyond the time when the hippocampus is required for memory. *The Journal of Neuroscience: The Official Journal of the Society for Neuroscience, 24*, 7477–7481.

Leuner, B., Gould, E., & Shors, T. J. (2006). Is there a link between adult neurogenesis and learning? *Hippocampus, 16*, 216–224.

Leuner, B., Caponiti, J. M., & Gould, E. (2012). Oxytocin stimulates adult neurogenesis even under conditions of stress and elevated glucocorticoids. *Hippocampus, 22*, 861–868.

Liebmann, M., Stahr, A., Guenther, M., Witte, O. W., & Frahm, C. (2013). Astrocytic Cx43 and Cx30 differentially modulate adult neurogenesis in mice. *Neuroscience Letters, 545*, 40–45.

Linnarsson, S., Willson, C. A., & Ernfors, P. (2000). Cell death in regenerating populations of neurons in BDNF mutant mice. *Brain Research Molecular Brain Research, 75*, 61–69.

Lisman, J. (2011). Formation of the non-functional and functional pools of granule cells in the dentate gyrus: role of neurogenesis, LTP and LTD. *The Journal of Physiology, 589*, 1905–1909.

Lois, C., & Alvarez-Buylla, A. (1994). Long-distance neuronal migration in the adult mammalian brain. *Science, 264*, 1145–1148.

Lomo, T. (1971). Potentiation of monosynaptic EPSPs in the perforant path-dentate granule cell synapse. *Experimental Brain Research, 12*, 46–63.

Lowenstein, D. H., & Arsenault, L. (1996). The effects of growth factors on the survival and differentiation of cultured dentate gyrus neurons. *The Journal of Neuroscience: The Official Journal of the Society for Neuroscience, 16*, 1759–1769.

Marin-Burgin, A., Mongiat, L. A., Pardi, M. B., & Schinder, A. F. (2012). Unique processing during a period of high excitation/inhibition balance in adult-born neurons. *Science, 335*, 1238–1242.

Markwardt, S. J., Wadiche, J. I., & Overstreet-Wadiche, L. S. (2009). Input-specific GABAergic signaling to newborn neurons in adult dentate gyrus. *The Journal of Neuroscience: The Official Journal of the Society for Neuroscience, 29*, 15063–15072.

McEwen, B. S., Cameron, H., Chao, H. M., Gould, E., Luine, V., Magarinos, A. M., et al. (1994). Resolving a mystery: progress in understanding the function of adrenal steroid receptors in hippocampus. *Progress in Brain Research, 100*, 149–155.

Minichiello, L., & Klein, R. (1996). TrkB and TrkC neurotrophin receptors cooperate in promoting survival of hippocampal and cerebellar granule neurons. *Genes and Development, 10*, 2849–2858.

Mongiat, L. A., Esposito, M. S., Lombardi, G., & Schinder, A. F. (2009). Reliable activation of immature neurons in the adult hippocampus. *PloS One, 4*, e5320.

Morrens, J., Van Den Broeck, W., & Kempermann, G. (2012). Glial cells in adult neurogenesis. *Glia, 60*, 159–174.

Morris, R. G., Hagan, J. J., & Rawlins, J. N. (1986). Allocentric spatial learning by hippocampectomised rats: a further test of the "spatial mapping" and "working memory" theories of hippocampal function. *The Quarterly Journal of Experimental Psychology B, Comparative and Physiological Psychology, 38*, 365–395.

Moser, M. B., Moser, E. I., Forrest, E., Andersen, P., & Morris, R. G. (1995). Spatial learning with a minislab in the dorsal hippocampus. *Proceedings of the National Academy of Sciences of the United States of America, 92*, 9697–9701.

Muller, S., Chakrapani, B. P., Schwegler, H., Hofmann, H. D., & Kirsch, M. (2009). Neurogenesis in the dentate gyrus depends on ciliary neurotrophic factor and signal transducer and activator of transcription 3 signaling. *Stem Cells, 27*, 431–441.

Namba, T., Mochizuki, H., Onodera, M., Mizuno, Y., Namiki, H., & Seki, T. (2005). The fate of neural progenitor cells expressing astrocytic and radial glial markers in the postnatal rat dentate gyrus. *The European Journal of Neuroscience, 22*, 1928–1941.

Orchinik, M., Weiland, N. G., & McEwen, B. S. (1994). Adrenalectomy selectively regulates $GABA_A$ receptor subunit expression in the hippocampus. *Molecular and Cellular Neurosciences, 5*, 451–458.

Overstreet Wadiche, L., Bromberg, D. A., Bensen, A. L., & Westbrook, G. L. (2005). GABAergic signaling to newborn neurons in dentate gyrus. *Journal of Neurophysiology, 94*, 4528–4532.

Paoletti, P., Bellone, C., & Zhou, Q. (2013). NMDA receptor subunit diversity: impact on receptor properties, synaptic plasticity and disease. *Nature Reviews Neuroscience, 14*, 383–400.

Paulsen, O., & Sejnowski, T. J. (2000). Natural patterns of activity and long-term synaptic plasticity. *Current Opinion in Neurobiology, 10*, 172–179.

Peissner, W., Kocher, M., Treuer, H., & Gillardon, F. (1999). Ionizing radiation-induced apoptosis of proliferating stem cells in the dentate gyrus of the adult rat hippocampus. *Brain Research Molecular Brain Research, 71*, 61–68.

Piatti, V. C., Davies-Sala, M. G., Esposito, M. S., Mongiat, L. A., Trinchero, M. F., & Schinder, A. F. (2011). The timing for neuronal maturation in the adult hippocampus is modulated by local network activity. *The Journal of Neuroscience: The Official Journal of the Society for Neuroscience, 31*, 7715–7728.

Raghavachari, S., Kahana, M. J., Rizzuto, D. S., Caplan, J. B., Kirschen, M. P., Bourgeois, B., et al. (2001). Gating of human theta oscillations by a working memory task. *The Journal of Neuroscience: The Official Journal of the Society for Neuroscience, 21*, 3175–3183.

Reagan, L. P., & McEwen, B. S. (1997). Controversies surrounding glucocorticoid-mediated cell death in the hippocampus. *Journal of Chemical Neuroanatomy, 13*, 149–167.

Rizzino, A. (2009). Sox2 and Oct-3/4: a versatile pair of master regulators that orchestrate the self-renewal and pluripotency of embryonic stem cells. *Wiley Interdisciplinary Reviews Systems Biology and Medicine, 1*, 228–236.

Rodenas-Ruano, A., Chavez, A. E., Cossio, M. J., Castillo, P. E., & Zukin, R. S. (2012). REST-dependent epigenetic remodeling promotes the developmental switch in synaptic NMDA receptors. *Nature Neuroscience, 15*, 1382–1390.

Sahay, A., Scobie, K. N., Hill, A. S., O'Carroll, C. M., Kheirbek, M. A., Burghardt, N. S., et al. (2011). Increasing adult hippocampal neurogenesis is sufficient to improve pattern separation. *Nature, 472*, 466–470.

Sauer, J. F., Struber, M., & Bartos, M. (2012). Interneurons provide circuit-specific depolarization and hyperpolarization. *The Journal of Neuroscience: The Official Journal of the Society for Neuroscience, 32*, 4224–4229.

Schmidt-Hieber, C., Jonas, P., & Bischofberger, J. (2004). Enhanced synaptic plasticity in newly generated granule cells of the adult hippocampus. *Nature, 429*, 184–187.

Seki, T., & Arai, Y. (1993). Distribution and possible roles of the highly polysialylated neural cell adhesion molecule (NCAM-H) in the developing and adult central nervous system. *Neuroscience Research, 17*, 265–290.

Seri, B., Garcia-Verdugo, J. M., Collado-Morente, L., McEwen, B. S., & Alvarez-Buylla, A. (2004). Cell types, lineage, and architecture of the germinal zone in the adult dentate gyrus. *The Journal of Comparative Neurology, 478*, 359–378.

Shapiro, L. A., & Ribak, C. E. (2005). Integration of newly born dentate granule cells into adult brains: hypotheses based on normal and epileptic rodents. *Brain Research. Brain Research Reviews, 48*, 43–56.

Shimazu, K., Zhao, M., Sakata, K., Akbarian, S., Bates, B., Jaenisch, R., et al. (2006). NT-3 facilitates hippocampal plasticity and learning and memory by regulating neurogenesis. *Learning and Memory, 13*, 307–315.

Shors, T. J., Townsend, D. A., Zhao, M., Kozorovitskiy, Y., & Gould, E. (2002). Neurogenesis may relate to some but not all types of hippocampal-dependent learning. *Hippocampus, 12*, 578–584.

Sierra, A., Encinas, J. M., Deudero, J. J., Chancey, J. H., Enikolopov, G., Overstreet-Wadiche, L. S., et al. (2010). Microglia shape adult hippocampal neurogenesis through apoptosis-coupled phagocytosis. *Cell Stem Cell, 7*, 483–495.

Snyder, E. M., Philpot, B. D., Huber, K. M., Dong, X., Fallon, J. R., & Bear, M. F. (2001). Internalization of ionotropic glutamate receptors in response to mGluR activation. *Nature Neuroscience, 4*, 1079–1085.

Snyder, E. M., Colledge, M., Crozier, R. A., Chen, W. S., Scott, J. D., & Bear, M. F. (2005). Role for A kinase-anchoring proteins (AKAPS) in glutamate receptor trafficking and long term synaptic depression. *The Journal of Biological Chemistry, 280*, 16962–16968.

Snyder, J. S., Clifford, M. A., Jeurling, S. I., & Cameron, H. A. (2012). Complementary activation of hippocampal-cortical subregions and immature neurons following chronic training in single and multiple context versions of the water maze. *Behavioural Brain Research, 227*, 330–339.

Snyder, J. S., Ferrante, S. C., & Cameron, H. A. (2012). Late maturation of adult-born neurons in the temporal dentate gyrus. *PloS One, 7*, e48757.

Soltesz, I., Smetters, D. K., & Mody, I. (1995). Tonic inhibition originates from synapses close to the soma. *Neuron, 14*, 1273–1283.

Song, H., Stevens, C. F., & Gage, F. H. (2002). Astroglia induce neurogenesis from adult neural stem cells. *Nature, 417*, 39–44.

Takigawa, T., & Alzheimer, C. (2002). Phasic and tonic attenuation of EPSPs by inward rectifier K+ channels in rat hippocampal pyramidal cells. *The Journal of Physiology, 539*, 67–75.

Tanaka, T., Serneo, F. F., Higgins, C., Gambello, M. J., Wynshaw-Boris, A., & Gleeson, J. G. (2004). Lis1 and doublecortin function with dynein to mediate coupling of the nucleus to the centrosome in neuronal migration. *The Journal of Cell Biology, 165*, 709–721.

Tanapat, P., Hastings, N. B., Reeves, A. J., & Gould, E. (1999). Estrogen stimulates a transient increase in the number of new neurons in the dentate gyrus of the adult female rat. *The Journal of neuroscience: The Official Journal of the Society for Neuroscience, 19*, 5792–5801.

Temprana, S. G., Mongiat, L. A., Yang, S. M., Trinchero, M. F., Alvarez, D. D., Kropff, E., et al. (2015). Delayed coupling to feedback inhibition during a critical period for the integration of adult-born granule cells. *Neuron, 85*, 116–130.

Toriyama, M., Mizuno, N., Fukami, T., Iguchi, T., Toriyama, M., Tago, K., et al. (2012). Phosphorylation of doublecortin by protein kinase A orchestrates microtubule and actin dynamics to promote neuronal progenitor cell migration. *The Journal of Biological Chemistry, 287*, 12691–12702.

Tozuka, Y., Fukuda, S., Namba, T., Seki, T., & Hisatsune, T. (2005). GABAergic excitation promotes neuronal differentiation in adult hippocampal progenitor cells. *Neuron, 47*, 803–815.

Trivedi, M. A., & Coover, G. D. (2004). Lesions of the ventral hippocampus, but not the dorsal hippocampus, impair conditioned fear expression and inhibitory avoidance on the elevated T-maze. *Neurobiology of Learning and Memory, 81*, 172–184.

Tsai, L. H., & Gleeson, J. G. (2005). Nucleokinesis in neuronal migration. *Neuron, 46*, 383–388.

van Praag, H., Christie, B. R., Sejnowski, T. J., & Gage, F. H. (1999). Running enhances neurogenesis, learning, and long-term potentiation in mice. *Proceedings of the National Academy of Sciences of the United States of America, 96*, 13427–13431.
van Praag, H., Kempermann, G., & Gage, F. H. (1999). Running increases cell proliferation and neurogenesis in the adult mouse dentate gyrus. *Nature Neuroscience, 2*, 266–270.
van Praag, H., Schinder, A. F., Christie, B. R., Toni, N., Palmer, T. D., & Gage, F. H. (2002). Functional neurogenesis in the adult hippocampus. *Nature, 415*, 1030–1034.
van Praag, H., Shubert, T., Zhao, C., & Gage, F. H. (2005). Exercise enhances learning and hippocampal neurogenesis in aged mice. *The Journal of Neuroscience: The Official Journal of the Society for Neuroscience, 25*, 8680–8685.
Vanderwolf, C. H. (1969). Hippocampal electrical activity and voluntary movement in the rat. *Electroencephalography and Clinical Neurophysiology, 26*, 407–418.
Veyrac, A., Gros, A., Bruel-Jungerman, E., Rochefort, C., Kleine Borgmann, F. B., Jessberger, S., et al. (2013). Zif268/egr1 gene controls the selection, maturation and functional integration of adult hippocampal newborn neurons by learning. *Proceedings of the National Academy of Sciences of the United States of America, 110*, 7062–7067.
Vivar, C., Traub, R. D., & Gutierrez, R. (2012). Mixed electrical-chemical transmission between hippocampal mossy fibers and pyramidal cells. *The European Journal of Neuroscience, 35*, 76–82.
Wang, D. D., & Kriegstein, A. R. (2008). GABA regulates excitatory synapse formation in the neocortex via NMDA receptor activation. *The Journal of Neuroscience: The Official Journal of the Society for Neuroscience, 28*, 5547–5558.
Wang, S., Scott, B. W., & Wojtowicz, J. M. (2000). Heterogenous properties of dentate granule neurons in the adult rat. *Journal of Neurobiology, 42*, 248–257.
Yang, P., Arnold, S. A., Habas, A., Hetman, M., & Hagg, T. (2008). Ciliary neurotrophic factor mediates dopamine D2 receptor-induced CNS neurogenesis in adult mice. *The Journal of Neuroscience: The Official Journal of the Society for Neuroscience, 28*, 2231–2241.
Yau, S. Y., & So, K. F. (2014). Adult neurogenesis and dendritic remodeling in hippocampal plasticity: which one is more important? *Cell Transplantation, 23*, 471–479.
Yeckel, M. F., & Berger, T. W. (1990). Feedforward excitation of the hippocampus by afferents from the entorhinal cortex: redefinition of the role of the trisynaptic pathway. *Proceedings of the National Academy of Sciences of the United States of America, 87*, 5832–5836.
Zhao, C., Teng, E. M., Summers, R. G., Jr., Ming, G. L., & Gage, F. H. (2006). Distinct morphological stages of dentate granule neuron maturation in the adult mouse hippocampus. *The Journal of Neuroscience: The Official Journal of the Society for Neuroscience, 26*, 3–11.
Zhu, Y., Li, H., Zhou, L., Wu, J. Y., & Rao, Y. (1999). Cellular and molecular guidance of GABAergic neuronal migration from an extracortical origin to the neocortex. *Neuron, 23*, 473–485.

CHAPTER

3

Cellular and Molecular Regulation

D.C. Lie

Friedrich-Alexander University Erlangen–Nuremberg, Erlangen, Germany

S. Jessberger

University of Zurich, Zurich, Switzerland

INTRODUCTION

The adult mammalian brain is capable of responding to changing environments and experiences. However, it has been long thought that these functional and structural changes are based on modifications of the connectivity between neurons generated during embryonic or early postnatal neurogenesis. Even though Altman and Das (1965) provided the first evidence in the early 1960s, the possibility that even in the adult mammalian brain new neurons are generated by neural stem/progenitor cells (NSPCs) was largely rejected by the community due to the hypothesis that the adult brain is too complex to support the generation and more importantly the meaningful integration of newborn neurons. However, technical advances in the mid-1990s that were largely based on the use of thymidine analogues such as bromodeoxyuridine (BrdU) to label proliferating cells and their progeny together with confocal microscopy to unambiguously phenotype newborn cells showed persistent neurogenesis in the adult mammalian brain (Kuhn & Cooper-Kuhn, 2007). However, this process, called adult neurogenesis, is not widespread but rather restricted to distinct brain areas. Within the mouse brain and rat brain, which are the most commonly studied species in the field of adult neurogenesis, NSPCs persist in two main neurogenic regions: the subventricular zone (SVZ) lining the lateral ventricles where NSPCs divide and give rise to newborn cells that migrate through the rostral migratory

Adult Neurogenesis in the Hippocampus
http://dx.doi.org/10.1016/B978-0-12-801977-1.00003-9

stream toward the olfactory bulb (OB) where newborn cells differentiate into distinct types of olfactory neurons, and the hippocampal dentate gyrus (DG) where only one subtype of glutamatergic, excitatory neurons, dentate granule cells, is generated (Ming & Song, 2011). In contrast to the rodent brain, SVZ/OB neurogenesis seems to be largely absent in the adult human brain (Bergmann et al., 2012; Curtis et al., 2007; Sanai, Berger, Garcia-Verdugo, & Alvarez-Buylla, 2007; Sanai et al., 2004); however, levels of neurogenesis remain substantial throughout life also in the human DG (Eriksson et al., 1998; Spalding et al., 2013). This review will focus on hippocampal neurogenesis and the cellular and molecular mechanisms that govern the life-long development of newborn neurons in the adult mammalian DG (Fig. 3.1).

From 1995 to 2015 tremendous efforts have been made to characterize the biology and the functional significance of adult neurogenesis for brain physiology and its role in a number of neuropsychiatric diseases.

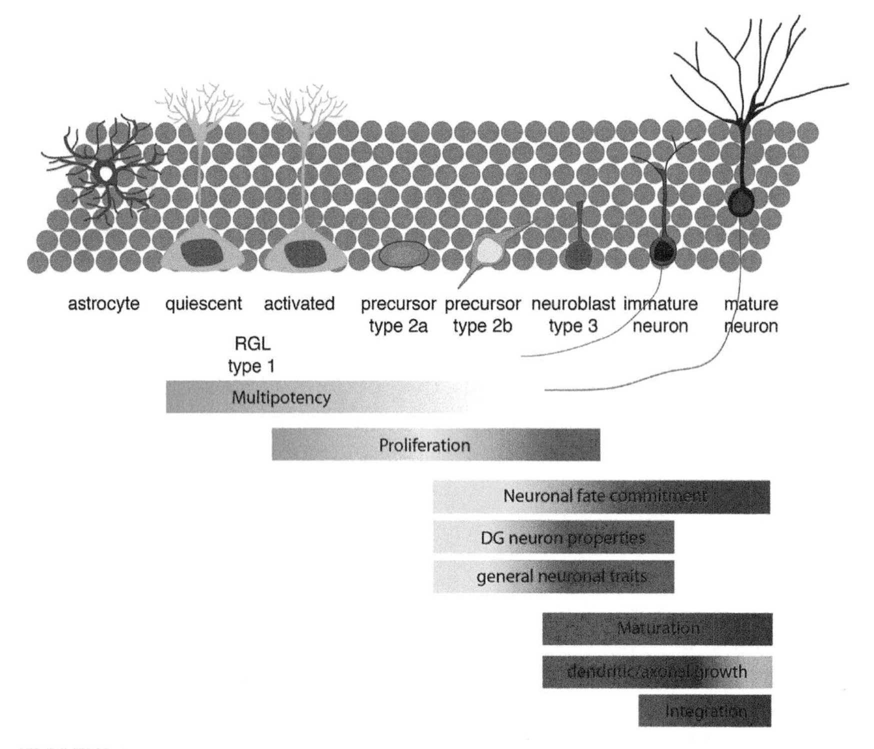

FIGURE 3.1 Neurogenesis in the dentate gyrus (DG) of the adult hippocampus. Quiescent and activated type 1 radial NSPCs (green) within the subgranular zone of the DG are assumed to be multipotent cells; multipotency may be gradually lost with differentiation. Later in the development of NSPC-derived progeny fate commitment and neuronal integration are regulated by a variety of intrinsic and extrinsic mechanisms.

Our knowledge regarding the neurogenic process in the adult brain is constantly increasing. The field is currently on the verge of relating its findings of physiology and disease from rodent models to the human brain. In the following sections we will summarize recent advances aiming to understand the basic mechanisms regulating NSPC activity and subsequent neuronal integration into the adult brain.

CELLULAR IDENTITY OF ADULT HIPPOCAMPAL NEURAL STEM/PROGENITOR CELLS

NSPCs progress through distinct developmental stages before neurons are integrated into the preexisting DG circuit. It is currently believed that radial glia-like (RGL) cells that are also called type 1 cells represent the bona fide neural stem cell in the adult DG (Kempermann, Jessberger, Steiner, & Kronenberg, 2004). These cells share certain cellular and molecular characteristics with astrocytes, such as endothelial endfeet and expression of glial fibrillary acidic protein (Seri, Garcia-Verdugo, Collado-Morente, McEwen, & Alvarez-Buylla, 2004). The vast majority of RGLs do not divide under physiological conditions but rather are kept in a quiescent state. Once activated RGLs seem to divide asymmetrically and give rise to more proliferative nonradial glia-like cells (also called type 2 cells). However, it remains controversial if RGLs have the capacity to asymmetrically divide over extended times (as suggested by clonal lineage tracing of RGLs in the adult DG) or if they divide for 3–4 times before they terminally differentiate into glial cells (as suggested by population lineage tracings) (Bonaguidi et al., 2011; Encinas et al., 2011). In other words: does the adult DG harbor a pool of NSPCs with extensive or unlimited self-renewal capacity or is the ability to expand and divide largely restricted to few cell divisions? Obviously, this is a conceptually important question to the field that currently remains unanswered. At this time, it seems plausible that the answer to this question will require longitudinal observation of labeled NSPCs over time within their niche using advanced deep tissue imaging approaches of individual clones. Despite the uncertainty regarding the cellular properties and identity of bona fide NSPCs, substantial progress has been made in characterizing the intrinsic and extrinsic signaling cues and environmental challenges that keep NSPCs in a quiescent state. This is important, similar to other somatic stem cell niches, for preventing premature exhaustion of the stem cell pool. Not surprisingly, key signaling pathways that have been implicated during embryonic development are also critically involved in maintaining adult NSPCs in a quiescent state. Among others the Notch signaling pathway has been shown to be key for NSPC quiescence as deletion of the canonical Notch signal leads to a burst of neurogenesis followed by exhaustion of the stem cell pool (Ables et al., 2010; Breunig, Silbereis, Vaccarino, Sestan, & Rakic, 2007; Ehm et al., 2010; Lugert et al., 2010). Similarly, BMP activity is critical for keeping NSPCs in a quiescent state as deletion of BMP receptor 1A that is highly expressed on RGLs leads to ectopic RGL proliferation (Mira et al., 2010). In addition to these signaling hubs, and again similar to other somatic stem cell systems, a critical role for cellular metabolism to maintain cells in a quiescent state has been recently identified. Specifically, the inhibition of the levels of de novo lipogenesis through the regulatory protein Spot14 keeps NSPCs in a nondividing state (Knobloch et al., 2013). Furthermore, chromatin modifiers and epigenetic regulators have been implicated in keeping NSPCs quiescent (Amador-Arjona et al., 2015; Jones et al., 2015; Ma et al., 2010; Raposo et al., 2015; Webb & Brunet, 2013; Zhang et al., 2013, 2014). One of the key challenges will be to analyze how these cellular regulators are integrated on a systems level, ie, if RGL proliferation is activated, for example, by epileptic activity that enhances RGL divisions. Interestingly, recent data clearly indicated that NSPCs can either directly or indirectly sense the activity of surrounding neural circuits through GABA-mediated signaling and potentially other neurotransmitters such as serotonin and glutamate and translate these signals into modified proliferative behavior (Alenina & Klempin, 2015; Klempin et al., 2013; Song et al., 2013; Song, Zhong, et al., 2012; Walker et al., 2008). Be that as it may, future experiments using advanced in vivo imaging in combination with a variety of transgenic labels of different presumed NSPC populations (or states) are required to truly identify the cellular identity of quiescent NSPCs in the adult hippocampus. These descriptive data are needed in parallel with genetic manipulations of pathways implicated in NSPC behavior to complete our knowledge underlying NSPC quiescence in the adult hippocampus.

MECHANISMS OF STEM CELL ACTIVATION AND PROLIFERATION

It is currently hypothesized that NSPCs need to exit a quiescent state to initiate cell proliferation, as outlined above. Many roads apparently lead to the activation of NSPC proliferation. Two key mechanisms need to happen: the removal of quiescence-inducing signals and the upregulation of proproliferative signals. Again, canonical Notch- and BMP-associated mechanisms appear to be key players that need to be downregulated to allow NSPCs to enter the cell cycle (Ables et al., 2010; Bonaguidi et al., 2008; Breunig et al., 2007; Ehm et al., 2010; Lugert et al., 2010; Mira et al., 2010). In combination with the induced expression of a number of transcription factors that regulate proliferation and/or cellular commitment, among others TLX and Ascl1 (Andersen et al., 2014; Elmi et al., 2010; Niu, Zou, Shen, & Zhang, 2011; Raposo et al., 2015; Shi et al., 2004; Zhang, Zou, He, Gage, & Evans, 2008), NSPCs appear to transform from a type 1 state, presumably dividing largely asymmetrically, to a more proliferative type 2 cell that is believed to expand the pool of newborn cells through symmetric cell divisions (again, this is not formally proven and needs to be confirmed using, for example, live-imaging approaches). Besides intrinsic transcriptional and metabolic adjustments, it is plausible that activation of NSPCs is controlled by extrinsic niche- or microenvironment-mediated mechanisms; for example, circuit activity is translated into altered proliferative behavior of NSPCs that may be regulated either by direct connections between neural cells and NSPCs or through spillover of neurotransmitters onto receptors decorating NSPC processes (Alenina & Klempin, 2015; Klempin et al., 2013; Moss & Toni, 2013; Song et al., 2013; Song, Zhong, et al., 2012; Walker et al., 2008). However, it remains largely unknown how these distinct intrinsic and extrinsic mechanisms converge on signaling hubs and/or if they rather function in parallel. Addressing these questions is complicated by the fact that experiments aiming to understand the cross talk between different regulators will preferentially need to be conducted in a setting that at least resembles the in vivo situation or are performed directly within the hippocampus.

MECHANISMS OF NEURONAL FATE DETERMINATION

Neuronal fate determination describes the commitment of a proliferating precursor to the neuronal lineage. In current models of the adult hippocampal neurogenic lineage, the neuronal fate determination is equated with the expression of the neuron-specific microtubule-associated protein doublecortin in late type 2 progenitor cells (type 2b cells) (Kempermann et al., 2004).

Similar to stem cell quiescence and activation, neuronal fate determination is controlled by the interplay of niche-derived signals and cell-intrinsic regulatory networks. Regulatory networks in fate determination are expected to terminate programs maintaining multipotency and to initiate neuron-specific gene expression programs. With regard to niche-derived signals, Notch- and Wnt-signaling pathway activities are central to the decision between multipotency and neuronal fate commitment. Canonical Notch signaling maintains Sox2 expression, which in turn activates multipotency-associated genes and represses neuronal differentiation genes (Ehm et al., 2010). Wnt/β-catenin signaling, on the contrary, drives the dentate granule neuron-specific transcription factor combination of Prox1 and NeuroD1, which each by itself potently drives in vitro neuronal differentiation of adult neural stem cells (Karalay et al., 2011; Kuwabara et al., 2009; Lie et al., 2005).

The regulation of NeuroD1 expression provides mechanistic insight into the opposing actions of the Notch- and Wnt-signaling pathways in neuronal fate determination: NeuroD1 expression is controlled by regulatory elements consisting of overlapping Sox2-binding sites and binding sites for the T-cell factor/lymphoid enhancer factor (TCF/LEF) transcription factors, ie, the transcriptional effectors of Wnt/β-catenin signaling. The Notch target Sox2 occupies these Sox/LEF-binding sites in multipotent precursor cells to repress NeuroD1 expression. Induction of neuronal differentiation by canonical Wnt signaling—potentially assisted by downregulation of Notch-dependent Sox2 expression—is paralleled by replacement of Sox2 by a β-catenin-containing transcriptional activator complex and expression of NeuroD1. Interestingly Sox/LEF-binding sites appear to be widely distributed, raising the possibility that the Sox2 to β-catenin/TCF/LEF switch may be a general mechanism for expressing the dentate granule neuron-specific program (Kuwabara et al., 2009).

The antagonism between the Notch/Wnt signaling and their transcriptional downstream targets represents an elegant model for explaining the lineage progression from a multipotent to a neuronally committed precursor cell. Current data, however, indicate that the regulatory network is far more complex. GABAergic signaling and Ca^{2+} signaling were found to promote NeuroD1 expression in neural precursors and neuronal differentiation, suggesting that neural network activity modulates the neuronal fate determination process (Deisseroth et al., 2004; Quadrato et al., 2014; Tozuka, Fukuda, Namba, Seki, & Hisatsune, 2005). Shutdown of the Notch/Sox2-dependent multipotency program is assisted by transcriptional repression of Sox2 by the SoxB2 transcription factor Sox21 (Matsuda et al., 2012) and the T-box-containing transcription factor Tbr2 (Hodge et al., 2008, 2012), which both show peak expression shortly before the onset of doublecortin expression. Moreover, expression of neuronal programs requires not only the shutdown of the Sox2 multipotency program but also relief of FOXO and REST/NRSF transcription factor-mediated repression of proneurogenic gene expression programs (Gao et al., 2011; Webb et al., 2013). Current data suggest that while Prox1 and NeuroD1 drive expression of genes that are specific for dentate granule neurons, they only play a minor function in the establishment of general neuronal traits. Expression of such traits is likely to be regulated by transcription factors of the SoxC family, ie, Sox4 and Sox11, which are initiated in parallel to NeuroD1 and Prox1 expression in type 2b cells (Haslinger, Schwarz, Covic, & Chichung Lie, 2009; Mu et al., 2012). Chromatin immunoprecipitation(ChIP)-Seq analyses as well as functional analyses demonstrate that SoxC directly controls the expression of structural proteins, which are expressed by virtually all developing neurons in the embryonic and adult CNS (Bergsland et al., 2011). Moreover, combined ablation of Sox4 and Sox11 results in failure to generate DCX-positive immature neurons (Mu et al., 2012). The regulation of SoxCs adds another layer of complexity to the mechanisms controlling neuronal fate determination as current data indicate that SoxC expression is under the tight control of specific epigenetic regulators such as the Chromodomain protein ChD7 and the BAF complex (Feng et al., 2013; Ninkovic et al., 2013). In the future, proteomic approaches, single-cell transcriptomic and epigenetic analyses, and in vivo visualization of pathway activities will be crucial for fully defining the components of the regulatory network and for understanding how these components coordinately control the adoption of a neuronal fate.

Based on marker expression and morphological data it is assumed that adult hippocampal neurogenesis generates only one neuronal subtype. A recent study, however, revealed that two functionally distinct populations of dentate granule neurons are generated, raising the question of whether regulatory mechanisms for neuronal fate determination differ between these populations (Brunner et al., 2014).

Finally, a fundamental open question is the validity of the current concept of neuronal fate determination, which comprises irreversibility of fate and a stage in development from which neuronal properties are exclusively expressed. The adult neurogenic lineage is considered unidirectional based on lineage tracing of radial glia-like stem cells. Stringent fate mapping studies of accepted neuronally fate-determined cells on the clonal level have not been conducted, leaving the possibility open that cells including "neuronally fate-determined cells" move along the different developmental stages in both directions. TUJ1- and DCX-expressing cells derived from adult neurogenic regions adopt a glial fate on transplantation into nonneurogenic areas (Seidenfaden, Desoeuvre, Bosio, Virard, & Cremer, 2006), calling into question the concept of irreversibility and/or the validity of markers that are being used to unequivocally identify a cell as a neuron. To further complicate this issue, transcripts of "exclusively" neuronal factors are already expressed in stem cells, which has led to the suggestion that stem cells are already primed or biased to differentiate into neurons (Beckervordersandforth et al., 2010). It will be interesting to determine whether the present static concept of neuronal fate determination in neurogenesis requires replacement by a fluid concept, in which the neuronal identity is stabilized over time.

MECHANISMS OF NEURONAL MATURATION AND FUNCTIONAL INTEGRATION

The developmental timeline of neuronal maturation and functional integration has been extensively analyzed in adult mice. During the first three weeks new neurons undergo rapid axonal and dendritic growth to project axons into the CA3 area and a dendritic tree with complex arborization in the molecular layer (Sun et al., 2013; Zhao, Teng, Summers, Ming, & Gage, 2006).

GABAergic synaptic inputs are detected as early as 1 week after neuronal birth (Ge et al., 2006). The GABAergic input is initially depolarizing but is becoming hyperpolarizing in week 3. Around the same time the first glutamatergic synaptic inputs and mossy fiber synaptic outputs to hilar and CA3 neurons are observed (Esposito et al., 2005; Ge et al., 2006; Toni et al., 2008; Zhao et al., 2006). The synaptic integration and neurophysiological properties of newborn neurons are refined over the following weeks. During weeks 4–6 neurons exhibit enhanced synaptic plasticity compared to older DG neurons, mediated by the expression of NR2B-containing NMDA receptors and the protracted development of GABAergic inhibition (Ge, Pradhan, Ming, & Song, 2007; Marin-Burgin, Mongiat, Pardi, & Schinder, 2012; Schmidt-Hieber, Jonas, & Bischofberger, 2004).

To date, hippocampal network activity is hypothesized to be the main driver for neuronal maturation and functional integration. During the early maturation phase network activity appears to be relayed to the developing neuron by depolarizing GABAergic input (Ge et al., 2006; Ge, Pradhan, et al., 2007). The depolarizing input then activates the transcription factor CREB, which results in the expression of maturation-associated gene expression programs. Consistent with this pathway, ablation of GABA-dependent depolarization and of CREB signaling severely impairs dendrite growth and delays synaptic integration (Ge et al., 2006; Jagasia et al., 2009). Glutamate and NMDAR-mediated signaling constitute prime candidates that drive maturation and synaptic integration during later phases of neurogenesis. Ablation of the NR2B subunit, which confers heightened excitability during weeks 4–6, reduces dendrite complexity, suggesting that NR2B-containing NMDARs control the limited dendritic growth after week 3 (Ge et al., 2007; Kheirbek, Tannenholz, & Hen, 2012). In addition, NMDAR-dependent signaling has also been implied in synaptogenesis and integration-dependent survival of newborn neurons (Tashiro, Makino, & Gage, 2007; Tashiro, Sandler, Toni, Zhao, & Gage, 2006; Toni et al., 2007).

There are increasing data implicating Wnts in the control not only of neuronal fate determination but also of maturation and functional integration. A recent study demonstrated that the Wnt-dependent planar cell polarity pathway controls dendritic orientation and patterning of adult-generated DG neurons (Schafer et al., 2015). The secreted Wnt inhibitors Dkk1 and Sfrp3 are expressed in the dentate gyrus and inhibit multiple steps in neurogenesis including proliferation, dendrite development, and synaptogenesis. Intriguingly, Dkk1 is increasingly expressed in the aged hippocampus, which provides an explanation for the age-associated decline in adult neurogenesis (Seib et al., 2013), whereas Sfsrp3 expression is negatively regulated by hippocampal network activity, which links Wnt signaling to activity-dependent proliferation and maturation processes (Jang et al., 2013).

As described above GABA signaling-induced activity of the transcription factor CREB is a key regulator of early maturation. Whether transcriptional regulators take also a center stage in late maturation steps is less well defined. There are, however, two promising candidates, namely corticosteroid receptors and the zinc finger transcription factor Krüppel-like factor 9 (Klf9). Stress and corticosteroids have received major attention as regulators of adult neurogenesis (Gould & Tanapat, 1999; Karten, Olariu, & Cameron, 2005; Snyder, Soumier, Brewer, Pickel, & Cameron, 2011). The fact that corticoid receptors function as transcription factors on ligand binding has been overlooked and their transcriptional targets in adult neurogenesis have not been systematically investigated. Recently, knockdown of corticosteroid receptors from newborn neurons was shown to produce defects in dendritogenesis and synaptogenesis (Fitzsimons et al., 2013), suggesting that corticosteroid-controlled transcriptional programs participate in the regulation of functional integration. Klf9 starts to be expressed around 3 weeks after the birth of the new DG neuron. Its expression thus coincides with the time when adult DG neurons begin to receive functional inputs (Scobie et al., 2009). Adult-generated neurons in Klf9 knockout mice show impaired formation of higher order dendritic branch points and severely impaired survival, suggesting that Klf9 may be essential for the expression of genetic programs controlling synaptic integration and integration-dependent survival (Scobie et al., 2009; Tashiro et al., 2006).

Owing to the need to restore ion gradients dissipated by the repeated generation of postsynaptic potentials and action potentials, and to the need to maintain the neurotransmitter cycle, functionally integrated neurons are among the cells with the highest energy needs (Alle, Roth, & Geiger, 2009; Attwell & Laughlin, 2001). Consequently, metabolism of the adult-born neuron needs to be adapted in the course of maturation and functional integration. The present studies have focused on mitochondria given their central role in neuronal metabolism. The mitochondrial compartment undergoes developmental stage-dependent changes in mass, morphology, and distribution, indicating that mitochondria-dependent metabolism is adapted to the changing needs of the maturing neuron. Notably, in vivo functional data strongly indicate that mitochondrial biogenesis and dendritic distribution impact on the speed of synaptogenesis and dendritic growth, and on maintenance of synapses (Cheng et al., 2012; Oruganty-Das, Ng, Udagawa, Goh, & Richter, 2012; Steib, Schaffner, Jagasia, Ebert, & Lie, 2014). Hence, adaptation of mitochondrial metabolism—rather than being a passive response to cellular function—may represent an important driving force for maturation.

The mechanism through which mitochondria-dependent metabolism controls maturation has not been elucidated. Moreover, the contribution of mitochondria-independent metabolic pathways to maturation has not been addressed. Finally it will be important to understand how coordination between adaptation of metabolic circuits with signaling pathways and genetic programs is achieved. Here, metabolic circuits may not only be positioned downstream of genetic regulatory circuits but could also directly influence regulatory pathways, eg, by generating substrates for posttranslational modifications.

CONCLUSIONS

It is now accepted that NSPCs persist throughout life and generate new neurons in the mammalian brain, including the hippocampal DG of humans. Adult neurogenesis is a highly dynamic process that dramatically changes hippocampal circuit connectivity in response to a variety of experiences and stimuli (Song, Christian, Ming, & Song, 2012). Given the association of decreased or altered hippocampal neurogenesis with a number of neuropsychiatric diseases, among others major depression and epilepsy, there may also be translational relevance to current efforts to understand the molecular and cellular mechanisms underlying the neurogenic process in the adult mammalian brain (Braun & Jessberger, 2014; Parent, 2007; Sahay & Hen, 2008; Winner, Kohl, & Gage, 2011). Despite substantial progress that has been made in the field from 1995 to 2015, many fundamental questions remain open. It is still not solved if single NSPCs truly show self-renewal over time and if individual NSPCs show repeated cycles of quiescence and activation. Furthermore, we are only at the beginning of understanding if there is—and if so why—heterogeneity in NSPC populations in the adult DG. Whereas the list of functionally significant signals/pathways that regulate distinct developmental steps in the course of adult neurogenesis is constantly growing, there are little data available on how these signals may converge on central signaling hubs and how they are orchestrated to eventually allow for a meaningful neurogenic response. Similarly, we are only at the beginning of understanding circuit effects on the integration of newborn granule cells. At which developmental stage are newborn neurons functionally important? Do they switch their computational function with advancing maturity? How do they contribute to physiological DG-dependent learning and memory? Is altered neurogenesis under pathological conditions a mere bystander or causally linked to disease development and/or progression? These are all at least partially unanswered questions that are currently addressed using a number of novel tools such as defined trangenesis-based genetic targeting, in vivo imaging, and optogenetic approaches. Finally, the field is currently on the verge of translating its findings that are largely based on rodent data or human in vitro approaches to human physiology and disease. Substantial progress has been made recently by the development of novel detection approaches of adult-born neurons in the human DG and by the establishment of innovative noninvasive imaging tools. However, much more evidence is required to truly relate the neurogenic process also in the human brain to physiological brain function and disease.

References

Ables, J. L., Decarolis, N. A., Johnson, M. A., Rivera, P. D., Gao, Z., Cooper, D. C., et al. (2010). Notch1 is required for maintenance of the reservoir of adult hippocampal stem cells. *The Journal of Neuroscience, 30*, 10484–10492.

Alenina, N., & Klempin, F. (2015). The role of serotonin in adult hippocampal neurogenesis. *Behavioural Brain Research, 277*, 49–57.

Alle, H., Roth, A., & Geiger, J. R. (2009). Energy-efficient action potentials in hippocampal mossy fibers. *Science, 325*, 1405–1408.

Altman, J., & Das, G. D. (1965). Autoradiographic and histological evidence of postnatal hippocampal neurogenesis in rats. *The Journal of Comparative Neurology, 124*, 319–335.

Amador-Arjona, A., Cimadamore, F., Huang, C. T., Wright, R., Lewis, S., Gage, F. H., et al. (2015). SOX2 primes the epigenetic landscape in neural precursors enabling proper gene activation during hippocampal neurogenesis. *Proceedings of the National Academy of Sciences of the United States of America, 112*, E1936–E1945.

Andersen, J., Urban, N., Achimastou, A., Ito, A., Simic, M., Ullom, K., et al. (2014). A transcriptional mechanism integrating inputs from extracellular signals to activate hippocampal stem cells. *Neuron, 83*, 1085–1097.

Attwell, D., & Laughlin, S. B. (2001). An energy budget for signaling in the grey matter of the brain. *Journal of Cerebral Blood Flow and Metabolism, 21*, 1133–1145.

Beckervordersandforth, R., Tripathi, P., Ninkovic, J., Bayam, E., Lepier, A., Stempfhuber, B., et al. (2010). In vivo fate mapping and expression analysis reveals molecular hallmarks of prospectively isolated adult neural stem cells. *Cell Stem Cell, 7*, 744–758.

Bergmann, O., Liebl, J., Bernard, S., Alkass, K., Yeung, M. S., Steier, P., et al. (2012). The age of olfactory bulb neurons in humans. *Neuron, 74*, 634–639.

Bergsland, M., Ramskold, D., Zaouter, C., Klum, S., Sandberg, R., & Muhr, J. (2011). Sequentially acting Sox transcription factors in neural lineage development. *Genes & development, 25*, 2453–2464.

Bonaguidi, M. A., Peng, C. Y., McGuire, T., Falciglia, G., Gobeske, K. T., Czeisler, C., et al. (2008). Noggin expands neural stem cells in the adult hippocampus. *The Journal of Neuroscience, 28*, 9194–9204.

Bonaguidi, M. A., Wheeler, M. A., Shapiro, J. S., Stadel, R. P., Sun, G. J., Ming, G. L., et al. (2011). In vivo clonal analysis reveals self-renewing and multipotent adult neural stem cell characteristics. *Cell, 145*, 1142–1155.

Braun, S. M., & Jessberger, S. (2014). Adult neurogenesis and its role in neuropsychiatric disease, brain repair and normal brain function. *Neuropathology and Appllied Neurobiology, 40*, 3–12.

Breunig, J. J., Silbereis, J., Vaccarino, F. M., Sestan, N., & Rakic, P. (2007). Notch regulates cell fate and dendrite morphology of newborn neurons in the postnatal dentate gyrus. *Proceedings of the National Academy of Sciences of the United States of America, 104*, 20558–20563.

Brunner, J., Neubrandt, M., Van-Weert, S., Andrasi, T., Kleine Borgmann, F. B., Jessberger, S., et al. (2014). Adult-born granule cells mature through two functionally distinct states. *Elife, 3*, e03104.

Cheng, A., Wan, R., Yang, J. L., Kamimura, N., Son, T. G., Ouyang, X., et al. (2012). Involvement of PGC-1alpha in the formation and maintenance of neuronal dendritic spines. *Nature Communications, 3*, 1250.

Curtis, M. A., Kam, M., Nannmark, U., Anderson, M. F., Axell, M. Z., Wikkelso, C., et al. (2007). Human neuroblasts migrate to the olfactory bulb via a lateral ventricular extension. *Science, 315*, 1243–1249.

Deisseroth, K., Singla, S., Toda, H., Monje, M., Palmer, T. D., & Malenka, R. C. (2004). Excitation-neurogenesis coupling in adult neural stem/progenitor cells. *Neuron, 42*, 535–552.

Ehm, O., Goritz, C., Covic, M., Schaffner, I., Schwarz, T. J., Karaca, E., et al. (2010). RBPJkappa-dependent signaling is essential for long-term maintenance of neural stem cells in the adult hippocampus. *The Journal of Neuroscience, 30*, 13794–13807.

Elmi, M., Matsumoto, Y., Zeng, Z. J., Lakshminarasimhan, P., Yang, W., Uemura, A., et al. (2010). TLX activates MASH1 for induction of neuronal lineage commitment of adult hippocampal neuroprogenitors. *Molecular and Cellular Neurosciences, 45*, 121–131.

Encinas, J. M., Michurina, T. V., Peunova, N., Park, J. H., Tordo, J., Peterson, D. A., et al. (2011). Division-coupled astrocytic differentiation and age-related depletion of neural stem cells in the adult hippocampus. *Cell Stem Cell, 8*, 566–579.

Eriksson, P. S., Perfilieva, E., Bjork-Eriksson, T., Alborn, A. M., Nordborg, C., Peterson, D. A., et al. (1998). Neurogenesis in the adult human hippocampus. *Nature Medicine, 4*, 1313–1317.

Esposito, M. S., Piatti, V. C., Laplagne, D. A., Morgenstern, N. A., Ferrari, C. C., Pitossi, F. J., et al. (2005). Neuronal differentiation in the adult hippocampus recapitulates embryonic development. *The Journal of Neuroscience, 25*, 10074–10086.

Feng, W., Khan, M. A., Bellvis, P., Zhu, Z., Bernhardt, O., Herold-Mende, C., et al. (2013). The chromatin remodeler CHD7 regulates adult neurogenesis via activation of SoxC transcription factors. *Cell Stem Cell, 13*, 62–72.

Fitzsimons, C. P., van Hooijdonk, L. W., Schouten, M., Zalachoras, I., Brinks, V., Zheng, T., et al. (2013). Knockdown of the glucocorticoid receptor alters functional integration of newborn neurons in the adult hippocampus and impairs fear-motivated behavior. *Molecular Psychiatry, 18*, 993–1005.

Gao, Z., Ure, K., Ding, P., Nashaat, M., Yuan, L., Ma, J., et al. (2011). The master negative regulator REST/NRSF controls adult neurogenesis by restraining the neurogenic program in quiescent stem cells. *The Journal of Neuroscience, 31*, 9772–9786.

Ge, S., Goh, E. L., Sailor, K. A., Kitabatake, Y., Ming, G. L., & Song, H. (2006). GABA regulates synaptic integration of newly generated neurons in the adult brain. *Nature, 439*, 589–593.

Ge, S., Pradhan, D. A., Ming, G. L., & Song, H. (2007). GABA sets the tempo for activity-dependent adult neurogenesis. *Trends in Neurosciences, 30*, 1–8.

Ge, S., Yang, C. H., Hsu, K. S., Ming, G. L., & Song, H. (2007). A critical period for enhanced synaptic plasticity in newly generated neurons of the adult brain. *Neuron, 54*, 559–566.

Gould, E., & Tanapat, P. (1999). Stress and hippocampal neurogenesis. *Biological Psychiatry, 46*, 1472–1479.

Haslinger, A., Schwarz, T. J., Covic, M., & Chichung Lie, D. (2009). Expression of Sox11 in adult neurogenic niches suggests a stage-specific role in adult neurogenesis. *The European Journal of Neuroscience, 29*, 2103–2114.

Hodge, R. D., Kowalczyk, T. D., Wolf, S. A., Encinas, J. M., Rippey, C., Enikolopov, G., et al. (2008). Intermediate progenitors in adult hippocampal neurogenesis: Tbr2 expression and coordinate regulation of neuronal output. *The Journal of Neuroscience, 28*, 3707–3717.

Hodge, R. D., Nelson, B. R., Kahoud, R. J., Yang, R., Mussar, K. E., Reiner, S. L., et al. (2012). Tbr2 is essential for hippocampal lineage progression from neural stem cells to intermediate progenitors and neurons. *The Journal of Neuroscience, 32*, 6275–6287.

Jagasia, R., Steib, K., Englberger, E., Herold, S., Faus-Kessler, T., Saxe, M., et al. (2009). GABA-cAMP response element-binding protein signaling regulates maturation and survival of newly generated neurons in the adult hippocampus. *The Journal of Neuroscience, 29*, 7966–7977.

Jang, M. H., Bonaguidi, M. A., Kitabatake, Y., Sun, J., Song, J., Kang, E., et al. (2013). Secreted frizzled-related protein 3 regulates activity-dependent adult hippocampal neurogenesis. *Cell Stem Cell, 12*, 215–223.

Jones, K. M., Saric, N., Russell, J. P., Andoniadou, C. L., Scambler, P. J., & Basson, M. A. (2015). CHD7 maintains neural stem cell quiescence and prevents premature stem cell depletion in the adult hippocampus. *Stem Cells, 33*, 196–210.

Karalay, O., Doberauer, K., Vadodaria, K. C., Knobloch, M., Berti, L., Miquelajauregui, A., et al. (2011). Prospero-related homeobox 1 gene (Prox1) is regulated by canonical Wnt signaling and has a stage-specific role in adult hippocampal neurogenesis. *Proceedings of the National Academy of Sciences of the United States of America, 108*, 5807–5812.

Karten, Y. J., Olariu, A., & Cameron, H. A. (2005). Stress in early life inhibits neurogenesis in adulthood. *Trends in Neurosciences, 28*, 171–172.

Kempermann, G., Jessberger, S., Steiner, B., & Kronenberg, G. (2004). Milestones of neuronal development in the adult hippocampus. *Trends in Neurosciences, 27*, 447–452.

Kheirbek, M. A., Tannenholz, L., & Hen, R. (2012). NR2B-dependent plasticity of adult-born granule cells is necessary for context discrimination. *The Journal of Neuroscience, 32*, 8696–8702.

Klempin, F., Beis, D., Mosienko, V., Kempermann, G., Bader, M., & Alenina, N. (2013). Serotonin is required for exercise-induced adult hippocampal neurogenesis. *The Journal of Neuroscience, 33*, 8270–8275.

Knobloch, M., Braun, S. M., Zurkirchen, L., von Schoultz, C., Zamboni, N., Arauzo-Bravo, M. J., et al. (2013). Metabolic control of adult neural stem cell activity by Fasn-dependent lipogenesis. *Nature, 493*, 226–230.

Kuhn, H. G., & Cooper-Kuhn, C. M. (2007). Bromodeoxyuridine and the detection of neurogenesis. *Current Pharmaceutical Biotechnology, 8*, 127–131.

Kuwabara, T., Hsieh, J., Muotri, A., Yeo, G., Warashina, M., Lie, D. C., et al. (2009). Wnt-mediated activation of NeuroD1 and retro-elements during adult neurogenesis. *Nature Neuroscience, 12*, 1097–1105.

Lie, D. C., Colamarino, S. A., Song, H. J., Desire, L., Mira, H., Consiglio, A., et al. (2005). Wnt signalling regulates adult hippocampal neurogenesis. *Nature, 437*, 1370–1375.

Lugert, S., Basak, O., Knuckles, P., Haussler, U., Fabel, K., Gotz, M., et al. (2010). Quiescent and active hippocampal neural stem cells with distinct morphologies respond selectively to physiological and pathological stimuli and aging. *Cell Stem Cell, 6*, 445–456.

Ma, D. K., Marchetto, M. C., Guo, J. U., Ming, G. L., Gage, F. H., & Song, H. (2010). Epigenetic choreographers of neurogenesis in the adult mammalian brain. *Nature Neuroscience, 13*, 1338–1344.

Marin-Burgin, A., Mongiat, L. A., Pardi, M. B., & Schinder, A. F. (2012). Unique processing during a period of high excitation/inhibition balance in adult-born neurons. *Science, 335*, 1238–1242.

Matsuda, S., Kuwako, K., Okano, H. J., Tsutsumi, S., Aburatani, H., Saga, Y., et al. (2012). Sox21 promotes hippocampal adult neurogenesis via the transcriptional repression of the Hes5 gene. *The Journal of Neuroscience, 32*, 12543–12557.

Ming, G. L., & Song, H. (2011). Adult neurogenesis in the mammalian brain: significant answers and significant questions. *Neuron, 70*, 687–702.

Mira, H., Andreu, Z., Suh, H., Lie, D. C., Jessberger, S., Consiglio, A., et al. (2010). Signaling through BMPR-IA regulates quiescence and long-term activity of neural stem cells in the adult hippocampus. *Cell Stem Cell, 7*, 78–89.

Moss, J., & Toni, N. (2013). A circuit-based gatekeeper for adult neural stem cell proliferation: parvalbumin-expressing interneurons of the dentate gyrus control the activation and proliferation of quiescent adult neural stem cells. *Bioessays, 35*, 28–33.

Mu, L., Berti, L., Masserdotti, G., Covic, M., Michaelidis, T. M., Doberauer, K., et al. (2012). SoxC transcription factors are required for neuronal differentiation in adult hippocampal neurogenesis. *The Journal of Neuroscience, 32*, 3067–3080.

Ninkovic, J., Steiner-Mezzadri, A., Jawerka, M., Akinci, U., Masserdotti, G., Petricca, S., et al. (2013). The BAF complex interacts with Pax6 in adult neural progenitors to establish a neurogenic cross-regulatory transcriptional network. *Cell Stem Cell, 13*, 403–418.

Niu, W., Zou, Y., Shen, C., & Zhang, C. L. (2011). Activation of postnatal neural stem cells requires nuclear receptor TLX. *The Journal of Neuroscience*, *31*, 13816–13828.

Oruganty-Das, A., Ng, T., Udagawa, T., Goh, E. L., & Richter, J. D. (2012). Translational control of mitochondrial energy production mediates neuron morphogenesis. *Cell Metabolism*, *16*, 789–800.

Parent, J. M. (2007). Adult neurogenesis in the intact and epileptic dentate gyrus. *Progress in Brain Research*, *163*, 529–540.

Quadrato, G., Elnaggar, M. Y., Duman, C., Sabino, A., Forsberg, K., & Di Giovanni, S. (2014). Modulation of GABAA receptor signaling increases neurogenesis and suppresses anxiety through NFATc4. *The Journal of Neuroscience*, *34*, 8630–8645.

Raposo, A. A., Vasconcelos, F. F., Drechsel, D., Marie, C., Johnston, C., Dolle, D., et al. (2015). Ascl1 coordinately regulates gene expression and the chromatin landscape during neurogenesis. *Cell Reports*, *10*(9), 1544–1556.

Sahay, A., & Hen, R. (2008). Hippocampal neurogenesis and depression. *Novartis Foundation Symposium*, *289*, 152–160 discussion 160-154, 193–155.

Sanai, N., Berger, M. S., Garcia-Verdugo, J. M., & Alvarez-Buylla, A. (2007). Comment on "Human neuroblasts migrate to the olfactory bulb via a lateral ventricular extension". *Science*, *318*, 393 author reply 393.

Sanai, N., Tramontin, A. D., Quinones-Hinojosa, A., Barbaro, N. M., Gupta, N., Kunwar, S., et al. (2004). Unique astrocyte ribbon in adult human brain contains neural stem cells but lacks chain migration. *Nature*, *427*, 740–744.

Schafer, S. T., Han, J., Pena, M., von Bohlen Und Halbach, O., Peters, J., & Gage, F. H. (2015). The Wnt adaptor protein ATP6AP2 regulates multiple stages of adult hippocampal neurogenesis. *The Journal of Neuroscience*, *35*, 4983–4998.

Schmidt-Hieber, C., Jonas, P., & Bischofberger, J. (2004). Enhanced synaptic plasticity in newly generated granule cells of the adult hippocampus. *Nature*, *429*, 184–187.

Scobie, K. N., Hall, B. J., Wilke, S. A., Klemenhagen, K. C., Fujii-Kuriyama, Y., Ghosh, A., et al. (2009). Kruppel-like factor 9 is necessary for late-phase neuronal maturation in the developing dentate gyrus and during adult hippocampal neurogenesis. *The Journal of Neuroscience*, *29*, 9875–9887.

Seib, D. R., Corsini, N. S., Ellwanger, K., Plaas, C., Mateos, A., Pitzer, C., et al. (2013). Loss of Dickkopf-1 restores neurogenesis in old age and counteracts cognitive decline. *Cell Stem Cell*, *12*, 204–214.

Seidenfaden, R., Desoeuvre, A., Bosio, A., Virard, I., & Cremer, H. (2006). Glial conversion of SVZ-derived committed neuronal precursors after ectopic grafting into the adult brain. *Molecular and Cellular Neurosciences*, *32*, 187–198.

Seri, B., Garcia-Verdugo, J. M., Collado-Morente, L., McEwen, B. S., & Alvarez-Buylla, A. (2004). Cell types, lineage, and architecture of the germinal zone in the adult dentate gyrus. *The Journal of Comparative Neurology*, *478*, 359–378.

Shi, Y., Chichung Lie, D., Taupin, P., Nakashima, K., Ray, J., Yu, R. T., et al. (2004). Expression and function of orphan nuclear receptor TLX in adult neural stem cells. *Nature*, *427*, 78–83.

Snyder, J. S., Soumier, A., Brewer, M., Pickel, J., & Cameron, H. A. (2011). Adult hippocampal neurogenesis buffers stress responses and depressive behaviour. *Nature*, *476*, 458–461.

Song, J., Christian, K., Ming, G. L., & Song, H. (2012). Modification of hippocampal circuitry by adult neurogenesis. *Developmental Neurobiology*, *72*(7), 1032–1043.

Song, J., Sun, J., Moss, J., Wen, Z., Sun, G. J., Hsu, D., et al. (2013). Parvalbumin interneurons mediate neuronal circuitry-neurogenesis coupling in the adult hippocampus. *Nature Neuroscience*, *16*, 1728–1730.

Song, J., Zhong, C., Bonaguidi, M. A., Sun, G. J., Hsu, D., Gu, Y., et al. (2012). Neuronal circuitry mechanism regulating adult quiescent neural stem-cell fate decision. *Nature*, *489*, 150–154.

Spalding, K. L., Bergmann, O., Alkass, K., Bernard, S., Salehpour, M., Huttner, H. B., et al. (2013). Dynamics of hippocampal neurogenesis in adult humans. *Cell*, *153*, 1219–1227.

Steib, K., Schaffner, I., Jagasia, R., Ebert, B., & Lie, D. C. (2014). Mitochondria modify exercise-induced development of stem cell-derived neurons in the adult brain. *The Journal of Neuroscience, 34,* 6624–6633.

Sun, G. J., Sailor, K. A., Mahmood, Q. A., Chavali, N., Christian, K. M., Song, H., et al. (2013). Seamless reconstruction of intact adult-born neurons by serial end-block imaging reveals complex axonal guidance and development in the adult hippocampus. *The Journal of Neuroscience, 33,* 11400–11411.

Tashiro, A., Makino, H., & Gage, F. H. (2007). Experience-specific functional modification of the dentate gyrus through adult neurogenesis: a critical period during an immature stage. *The Journal of Neuroscience, 27,* 3252–3259.

Tashiro, A., Sandler, V. M., Toni, N., Zhao, C., & Gage, F. H. (2006). NMDA-receptor-mediated, cell-specific integration of new neurons in adult dentate gyrus. *Nature, 442,* 929–933.

Toni, N., Laplagne, D. A., Zhao, C., Lombardi, G., Ribak, C. E., Gage, F. H., et al. (2008). Neurons born in the adult dentate gyrus form functional synapses with target cells. *Nature Neuroscience, 11,* 901–907.

Toni, N., Teng, E. M., Bushong, E. A., Aimone, J. B., Zhao, C., Consiglio, A., et al. (2007). Synapse formation on neurons born in the adult hippocampus. *Nature Neuroscience, 10,* 727–734.

Tozuka, Y., Fukuda, S., Namba, T., Seki, T., & Hisatsune, T. (2005). GABAergic excitation promotes neuronal differentiation in adult hippocampal progenitor cells. *Neuron, 47,* 803–815.

Walker, T. L., White, A., Black, D. M., Wallace, R. H., Sah, P., & Bartlett, P. F. (2008). Latent stem and progenitor cells in the hippocampus are activated by neural excitation. *The Journal of Neuroscience, 28,* 5240–5247.

Webb, A. E., & Brunet, A. (2013). FOXO flips the longevity SWItch. *Nature Cell Biology, 15,* 444–446.

Webb, A. E., Pollina, E. A., Vierbuchen, T., Urban, N., Ucar, D., Leeman, D. S., et al. (2013). FOXO3 shares common targets with ASCL1 genome-wide and inhibits ASCL1-dependent neurogenesis. *Cell Reports, 4,* 477–491.

Winner, B., Kohl, Z., & Gage, F. H. (2011). Neurodegenerative disease and adult neurogenesis. *The European Journal of Neuroscience, 33,* 1139–1151.

Zhang, R. R., Cui, Q. Y., Murai, K., Lim, Y. C., Smith, Z. D., Jin, S., et al. (2013). Tet1 regulates adult hippocampal neurogenesis and cognition. *Cell Stem Cell, 13,* 237–245.

Zhang, J., Ji, F., Liu, Y., Lei, X., Li, H., Ji, G., et al. (2014). Ezh2 regulates adult hippocampal neurogenesis and memory. *The Journal of Neuroscience, 34,* 5184–5199.

Zhang, C. L., Zou, Y., He, W., Gage, F. H., & Evans, R. M. (2008). A role for adult TLX-positive neural stem cells in learning and behaviour. *Nature, 451,* 1004–1007.

Zhao, C., Teng, E. M., Summers, R. G., Jr., Ming, G. L., & Gage, F. H. (2006). Distinct morphological stages of dentate granule neuron maturation in the adult mouse hippocampus. *The Journal of Neuroscience, 26,* 3–11.

CHAPTER

4

Learning and Memory

C.M. Merkley, J.M. Wojtowicz

University of Toronto, Toronto, ON, Canada

INTRODUCTION

In this chapter, we will address the role of adult neurogenesis (ANG) across the life span in learning and memory processes. We will outline behavioral, anatomical, and cellular studies that have shaped our understanding of the role of ANG in learning and memory. We will also discuss neurogenesis in the context of cognitive reserve in humans and recent findings using Alzheimer's disease (AD) model mice that are changing the way we think about neurogenesis across the life span and the importance of its regulation as an early tool to promote healthy aging.

Our understanding of the role of ANG in learning and memory begins with two fundamental properties of newly born neurons: hyperplasticity and turnover. Hyperplasticity exists in several forms and during several distinct time windows or, so-called, critical periods. Turnover, refers to the transition of each successive cohort of new neurons from one developmental state to the next. For a few months following each cell's birth these different states are represented by a distinct form of hyperplasticity such as altered excitation/inhibition balance, enhanced long-term potentiation, and enhanced ability to form dendritic spines. Each such state can occur transiently among neurons of certain ages but surrounded by neurons that are either younger or more mature. Thus, the hyperplastic neurons stand out from the rest of the surrounding neurons and may be preferentially used to alter the neuronal circuitry. However, as the neurons mature they assume a state of traditional plasticity, characteristic of the prenatally born neurons. This form of plasticity is still powerful but does not offer the sharp contrasts between neuronal populations afforded by the former.

http://dx.doi.org/10.1016/B978-0-12-801977-1.00004-0

Thus, ANG is a "living proof" of postnatal brain development and it is never at steady state. This may account for many apparently contradictory results from experiments that, in the past, may not have taken these different states into consideration. With improving understanding of the properties of ANG at the circuit level (Jonas & Lisman, 2014; Wojtowicz, 2012) there will come better understanding of the behavioral effects on learning and memory.

ADULT NEUROGENESIS ACROSS LIFE SPAN

The original description of a critical period in ANG came with a pioneering paper by Gould et al. (1999) who showed that a training task introduced during the second week of cellular life span can enhance the survival of a neuronal cohort in the dentate gyrus (DG). Although several reports supported these initial findings (Tashiro, Makino, & Gage, 2007; Trouche, Bontempi, Roullet, & Rampon, 2009), others did not or at least not entirely (Döbrössy et al., 2003; Snyder, Hong, Mcdonald, & Wojtowicz, 2005). For example, Snyder et al. (2005) did not observe enhanced survival effects, whereas Döbrössy et al. (2003) reported either enhanced or reduced survival of adult-born neurons that was dependent on the phase of the training that coincided with the birth of neurons. Other studies elaborated on this concept by showing that the enhanced survival of new neurons is correlated with subsequent improvement of spatial recognition (Tashiro et al., 2007) or spatial learning (Trouche et al., 2009). In particular, the outstanding feature of these survival effects was their specificity; i.e., in both studies the activity of the additional surviving neurons correlated with enhanced performance only on the tasks that were originally presented during the critical period.

Among the several regulatory mechanisms operating during this first developmental period, the neurotransmitter GABA stands out as particularly important, and serves as a dominant neurotransmitter in the DG. In general, GABA is inhibitory to the majority of mature granule neurons and its release is either regulated by phasic synaptic transmission via terminals of inhibitory interneurons (Overstreet Wadiche, Bromberg, Bensen, & Westbrook, 2005; Song et al., 2012, 2013) or tonically released into the extracellular space (Ge et al., 2006; Song et al., 2012). In addition, GABA serves as a regulatory trophic factor acting on the growing new neurons within the neurogenic zone of the DG. This action has been characterized in some detail and it differs depending on the developmental stage of target cells (Dieni, Chancey, & Overstreet-Wadiche, 2013).

The second critical period for ANG corresponds to 3–4 weeks after cellular birth and coincides with enhanced LTP in the afferent glutamatergic synapses contacting the new neurons. This enhancement is putatively caused by expression of an NR2B subtype of glutamate receptors present at the synapses (Ge, Yang, Hsu, Ming, & Song, 2007; Snyder, Kee, & Wojtowicz, 2001). This enhanced plasticity is not due to stronger synaptic input onto the adult-born/young neurons, but rather relates to the spiking properties of the young neurons (Mongiat, Esposito, Lombardi, & Schinder, 2009) and a low activation threshold (Marin-Burgin, Mongiat, Pardi, & Schinder, 2012). As such, this glutamate-dependent critical period follows the earlier one, in which GABA depolarization is a feature (see above), and hence it affects neurons that are more developed in terms of their synaptic connectivity. It is not clear whether this second critical period is separate from the first and whether it can support behavioral plasticity that is different from the GABA-mediated enhancement of neuronal survival. In fact, the GABA- and glutamate-dependent critical periods may overlap and there is evidence showing that GABA-dependent environmental exposure may regulate the functionality of the NMDA receptors on the young neurons (Chancey et al., 2013). In another study, Nakashiba et al. (2012) depleted newly born neurons at 3 weeks of age in transgenic mice and reported deficits in the tasks requiring pattern separation in spatial learning. This suggests that young neurons are being preferentially recruited to these types of behavioral tasks. On the other hand, Deng, Mayford, and Gage (2013) argue that pattern separation does not recruit the young neurons but rather a more mature population of granule neurons.

Evidence for yet another (third) critical period appears to encompass neuronal ages of a few months after birth (Deng et al., 2013; Lemaire et al., 2012; Tronel et al., 2010). This window of plasticity involves NMDA receptors, presumably mostly NR2A subunits, and is characterized by growth of dendrites and dendritic spines in response to spatial learning (Tronel et al., 2010). This phase, when neurons are months old and putatively mature, also involves GABAergic inhibition which is particularly important for controlling neuronal excitability and the output of mature granule neurons. Disinhibition of these inhibitory circuits allows unmasking of NMDA receptors and facilitation of plasticity (Dieni et al., 2013). A study by Lemaire et al. (2012) showed that mature, 4-month-old adult-born neurons, not developmentally generated neurons, are shaped by learning tasks with respect to their morphology. This is further evidence that fully mature, adult-born neurons are being recruited to spatial learning tasks, extending the idea that they still retain plasticity, even months after being generated (Lemaire et al., 2012).

This third critical period has extended our understanding of plasticity across the neuronal life span, into maturity. The notion that when neurons reach maturity, they are indistinguishable, electrophysiologically and morphologically, from other mature granule neurons, is being challenged by experiments such as the one by Ramirez-Amaya, Marrone, Gage, Worley, and Barnes (2006), who showed that even after 5 months, adult-born neurons are activated in response to a spatial experience, and importantly, a larger proportion of these 5-month-old neurons are activated than non-BrdU-labeled existing granule cells (Ramirez-Amaya et al., 2006). The caveat here is that there is no way to tell whether the other non-BrdU-activated neurons were also generated in adulthood; nevertheless, this work provides evidence that long into maturity, labeled adult-born neurons can be preferentially enlisted to certain tasks.

Given such wide coverage of the cellular life span by successive critical periods, new neurons are susceptible to physiological and behavioral stimuli for several months after their birth. Consequently, stimulation of ANG could involve several external factors occurring in appropriate sequence and in synchrony with the critical windows (Fig. 4.1).

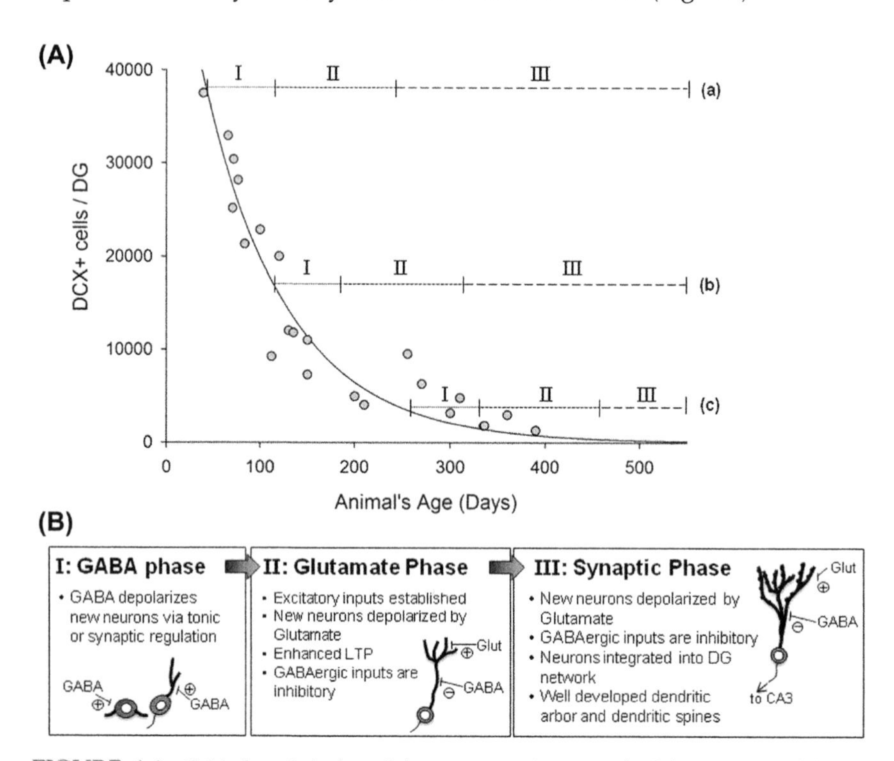

FIGURE 4.1 Critical periods for adult neurogenesis across the life span. (A) Although levels of adult neurogenesis (ANG), as delineated by numbers of doublecortin-positive (DCX+) cells, decline exponentially with age, it is still present to some extent in aged animals. Critical periods of neuronal development (shown as I, II, III) are characteristic of adult-born neurons born at different points throughout life and are defined in the panels below. (B) The first two critical periods, GABA phase (I) and Glutamate phase (II), are shown here as the same approximate length in animals that are adolescent (a), late adulthood (b), and middle age (c). However, this schematic depicts that neurons born later in life have a shorter Synaptic phase (III) than neurons born early on. The data points (filled circles) and the fitted exponential curve were obtained in numerous studies on rats in our laboratory.

A consequence of the multiple critical windows may be the differing functions of the newborn neurons at their different developmental stages. Tronel, Lemaire, Charrier, Montaron, and Abrous (2015), for example, have found that neurons born within the first week of the animal's birth are more likely to be involved in context discrimination 7 months later, in comparison to the neurons born at 2 months of age that are 5 months old during training on a behavioral task. This finding is seemingly at odds with a study of Nakashiba et al. (2012) who suggested that the younger cohorts (approx. 3 weeks old) of new neurons are involved in contextual pattern separation while the older neurons at 5–6 weeks are not. However, numerous methodological and species differences could account for the different results in these two studies.

The sequential critical periods may shed some light on the concept of "neuronal tagging," ie, a mechanism that allows a group of neurons to be marked by some yet undetermined molecular "stamp" which would later be recognized by incoming neuronal activity and enable that activity to be facilitated. This phenomenon would be analogous to synaptic tagging described previously in relation to the traditional synaptic plasticity such as long-term potentiation (Frey & Morris, 1997). As noted above, some studies have suggested that stimulation of neuronal survival during the second week after the cell's birth can facilitate learning several weeks or even months later (Tashiro et al., 2007; Trouche et al., 2009). Mechanisms mediating such cellular "memories" are of interest and relevant to studies that address a personal history of individuals exposed to stimulating environments early in life and then again in adulthood. Suppose a cohort of granule neurons is born during childhood or adolescence in an individual that is exposed to a particularly intense period of education or some other stimulating environment. Would this cohort be later selectively recruited when the individual encounters a similar environment?

A study by Alme et al. (2010) addressed a related question: whether neurons are selectively tagged or tuned to certain experiences and are subsequently activated on reinstatement of those experiences. To test this, Alme et al. (2010) exposed fully adult rats (10–12 months old) to four different or four identical environments over the course of approximately 3 months. One week after the last exposure, rats were placed in the same or different environments, and neuronal activity was measured in the DG. Their results showed that the activity on reexposure was dependent on the number of environments on reexposure and not on the original exposure(s) (Alme et al., 2010). Thus the adult-born neurons initially recruited were hypothesized to "opt for early retirement" and not for reuse on retrieval of old memories (Alme et al., 2010). The caveat of this study was the lack of precise identification of active neurons and thus, it was not at all clear whether the activity was measured in newly born or preexisting mature neurons.

The idea that neurons can be "tagged," and preferentially used later on with reintroduction of the same experience, is relevant to the concept of cognitive reserve, introduced by Stern and colleagues in the mid-1990s (Stern, 2002; Stern et al., 1995, 1994). The phenomenon of cognitive reserve, which is well supported by human studies, generally refers to some form of "memory" within the brain that is produced during childhood or early adulthood, but influences cognition much later in life (Stern, 2002). With the strong evidence that neurogenesis is closely linked with cognition and aging (Snyder et al., 2005; Lazarov, Mattson, Peterson, Pimplikar, & Van Praag, 2010), it is entirely reasonable to hypothesize that cognitive reserve could be related to ANG. In effect, a biological basis for cognitive reserve remains unknown, and neurogenesis is well poised as an underlying mechanism. Cohorts of neurons that are born during a period of increased activity, or challenge (albeit cognitive or physical), may be incorporated into neural networks that are able to withstand some form of neurological/pathological damage. Essentially, the capacity to withstand damage and retain cognitive capability into old age is an important aspect of the cognitive reserve concept. This aspect and the role of neurogenesis in cognitive reserve will be discussed further in the next section.

In conclusion, although various aspects of postnatal neurogenesis have been examined in numerous studies, there is a paucity of studies that address the changes along the life span of an animal in general, and moreover the effects of early life stimulation on neurogenesis and cognition later on. This gap in our knowledge needs to be filled to understand if and how the cognitive reserve is related to the cognitive decline in old age, particularly with neurodegenerative disorders such as AD.

EVIDENCE FOR COGNITIVE RESERVE

In the mid-1960s neuropathology characteristic of AD present in the brains of nondemented individuals was first reported (Tomlinson, Blessed, & Roth, 1968). From 1985 to 2015 numerous studies, including a number of community-based longitudinal studies, have substantiated the early work by Tomlinson et al. (1968), expanding on those initial observations and generating interest worldwide (Bennett et al., 2006; Crystal et al., 1988; Davis, Schmitt, Wekstein, & Markesbery, 1999; Green, Kaye, & Ball, 2000; Haroutunian et al., 1999; Hulette et al., 1998; Ince, 2001; Katzman et al., 1988; Knopman et al., 2003; Lim et al., 1999; Mortimer, 1997; Snowdon et al., 1996; Tomlinson et al., 1968, 1970). Included in these studies is the well-known work of Katzman et al. (1988), in which 10 elderly subjects with normal cognitive function showed pathology indicative of AD at autopsy. Only a few months later, Crystal et al. (1988) demonstrated similar findings, showing in their study that plaque pathology did not distinguish between demented and nondemented subjects, although some of these nondemented subjects may have fit criteria for mild cognitive impairment, which in recent years has been shown to be associated with AD pathology (Bennett, Schneider, Bienias, Evans, & Wilson, 2005; Guillozet, Weintraub, Mash, & Mesulam, 2003; Riley, Snowdon, & Markesbery, 2002). Overall, the discrepancy between the degree of brain damage and its manifestation in cognitive functioning/performance has led to the generation of the cognitive reserve hypothesis, proposed by Stern (2002) over a decade ago.

The concept of cognitive reserve refers to the ability of the brain to maintain cognitive functioning in the face of physical damage and disease pathology (Stern, 2002). Evidence in support of the cognitive reserve hypothesis lies in studies showing that early life influences such as education, physical activity, occupational attainment, linguistic ability, musical training, etc., may have an impact on later cognition, by reducing the risk of developing dementia or AD (Alexander et al., 1997; Bennett et al., 2012; Manly, Touradji, Tang, & Stern, 2003; Snowdon et al., 1996; Stern et al., 1994, 1995; Valenzuela & Sachdev, 2006; White-Schwoch, Woodruff Carr, Anderson, Strait, & Kraus, 2013). This may occur even in the face of physical brain damage that in other individuals manifests as severely impaired cognitive functioning (Katzman et al., 1988; Stern, 2002). These individual differences may be explained by some form of "active coping" within the brain of individuals with cognitive reserve that allows them to overcome loss of functionality associated with pathology (Stern, 2002).

THE NEUROGENIC RESERVE HYPOTHESIS

An intriguing and yet unanswered question is how early life experiences shape future brain functioning substantially enough that they can serve as a reserve in the face of damage? Further, what elements comprise the biological underpinning of this cognitive reserve and how does it alter cognitive functioning?

As a potential explanation for cognitive reserve, the neurogenic reserve hypothesis was proposed (Kempermann, 2008). This hypothesis states that activity-dependent increases in hippocampal neurogenesis may serve as a platform for which new neuron production throughout life can be sustained, and produces a potential for greater neural and cognitive efficiency across the life span and into old age (Kempermann, 2008). In other words, the neurogenic reserve represents the brain's compensatory potential in the face of neurodegenerative disease. Although physical activity and enrichment can increase neurogenesis and learning performance throughout life, including in old age (Bizon & Gallagher, 2003; Kempermann & Gage, 1999; Kempermann, Gast, & Gage, 2002; Kronenberg et al., 2006;

Kuhn, Dickinson-Anson, & Gage, 1996; Seki & Arai, 1995; Van Praag, Shubert, Zhao, & Gage, 2005), early life activity may be key for building an optimized neurogenic reserve and hippocampal network. Moreover, the survival of these new neurons may depend on sustained lifelong activity (Kempermann, 2008; Kempermann et al., 2002).

A recently published study provides some of the first evidence that activity in rodents in early life can exert long-lasting effects on neurogenesis with aging (Merkley, Jian, Mosa, Tan, & Wojtowicz, 2014). The authors used running to increase neurogenesis acutely during the adolescent period in rats, and neurogenesis was examined at various time points, up to 9 months later. Regression analysis on these data showed that while initial increases in neuronal differentiation and 1-week survival were brought down to control levels with increasing age, 4-week cell survival and maturation remained enhanced throughout the period of the study (up to 9 months later in 11-month-old rats) (Merkley et al., 2014). This is the first evidence in support of the idea that early life activity can have long-term effects on neurogenesis with age. Future studies expanding on these findings will help to elucidate how these neurogenic changes are related to cognitive changes with aging. In addition, these findings are relevant for work in humans, showing that early life activity results in a decreased risk of developing AD later in life (Valenzuela & Sachdev, 2006); however, whether these effects in humans are mediated by long-lasting changes in neurogenesis remains elusive. Fig. 4.2 shows the time course of neurogenic reserve superimposed on the time course of a person's life with possible changes in the trajectory of the brain's cognitive state. A specific relationship to ANG does not preclude other forms of plasticity of the brain.

PRESENCE OF EARLY NEUROGENIC CHANGES IN ALZHEIMER'S DISEASE MOUSE MODELS

Based on evidence that ANG declines with age (Epp, Barker, & Galea, 2009; Kuhn et al., 1996; Seki & Arai, 1995), and the well-known role for the hippocampus and adult-born neurons in learning and memory (Clelland et al., 2009; Snyder et al., 2005), many have speculated that changes in neurogenesis may mediate/underlie age-related cognitive decline and functional deficits characterizing AD (Lazarov et al., 2010; Ohm, 2007). In support of this contention, recent years have seen accumulating evidence that remarkable alterations in hippocampal neurogenesis (including cell proliferation and survival, neuronal differentiation and maturation) take place *before* the onset of neuropathological hallmarks of AD and cognitive impairment (Demars, Hu, Gadadhar, & Lazarov, 2010; Ekonomou et al., 2014; Hamilton et al., 2010; Krezymon et al., 2013).

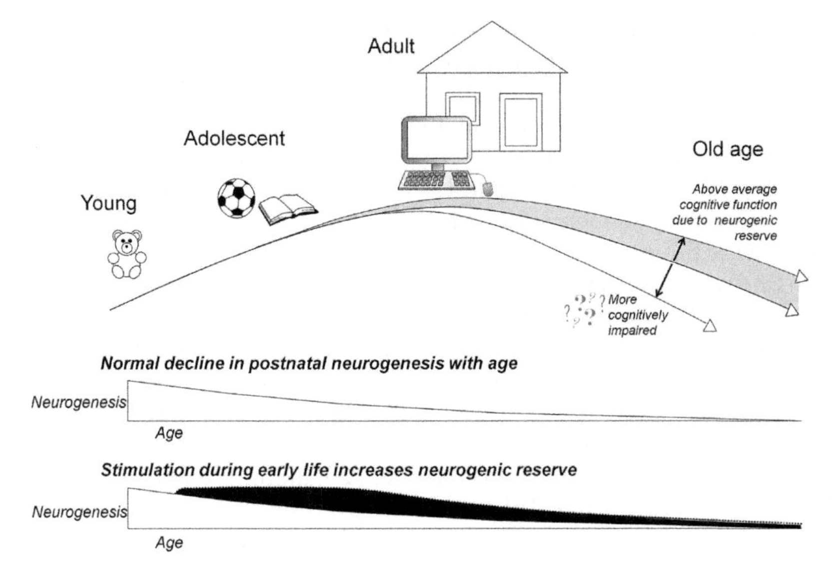

FIGURE 4.2 The neurogenic reserve hypothesis. The trajectory for cognitive functioning into old age is hypothesized to be modulated by early-life stimulation, which builds a neurogenic reserve. According to this hypothesis, activity-dependent increases in neurogenesis lead to a reserve pool of cells that can persist into old age and contribute to conserved cognitive function via enhanced cellular and functional plasticity. A lack of stimulation during adolescence may render the brain susceptible to declines in neurogenesis that might lead to a range of cognitive problems with aging, including an increased risk of dementia or Alzheimer's disease.

Using the APPswe/PS1ΔE9 mouse model of AD, Demars et al. (2010) showed a decrease in BrdU-labeled cells that was apparent at 2 months of age, well before the presence of amyloid deposition or memory impairments. Another study looking at very early stages of plaque and tangle formation in the 3xTg mouse model of AD showed a dramatic decrease in cell proliferation (80%) as well as a decrease in 3-week cell survival relative to nontransgenic (non-Tg) controls (Hamilton et al., 2010). However, in the latter study, cognitive impairments had likely already manifested, based on other characterizations of that mouse model (Billings, Oddo, Green, Mcgaugh, & Laferla, 2005; Hamilton et al., 2010). In contrast to the data in these studies suggesting a decrease in proliferation, Krezymon et al. (2013) have recently shown an increase in proliferation in 3-month-old Tg2576 mice relative to non-Tg controls. Interestingly, despite the increase in proliferation, 30-day cell survival was decreased in these mice, although these changes were not present when 5-month-old mice were examined (Krezymon et al., 2013). The literature is further convoluted by other studies that suggest increases in proliferation early on (Gan et al.,

2008; Kolecki et al., 2008; Lopez-Toledano & Shelanski, 2007; Mirochnic, Wolf, Staufenbiel, & Kempermann, 2009). It may be concluded that measures of proliferation are not useful in determining the rate of neurogenesis. First, the proliferation is only the first step in the long process of cell production and development. Second, proliferation appears to be the subject of feedback regulation and hence may increase to compensate for reduced levels of young neurons and not as a result of a direct pathological insult to the brain.

In line with this hypothesis, the data in AD mouse models for changes in neuronal differentiation using doublecortin (DCX) as a marker of immature neurons and neuroblasts appear to be fairly consistent in the literature. Three separate studies using different AD mouse models have shown a decrease in DCX-labeled cells early on in transgenic mice compared to controls (Beauquis, Vinuesa, Pomilio, Pavía, & Saravia, 2014; Demars et al., 2010; Hamilton et al., 2010). In two of these studies, the decreased neuronal differentiation preceded amyloid deposition and memory impairments (Beauquis et al., 2014; Demars et al., 2010), while in the third study, amyloid plaque and neurofibrillary tangles were at an early stage but cognitive impairments were already present (Hamilton et al., 2010).

Neuronal maturation in mouse models of AD has also been investigated in a few studies. Krezymon et al. (2013) showed not only that fewer neurons are surviving to 30 days in 3-month-old Tg2756 mice but also that significantly fewer of them express the mature neuronal marker NeuN. Therefore, long before neuropathology and cognitive deficits are present, neuronal maturation and survival are compromised. Another recent study looking at transgenic AD mice before amyloid deposition showed a decrease in mature neurons in the DG, using NeuN (Beauquis et al., 2014). Finally, in a third study investigating changes in numbers of mature neurons, Mirochnic et al. (2009) showed a decrease in Calretinin-labeled cells in young mice prior to the presence of amyloid plaques.

Altogether, because of the nature of these early neurogenic changes that take place prior to neuropathological and memory impairments, changes in neurogenesis may be more likely a contributor to, rather than a consequence of, neuronal and cognitive dysfunction (Demars et al., 2010; Hamilton et al., 2010; Krezymon et al., 2013). Furthermore, the data suggest that modulating neurogenesis might serve as an important tool to prevent or delay the development of age-related cognitive decline. Further research investigating early changes in neurogenesis in Alzheimer's mouse models will be important for our understanding of how changes in neurogenesis early on contribute to and may be an important indicator of impending AD manifestation later in life.

NEUROPATHOLOGY: NOT THE COMPLETE PICTURE

There are several lines of evidence suggesting that neuropathology is not the key contributor to learning and memory impairments in AD, and that neurogenesis is a worthwhile avenue of investigation as an underlying mechanism. First, the presence of changes in neurogenesis that take place prior to the onset of amyloid plaques strongly implicates that plaque formation occurs postliminary (downstream) of neurogenic alterations (Demars et al., 2010; Hamilton et al., 2010; Krezymon et al., 2013). Secondly, while all patients with AD show neuropathology indicating the presence of plaques, not all individuals with this pathology express AD and its characteristic cognitive deficits (Katzman et al., 1988; Tomlinson et al., 1968, 1970). Thirdly, removal of amyloid plaques in clinical studies using Alzheimer's patients has failed to reverse the cognitive symptoms of AD (Tayeb, Murray, Price, & Tarazi, 2013). Based on findings from the latter clinical studies, the presence of plaques appears to be only one facet and not a sole contributor to cognitive impairments. Therefore, movement toward interventions aimed at prevention rather than reversal may be more promising, namely, those that target neurogenesis through activity or exercise starting early in life.

PATTERN SEPARATION, NEUROGENESIS, AND ALZHEIMER'S DISEASE

In recent years, pattern separation has come into the spotlight as an important function of adult-born neurons in the DG. Pattern separation refers to the ability to distinguish between highly similar stimuli, transforming highly similar inputs to more dissimilar representations (Marr, 1971; Piatti, Ewell, & Leutgeb, 2013). Behavioral pattern separation tasks in rodents are dependent on the DG (Gilbert, Kesner, & Lee, 2001; Kesner, 2013; Morris, Churchwell, Kesner, & Gilbert, 2012) and specifically, adult-born neurons in the DG have been shown to play an important role (Clelland et al., 2009; Kheirbek, Tannenholz, & Hen, 2012; Nakashiba et al., 2012). Evidence implicating neurogenesis in rat studies shows that animals with reduced neurogenesis perform poorly on a pattern separation task (Clelland et al., 2009; Morris et al., 2012; Nakashiba et al., 2012), while increasing ANG (via exercise or genetic manipulations) leads to better performance on these tasks (Creer, Romberg, Saksida, Van Praag, & Bussey, 2010; Sahay et al., 2011). Pattern separation tasks conducted in human studies have also implicated the DG (Bakker, Kirwan, Miller, & Stark, 2008) and demonstrate age-related decline in performance (Toner, Pirogovsky, Kirwan, & Gilbert, 2009). In addition, two recent studies have demonstrated that individuals with AD are impaired on pattern separation tasks (Ally, Hussey, Ko, & Molitor, 2013; Wesnes, Annas, Basun, Edgar, & Blennow, 2014). It is tempting to speculate that impairments on performance of this task are a function of reduced neurogenesis, and although implicated, this is not directly shown. Thus, pattern separation ability could be rationally employed as a frontline task in studies of AD, and may be used to understand the role of ANG in disease progression.

CONCLUSIONS

In conclusion, there appears to be a disconnect between the studies on molecular/cellular mechanisms of neurogenesis in normal animals and the studies of early changes in pathological cases, such as AD mouse models, with the possible influence of early life experiences on pathology during aging. In this chapter, we have outlined the main stages of cell development as they relate to the animal's age and the cell's age. The two are not the same since the process of cellular development for a cell born in a mature animal is not exactly the same as for a cell born in the first few days after an animal's birth. Yet, in both cases cell development takes a long time and there is sound rationale for proposing that cell survival and its functional properties may be amenable to environmental/physiological influence at several "critical" periods along the developmental path. There are only a few studies addressing the mechanisms that could tag or render cells selectively tuned at early stages of their growth for later use. The neurogenic reserve hypothesis described in our chapter calls for precisely such studies and the field is ripe for examination of if and how the neurogenic reserve can compensate for cognitive decline in old age.

A convergence of studies investigating neurogenesis across the life span in response to early stimulation and those investigating neurogenic reserve may in the future help to explain the functional role of postnatal neurogenesis and help to clarify the enigmatic origin of dementia. As depicted in Fig. 4.2, the trajectory of cellular development for neurons born in adolescence when the numbers of potentially recruitable neurons are very high may depend on brain stimulation during that time, and may also be when brain stimulation is most effective. If this trajectory is altered by brain stimulation, an individual could experience a change in the quality of life and perseverance of cognitive function due to longevity of the generated neurons. Such a mechanism for neurogenic reserve does not preclude the importance of the brain's plasticity in adulthood, but as represented in Figs. 4.1 and 4.2, increased plasticity of new neurons born during adulthood may be inherently less effective than that of neurons born during adolescence, due to contraction of the critical phase and shorter lifetime of the generated neurons.

References

Alexander, G. E., Furey, M. L., Grady, C. L., Pietrini, P., Brady, D. R., Mentis, M. J., et al. (1997). Association of premorbid intellectual function with cerebral metabolism in Alzheimer's disease: implications for the cognitive reserve hypothesis. *American Journal of Psychiatry, 154*, 165–172.

Ally, B. A., Hussey, E. P., Ko, P. C., & Molitor, R. J. (2013). Pattern separation and pattern completion in Alzheimer's disease: evidence of rapid forgetting in amnestic mild cognitive impairment. *Hippocampus, 23*, 1246–1258. http://dx.doi.org/10.1002/hipo.22162.

Alme, C., Buzzetti, R., Marrone, D., Leutgeb, J. K., Chawla, M. K., Schaner, M. J., et al. (2010). Hippocampal granule cells opt for early retirement. *Hippocampus, 20*, 1109–1123.

Bakker, A., Kirwan, C. B., Miller, M., & Stark, C. E. (2008). Pattern separation in the human hippocampal CA3 and dentate gyrus. *Science, 319*, 1640–1642.

Beauquis, J., Vinuesa, A., Pomilio, C., Pavía, P., & Saravia, F. (2014). Hippocampal and cognitive alterations precede amyloid deposition in a mouse model for Alzheimer's disease. *Medicina (Buenos Aires), 74*, 282–286.

Bennett, D. A., Schneider, J. A., Bienias, J. L., Evans, D. A., & Wilson, R. S. (2005). Mild cognitive impairment is related to Alzheimer disease pathology and cerebral infarctions. *Neurology, 64*, 834–841.

Bennett, D. A., Schneider, J. A., Arvanitakis, Z., Kelly, J. F., Aggarwal, N. T., Shah, R. C., et al. (2006). Neuropathology of older persons without cognitive impairment from two community-based studies. *Neurology, 66*, 1837–1844. http://dx.doi.org/10.1212/01.wnl.0000219668.47116.e6.

Bennett, D. A., Schneider, J. A., Buchman, A. S., Barnes, L. L., Boyle, P. A., & Wilson, R. S. (2012). Overview and findings from the rush memory and aging Project. *Current Alzheimer Research, 9*, 646–663.

Billings, L. M., Oddo, S., Green, K. N., Mcgaugh, J. L., & Laferla, F. M. (2005). Intraneuronal Abeta causes the onset of early Alzheimer's disease-related cognitive deficits in transgenic mice. *Neuron, 45*, 675–688.

Bizon, J. L., & Gallagher, M. (2003). Production of new cells in the rat dentate gyrus over the lifespan: relation to cognitive decline. *European Journal of Neuroscience, 18*, 215–219.

Chancey, J. H., Adlaf, E. W., Sapp, M. C., Pugh, P. C., Wadiche, J. I., & Overstreet-Wadiche, L. S. (2013). GABA depolarization is required for experience-dependent synapse unsilencing in adult-born neurons. *Journal of Neuroscience, 33*, 6614–6622.

Clelland, C., Choi, M., Romberg, C., Clemenson, G., Fragniere, A., Tyers, P., et al. (2009). A functional role for adult hippocampal neurogenesis in spatial pattern separation. *Science, 325*, 210–213.

Creer, D. J., Romberg, C., Saksida, L. M., Van Praag, H., & Bussey, T. J. (2010). Running enhances spatial pattern separation in mice. *Proceedings of the National Academy of Sciences of the United States of America, 107*, 2367–2372.

Crystal, H., Dickson, D., Fuld, P., Masur, D., Scott, R., Mehler, M., et al. (1988). Clinicopathologic studies in dementia: nondemented subjects with pathologically confirmed Alzheimer's disease. *Neurology, 38*, 1682.

Davis, D. G., Schmitt, F. A., Wekstein, D. R., & Markesbery, W. R. (1999). Alzheimer neuropathologic alterations in aged cognitively normal subjects. *Journal of Neuropathology and Experimental Neurology, 58*, 376–388.

Demars, M., Hu, Y.-S., Gadadhar, A., & Lazarov, O. (2010). Impaired neurogenesis is an early event in the etiology of familial Alzheimer's disease in transgenic mice. *Journal of Neuroscience Research, 88*, 2103–2117. http://dx.doi.org/10.1002/jnr.22387.

Deng, W., Mayford, M., & Gage, F. H. (2013). Selection of distinct populations of dentate granule cells in response to inputs as a mechanism for pattern separation in mice. *eLife*, e00312.

Dieni, C. V., Chancey, J. H., & Overstreet-Wadiche, L. S. (2013). Dynamic functions of GABA signaling during granule cell maturation. *Frontiers in Neural Circuits, 6*.

Döbrössy, M. D., Drapeau, E., Aurousseau, C., Le Moal, M., Piazza, P. V., & Abrous, D. N. (2003). Differential effects of learning on neurogenesis: learning increases or decreases the number of newly born cells depending on their birth date. *Molecular Psychiatry, 8*, 974–982.

Ekonomou, A., Savva, G. M., Brayne, C., Forster, G., Francis, P. T., Johnson, M., et al. (2014). Stage-specific changes in neurogenic and Glial markers in Alzheimer's disease. *Biological Psychiatry, 77*(8), 711–719.

Epp, J. R., Barker, J. M., & Galea, L. A. (2009). Running wild: neurogenesis in the hippocampus across the lifespan in wild and laboratory-bred Norway rats. *Hippocampus, 19*, 1040–1049.

Frey, U., & Morris, R. G. M. (1997). Synaptic tagging and long-term potentiation. *Nature, 385*, 533–536.

Gan, L., Qiao, S., Lan, X., Chi, L., Luo, C., Lien, L., et al. (2008). Neurogenic responses to amyloid-beta plaques in the brain of Alzheimer's disease-like transgenic (pPDGF-APPSw,Ind) mice. *Neurobiology of Disease, 29*, 71–80.

Ge, S., Goh, E. L. K., Sailor, K. A., Kitabatake, Y., Ming, G.-L., & Song, H. (2006). GABA regulates synaptic integration of newly generated neurons in the adult brain. *Nature, 439*, 589–593. http://www.nature.com/nature/journal/v439/n7076/suppinfo/nature04404_S1.html.

Ge, S., Yang, C. H., Hsu, K. S., Ming, G. L., & Song, H. (2007). A critical period for enhanced synaptic plasticity in newly generated neurons of the adult brain. *Neuron, 54*, 559–566.

Gilbert, P. E., Kesner, R. P., & Lee, I. (2001). Dissociating hippocampal subregions: double dissociation between dentate gyrus and CA1. *Hippocampus, 11*, 626–636.

Gould, E., Beylin, A., Tanapat, P., Reeves, A., Shors, T.J. (1999). Learning enhances adult neurogenesis in the hippocampal formation. *Nature Neuroscience, 2*, 260–265.

Green, M. S., Kaye, J. A., & Ball, M. J. (2000). The Oregon brain aging study: neuropathology accompanying healthy aging in the oldest old. *Neurology, 54*, 105–113.

Guillozet, A. L., Weintraub, S., Mash, D. C., & Mesulam, M. M. (2003). Neurofibrillary tangles, amyloid, and memory in aging and mild cognitive impairment. *Archives of Neurology, 60*, 729–736.

Hamilton, L. K., Aumont, A., Julien, C., Vadnais, A., Calon, F., & Fernandes, K. J. L. (2010). Widespread deficits in adult neurogenesis precede plaque and tangle formation in the 3xTg mouse model of Alzheimer's disease. *European Journal of Neuroscience, 32*, 905–920.

Haroutunian, V., Purohit, D. P., Perl, D. P., Marin, D., Khan, K., Lantz, M., et al. (1999). Neurofibrillary tangles in nondemented elderly subjects and mild Alzheimer disease. *Archives of Neurology, 56*, 713–718.

Hulette, C. M., Welsh-Bohmer, K. A., Murray, M. G., Saunders, A. M., Mash, D. C., & Mcintyre, L. M. (1998). Neuropathological and neuropsychological changes in "normal" aging: evidence for preclinical Alzheimer disease in cognitively normal individuals. *Journal of Neuropathology and Experimental Neurology, 57*, 1168–1174.

Ince, P. (2001). Pathological correlates of late-onset dementia in a multicentre, community-based population in England and Wales. *The Lancet, 357*, 169–175.

Jonas, P., & Lisman, J. (2014). Structure, function, and plasticity of hippocampal dentate gyrus microcircuits. *Frontiers in Neural Circuits, 8*.

Katzman, R., Terry, R., Deteresa, R., Brown, T., Davies, P., Fuld, P., et al. (1988). Clinical, pathological, and neurochemical changes in dementia: a subgroup with preserved mental status and numerous neocortical plaques. *Annals of Neurology, 23*, 138–144.

Kempermann, G. (2008). The neurogenic reserve hypothesis: what is adult hippocampal neurogenesis good for? *Trends in Neurosciences, 31*, 163–169.

Kempermann, G., & Gage, F. H. (1999). Experience-dependent regulation of adult hippocampal neurogenesis: effects of long-term stimulation and stimulus withdrawal. *Hippocampus, 9*, 321–332.

Kempermann, G., Gast, D., & Gage, F. H. (2002). Neuroplasticity in old age: sustained fivefold induction of hippocampal neurogenesis by long-term environmental enrichment. *Annals of Neurology, 52*, 135–143.

Kesner, R. P. (2013). An analysis of the dentate gyrus function. *Behavioural Brain Research, 254*, 1–7.

Kheirbek, M. A., Tannenholz, L., & Hen, R. (2012). NR2B-dependent plasticity of adult-born granule cells is necessary for context discrimination. *Journal of Neuroscience, 32*, 8696–8702.

Knopman, D. S., Parisi, J. E., Salviati, A., Floriach-Robert, M., Boeve, B. F., Ivnik, R. J., et al. (2003). Neuropathology of cognitively normal elderly. *Journal of Neuropathology and Experimental Neurology, 62*, 1087–1095.

Kolecki, R., Lafauci, G., Rubenstein, R., Mazur-Kolecka, B., Kaczmarski, W., & Frackowiak, J. (2008). The effect of amyloidosis-beta and ageing on proliferation of neuronal progenitor cells in APP-transgenic mouse hippocampus and in culture. *Acta Neuropathologica, 116*, 419–424.

Krezymon, A., Richetin, K., Halley, H., Roybon, L., Lasalle, J.-M., Francès, B., et al. (2013). Modifications of hippocampal circuits and early disruption of adult neurogenesis in the Tg2576 mouse model of Alzheimer's disease. *PLoS One, 8*, e76497.

Kronenberg, G., Bick-Sander, A., Bunk, E., Wolf, C., Ehninger, D., & Kempermann, G. (2006). Physical exercise prevents age-related decline in precursor cell activity in the mouse dentate gyrus. *Neurobiology of Aging, 27*, 1505–1513.

Kuhn, H. G., Dickinson-Anson, H., & Gage, F. H. (1996). Neurogenesis in the dentate gyrus of the adult rat: age-related decrease of neuronal progenitor proliferation. *Journal of Neuroscience, 16*, 2027–2033.

Lazarov, O., Mattson, M., Peterson, D., Pimplikar, S., & Van Praag, H. (2010). When neurogenesis encounters aging and disease. *Trends in Neurosciences, 33*, 569–579.

Lemaire, V., Tronel, S., Montaron, M. F., Fabre, A., Dugast, E., & Abrous, D. N. (2012). Long-lasting plasticity of hippocampal adult-born neurons. *Journal of Neuroscience, 32*, 3101–3108.

Lim, A., Tsuang, D., Kukull, W., Nochlin, D., Leverenz, J., Mccormick, W., et al. (1999). Clinico-neuropathological correlation of Alzheimer's disease in a community-based case series. *Journal of the American Geriatrics Society, 47*, 564–569.

Lopez-Toledano, M. A., & Shelanski, M. L. (2007). Increased neurogenesis in young transgenic mice overexpressing human APP(Sw, Ind). *Journal of Alzheimer's Disease, 12*, 229–240.

Manly, J. J., Touradji, P., Tang, M. X., & Stern, Y. (2003). Literacy and memory decline among ethnically diverse elders. *Journal of Clinical and Experimental Neuropsychology, 25*, 680–690.

Marin-Burgin, A., Mongiat, L. A., Pardi, M. B., & Schinder, A. F. (2012). Unique processing during a period of high excitation/inhibition balance in adult-born neurons. *Science, 335*, 1238–1242.

Marr, D. (1971). Simple memory: a theory for archicortex. *Philosophical Transactions of the Royal Society of London. Series B, Biological Sciences, 262*, 23–81.

Merkley, C. M., Jian, C., Mosa, A., Tan, Y.-F., & Wojtowicz, J. M. (2014). Homeostatic regulation of adult hippocampal neurogenesis in aging rats: long-term effects of early exercise. *Frontiers in Neuroscience, 8*, 174. http://dx.doi.org/10.3389/fnins.2014.00174.

Mirochnic, S., Wolf, S., Staufenbiel, M., & Kempermann, G. (2009). Age effects on the regulation of adult hippocampal neurogenesis by physical activity and environmental enrichment in the APP23 mouse model of Alzheimer disease. *Hippocampus, 19*, 1008–1018.

Mongiat, L. A., Esposito, M. S., Lombardi, G., & Schinder, A. F. (2009). Reliable activation of immature neurons in the adult hippocampus. *PLoS One, 4*, e5320.

Morris, A. M., Churchwell, J. C., Kesner, R. P., & Gilbert, P. E. (2012). Selective lesions of the dentate gyrus produce disruptions in place learning for adjacent spatial locations. *Neurobiology of Learning and Memory, 97*, 326–331.

Mortimer, J. A. (1997). Brain reserve and the clinical expression of Alzheimer's disease. *Geriatrics, 52*(Suppl. 2), S50–S53.

Nakashiba, T., Cushman, J. D., Pelkey, K. A., Renaudineau, S., Buhl, D. L., Mchugh, T. J., et al. (2012). Young dentate granule cells mediate pattern separation, whereas old granule cells facilitate pattern completion. *Cell, 149*, 188–201.

Ohm, T. G. (2007). The dentate gyrus in Alzheimer's disease. In E. S. Helen (Ed.), *Progress in brain research* (pp. 723–740). Elsevier.

Overstreet Wadiche, L., Bromberg, D. A., Bensen, A. L., & Westbrook, G. L. (2005). GABAergic signaling to newborn neurons in dentate gyrus. *Journal of Neurophysiology, 94*, 4528–4532. http://dx.doi.org/10.1152/jn.00633.2005.

Piatti, V. C., Ewell, L. A., & Leutgeb, J. K. (2013). Neurogenesis in the dentate gyrus: carrying the message or dictating the tone. *Frontiers in Neuroscience, 7*.

Ramirez-Amaya, V., Marrone, D., Gage, F. H., Worley, P. F., & Barnes, C. A. (2006). Integration of new neurons into functional neural networks. *The Journal of Neuroscience, 26*, 12237–12241.

Riley, K. P., Snowdon, D. A., & Markesbery, W. R. (2002). Alzheimer's neurofibrillary pathology and the spectrum of cognitive function: findings from the Nun Study. *Annals of Neurology, 51*, 567–577.

Sahay, A., Scobie, K., Hill, A., O'carroll, C., Kheirbek, M., Burghardt, N., et al. (2011). Increasing adult hippocampal neurogenesis is sufficient to improve pattern separation. *Nature, 472*, 466–470.

Seki, T., & Arai, Y. (1995). Age-related production of new granule cells in the adult dentate gyrus. *Neuroreport, 6*, 2479–2482.

Snowdon, D. A., Kemper, S. J., Mortimer, J. A., Greiner, L. H., Wekstein, D. R., & Markesbery, W. R. (1996). Linguistic ability in early life and cognitive function and Alzheimer's disease in late life: findings from the Nun study. *Journal of the American Medical Association, 275*, 528–532.

Snyder, J., Kee, N., & Wojtowicz, J. (2001). Effects of adult neurogenesis on synaptic plasticity in the rat dentate gyrus. *Journal of Neurophysiology, 85*, 2423–2431.

Snyder, J. S., Hong, N. S., Mcdonald, R. J., & Wojtowicz, J. M. (2005). A role for adult neurogenesis in spatial long-term memory. *Neuroscience, 130*, 843–852.

Song, J., Zhong, C., Bonaguidi, M. A., Sun, G. J., Hsu, D., Gu, Y., et al. (2012). Neuronal circuitry mechanism regulating adult quiescent neural stem-cell fate decision. *Nature, 489*, 150–154. http://dx.doi.org/10.1038/nature11306.

Song, J., Sun, J., Moss, J., Wen, Z., Sun, G. J., Hsu, D., et al. (2013). Parvalbumin interneurons mediate neuronal circuitry-neurogenesis coupling in the adult hippocampus. *Nature Neuroscience, 16*, 1728–1730. http://dx.doi.org/10.1038/nn.3572.

Stern, Y., Gurland, B., Tatemichi, T. K., Tang, M. X., Wilder, D., & Mayeux, R. (1994). Influence of education and occupation on the incidence of Alzheimer's disease. *Journal of the American Medical Association, 271*, 1004–1010.

Stern, Y., Alexander, G. E., Prohovnik, I., Stricks, L., Link, B., Lennon, M. C., et al. (1995). Relationship between lifetime occupation and parietal flow: implications for a reserve against Alzheimer's disease pathology. *Neurology, 45*, 55–60.

Stern, Y. (2002). What is cognitive reserve? Theory and research application of the reserve concept. *Journal of the International Neuropsychological Society, 8*, 448–460.

Tashiro, A., Makino, H., & Gage, F. H. (2007). Experience-specific functional modification of the dentate gyrus through adult neurogenesis: a critical period during an immature stage. *Journal of Neuroscience, 27*, 3252–3259.

Tayeb, H. O., Murray, E. D., Price, B. H., & Tarazi, F. I. (2013). Bapineuzumab and solanezumab for Alzheimer's disease: is the 'amyloid cascade hypothesis' still alive? *Expert Opinion on Biological Therapy, 13*, 1075–1084.

Tomlinson, B. E., Blessed, G., & Roth, M. (1968). Observations on the brains of non-demented old people. *Journal of Neurological Sciences, 7*, 331–356. http://dx.doi.org/10.1016/0022-510X(68)90154-8.

Tomlinson, B. E., Blessed, G., & Roth, M. (1970). Observations on the brains of demented old people. *Journal of Neurological Sciences, 11*, 205–242.

Toner, C. K., Pirogovsky, E., Kirwan, C. B., & Gilbert, P. E. (2009). Visual object pattern separation deficits in nondemented older adults. *Learning and Memory, 16*, 338–342.

Tronel, S., Fabre, A., Charrier, V., Oliet, S. H. R., Gage, F. H., & Abrous, D. N. (2010). Spatial learning sculpts the dendritic arbor of adult-born hippocampal neurons. *Proceedings of the National Academy of Sciences of the United States of America, 107*, 7963–7968.

Tronel, S., Lemaire, V., Charrier, V., Montaron, M-F., Abrous, D.N. (2015). Influence of ontogenetic age on the role of dentate granule neurons. *Brain Structure and Function, 220*, 645–661.

Trouche, S., Bontempi, B., Roullet, P., & Rampon, C. (2009). Recruitment of adult-generated neurons into functional hippocampal networks contributes to updating and strengthening of spatial memory. *Proceedings of the National Academy of Sciences of the United States of America, 106*, 5919–5924.

Valenzuela, M. J., & Sachdev, P. (2006). Brain reserve and dementia: a systematic review. *Psychological Medicine, 36*, 441–454.

Van Praag, H., Shubert, T., Zhao, C., & Gage, F. H. (2005). Exercise enhances learning and hippocampal neurogenesis in aged mice. *Journal of Neuroscience, 25*, 8680–8685.

Wesnes, K. A., Annas, P., Basun, H., Edgar, C., & Blennow, K. (2014). Performance on a pattern separation task by Alzheimer's patients shows possible links between disrupted dentate gyrus activity and apolipoprotein E ∈4 status and cerebrospinal fluid amyloid-β42 levels. *Alzheimer's Research and Therapy, 6*, 20.

White-Schwoch, T., Woodruff Carr, K., Anderson, S., Strait, D. L., & Kraus, N. (2013). Older adults benefit from music training early in life: biological evidence for long-term training-driven plasticity. *Journal of Neuroscience, 33*, 17667–17674.

Wojtowicz, J. M. (2012). Adult neurogenesis. From circuits to models. *Behavioural Brain Research, 227*, 490–496.

CHAPTER

5

Physical Exercise

S.-Y. Yau, A. Patten*, Z. Sharp, B.R. Christie*

University of Victoria, Victoria, BC, Canada

INTRODUCTION

The brain is not a static entity; rather it can show both structural and functional changes in response to environmental and experiential demands. One example of brain plasticity is adult neurogenesis, which is a process of the production of new neurons in the adult brain and integration into the existing central nervous system (CNS) circuitry throughout the life span. Adult neurogenesis occurs in two major neurogenic areas of the brain: the subventricular zone and the subgranular zone (SGZ) in the dentate gyrus (DG) of the hippocampus. In the SGZ, progenitor cells create neuroblasts that migrate into the granule cell layer and differentiate into dentate granule cells (Cameron, Woolley, McEwen, & Gould, 1993). These new neurons can be incorporated into existing neuronal circuits (Gheusi & Lledo, 2007) and display action potentials and functional synaptic connections similar to those of DG granule cells born during the developmental period (van Praag et al., 2002). Adult hippocampal neurogenesis has been shown to be important for improving hippocampal-dependent learning tasks (Gould, Beylin, Tanapat, Reeves, & Shors, 1999; Kempermann & Gage, 2002a) and synaptic plasticity (van Praag, Christie, Sejnowski, & Gage, 1999), and it may be the cause of the therapeutic effects of antidepressants (Santarelli et al., 2003; Surget et al., 2008) and physical exercise (Bjornebekk, Mathe, & Brene, 2005; Yau et al., 2011), as well as regulation of the stress response (Snyder, Soumier, Brewer, Pickel, & Cameron, 2011).

Animal studies have demonstrated that physical exercise facilitates both structural and functional plasticity in the hippocampus in which voluntary wheel running not only enhances cell proliferation, neuronal differentiation,

*Both authors contributed equally to this chapter.

http://dx.doi.org/10.1016/B978-0-12-801977-1.00005-2

and survival (van Praag, Christie, et al., 1999; van Praag, Kempermann, & Gage, 1999) of newborn cells, but also improves synaptic plasticity (Farmer et al., 2004; Kronenberg et al., 2006; Liu, Zhao, Cai, Zhao, & Shi, 2011; van Praag, Christie, et al., 1999; Titterness, Wiebe, Kwasnica, Keyes, & Christie, 2011) and spatial learning (Fordyce & Wehner, 1993), in both rats and mice. New neurons may play an important role in synaptic plasticity and thus learning and memory (Deng, Aimone, & Gage, 2010), raising the possibility that exercise-elicited cognitive benefits in humans may be mediated by alterations in adult neurogenesis (Yau, Gil-Mohapel, Christie, & So, 2014).

Despite the fact that physical exercise has long been known to be beneficial for brain health, the duration, intensity, or type of physical exercise which should be considered as the most effective exercise prescription for improving cognitive function across the age spectrum remain unclear. Understanding the relationships between cognitive benefits from physical exercise and these variables, such as duration, intensity, and type of exercise training, will be critical for reliably producing benefits for both structural and functional plasticity, and hence for obtaining the most beneficial effects of exercise in the human brain.

In animal models, the intensity, type, and duration of exercise training and the animal strain, species, and age when used in experiments can all affect the benefits that exercise has on cognition (Kronenberg et al., 2006; Merritt & Rhodes, 2015; Wyss, Chambless, Kadish, & van Groen, 2000) as well as adult neurogenesis (Clark et al., 2011; Holmes, Galea, Mistlberger, & Kempermann, 2004; Kronenberg et al., 2006; Merritt & Rhodes, 2015). In this chapter, we discuss how these variables affect physical exercise-induced neurogenesis and the potential mechanisms behind how physical exercise induces neurogenesis based on findings from the animal studies. Discussion is also extended to human studies with information on recent approaches for measuring adult neurogenesis in the live human brain, followed by evidence showing exercise-enhanced cognition in humans and its dosage effects.

Exercise regimes range from voluntary wheel running (van Praag, Christie, et al., 1999; Yau et al., 2011) and voluntary resistance running (Lee, Inoue, et al., 2013) to forced exercise including treadmill running (Arida, Scorza, da Silva, Scorza, & Cavalheiro, 2004; Uda, Ishido, Kami, & Masuhara, 2006) and forced swimming (van Praag, Kempermann, et al., 1999). Both forms of physical exercise have consistently been shown to promote adult neurogenesis (van Praag, Christie, et al., 1999; van Praag, Kempermann, et al., 1999) and learning and memory (Ang, Dawe, Wong, Moochhala, & Ng, 2006; Fordyce & Wehner, 1993; van Praag, Shubert, Zhao, & Gage, 2005).

The beneficial effects of physical exercise are mostly reported in animal studies using voluntary wheel running. Voluntary running allows an animal to freely access a running wheel and to decide when, how often, and how much they want to run; whereas forced running has predetermined parameters to force the animals to run, even if they are not motivated to do so. The voluntary running regime is thought to be easily translated to interpret the effect of physical exercise in human conditions, because animals are allowed to choose how much and when to run. However, for quantifying and analyzing the beneficial effects of physical exercise, forced exercise (eg, treadmill running) is often a better choice to ensure a more accurate correlation between the amount of physical activity and the benefits on the outcome measured. For example, treadmill running has an advantage over voluntary wheel running whereby the researcher can precisely control the exercise speed, frequency, intensity, duration, and time of physical training.

Despite the fact that there is large variability in the model (voluntary to forced) and duration (few days to months) of exercise training used in different laboratory animals (mice to rats), exercise-induced neurogenesis has been consistently reported (Eadie, Redila, & Christie, 2005; Kronenberg et al., 2006; van Praag, Kempermann, et al., 1999; van Praag et al., 2002; van Praag et al., 2005; Yau et al., 2011, 2012). See Table 5.1 for a summary of running-induced hippocampal neurogenesis in C57BL/6 mice. However, the most beneficial type, intensity, and duration of physical exercise have yet to be established. Forced and voluntary exercise may exert differential effects on the brain and behavior (Arida et al., 2004; Leasure & Jones, 2008).

VOLUNTARY WHEEL RUNNING

Nocturnal animals exhibit high running activity during their active phase (dark period) and very little running activity during their inactive phase (light period). Rodents with access to a wheel usually run during their active phase (Mistlberger & Holmes, 2000). Mice vary with spontaneous daily running distances ranging from ~2 to 10 km/day across difference strains (Clark et al., 2011). Most of these studies were conducted using voluntary wheel running with different housing conditions, for example, individual housing with a running wheel (Stranahan, Khalil, & Gould, 2006; Yau et al., 2011), two animals per wheel (Kronenberg et al., 2006), three to four animals per wheel (van Praag, Kempermann, et al., 1999), or six animals with three wheels in a cage (Yau, Li, et al., 2014). Studies using voluntary wheel running mainly focus on the running distance, but often neglect the exercise intensity; however, this may be due to the difficulties in precisely controlling the intensity, duration, and time interval of running using this paradigm.

TABLE 5.1 Effects of Physical Exercise on Hippocampal Cell Proliferation and Neurogenesis in C57BL/6 Mice

Reference	Age	Gender	Housing	Type of exercise	Duration	Distance	Proliferation (fold of increase)	Neurogenesis (fold of increase)	Other factors?
van Praag, Kempermann et al. (1999)	3 months	Female	Group	Voluntary: Running wheel	12 and 42 days		BrdU: 1.48	BrdU 2.01	Enrichment, learning
	"	"	"	Forced: Swimming (at most 2 × 40 s per day)	"			NS decre	
van Praag et al. (2005)	3 months	Male	Individual	Voluntary: Running wheel	45 days	4.9 km/day	X	BrdU: 3.84	Age
	19 months	"	"	"	45 days	3.9 km/day	X	BrdU: 5.61	
Kronenberg et al. (2006)	74 days	Male	Group	Voluntary: Running wheel	3 days		Ki67: 1.29 // BrdU: 1.47	DCX: NS incre	Age
	"	"	"	"	10 days		Ki67: 1.86// BrdU: 1.45	DCX: 1.69	
	"	"	"	"	32 days		Ki67: 0// BrdU: NS	DCX: 2	
	12 months	"	"	"	10 days	3.4 km/day	BrdU: 2	DCX: 1.25 non	
	24 months	"	"	"	10 days	1.4 km/day	BrdU: 2	DCX: 1.8	
	9 months	"	"	"	6 months	1.5 km/day	BrdU: 2.5	DCX: 2	
Wu et al. (2008)	9.5 months	Male	?	Forced: Treadmill 10 m/min for 20 min/day			BrdU: 2.14 (mix?)	DCX: 2.22	

	13.5 months	"	"	with increasing 10 min/day until 60 min/day (~70% maximal oxygen consumption)			BrdU: 1.67 (mix?)	DCX: 3.33	
Van der Borght et al. (2009)	8 weeks	Male	Individual	Voluntary: Running wheel	1	5–6 km/day	Ki67: 0	DCX: 0	Duration/deprivation
	"	"	"	"	3	5–6 km/day	Ki67: 1.27	DCX: 0	
	"	"	"	"	10	13 km/day near end	Ki67: 1.6	DCX: 1.25	
	"	"	"	"	10 (removed; 1 day later)		Returned to control	Remained inc	
	"	"	"	"	10 (removed; 6 days later)		Returned to control	Remained inc	
van Praag, Christie BR et al. (1999)	3 months	Female	Group	Voluntary: Running Wheel	2–4 months	4.78 km/day		2.66	
Snyder et al. (2009)	7 weeks	Male	Individual	Voluntary: ?	12 days		PCNA: 1.62–1.72	BrdU: 1.6	Stress
					19 days		0?	BrdU: 1.6	
Rei et al. (2011)	"Adult"	Male	?	Voluntary: ?	6 weeks		BrdU: 1.25	X	PGRN
Garrett et al. (2012)	8 weeks	Male	Individual	Voluntary: Running saucer/dish	14 days		Ki67: 1.4	BrdU??	

Continued

TABLE 5.1 Effects of Physical Exercise on Hippocampal Cell Proliferation and Neurogenesis in C57BL/6 Mice—cont'd

Reference	Age	Gender	Housing	Type of exercise	Duration	Distance	Proliferation (fold of increase)	Neurogenesis (fold of increase)	Other factors?
					28 days		NS		
Fuss et al. (2009)	8 weeks + 10 days	Male	Individual	Voluntary: ?	6 weeks		Ki67: 1.14 (NS)		
Kannangara et al. (2011)	2 months (+7 days)	Male	Individual	Voluntary: ?	12 days		BrdU: 1.5		Social and stress
			Group	Voluntary: ?	"		BrdU: 2		
Holmes et al. (2004)	10 weeks	Male	Individual	Voluntary: Restricted access 1 or 3 h/day	7 days/ 28 days		BrdU: 2 (3 h/ day)	BrdU: 2 (3 h/ day)	Circadian
Mustroph et al. (2012)	7 weeks	Male	Individual	Voluntary: ?	32 days		X	BrdU: 2	Enrichment
Clark et al. (2008)	154 days	Both	Individual	Voluntary: ?	40 days		X	BrdU: 4	Ablation/ behavior
Pereira et al. (2007)	7 weeks	Unknown	Unknown	Voluntary: ?	2 weeks		X	BrdU: 1.5	Cerebral blood volume
Klempin et al. (2013)	P42	Unknown	Individual	Voluntary: ?	6 days		BrdU: 2.4		Serotonin depletion
	P80						BrdU: 2.3		
	1 year						BrdU: 2		
Van der Borght (2006)	6–8 weeks	Male	Unknown	Voluntary: ?	9 days		Ki67: 1.2		Circadian

Kitamura et al. (2003)	42 days	Unknown	Group	Voluntary: ?	21 days		BrdU: 1.4	BrdU: 2	NR2A KO
Klaus et al. (2012)	12 weeks	Female	Individual	Voluntary: ?	2 weeks		Ki67: 1.3	DCX: 1.7	House mice
				Run for food			0?	DCX: 1.4	
Kobilo et al. (2011)	5 weeks	Female	Group	Voluntary: 10 mice/10 wheels	12 or 42 days		BrdU: 1.6	BrdU: 2.3	Enrichment
Kannangara et al. (2011)	2 month	Female	Individual	Voluntary: ?	11 days		BrdU: 2		Age and social
			Group				?		
	18 month		Individual				?		
			Group				?		
Naylor et al. (2008)	8 weeks	Male	Individual	Voluntary: Running wheel	30 days		Sox-2: 1.16	BrdU: 1.96	Irradiation
Marlatt et al. (2012)	17 months	Female	Individual	Voluntary: Running wheel	8 months	4 km/day		BrdU: 6	BDNF, MWM
Li et al. (2013)	3 months	Male	Group	Forced treadmill running 10 min at speed of 10 m/min	8 or 29 days	~600 m/day		BrdU: 2–2.5 (acute)	Varied duration or daily distance
								BrdU: 0.5–2 (chronic)	
Fisher et al. (2014)	10 weeks	Female	Individual	Voluntary: Running wheel	5 days		CIdU: 1.56		

BrdU, 5-bromo-2′-deoxyuridine; *CIdU*, chlorodeoxyuridine; *DCX*, doublecortin; *NS*, nonsignificant; *Incre*, increase; *Decre*, decrease; ?, unknown.

Dosage (Intensity and Duration) Effects

Because it is hard to control the amount of running each animal does using the voluntary running regime, the minimum amount of voluntary running or the duration that is required for achieving a significant increase in neurogenesis or an optimal increase in neurogenesis remains unknown. Kronenberg et al. (2006) investigated changes of cell proliferation and neurogenesis with different durations of voluntary wheel running on days 0, 3, 10, and 32 in C57BL/6 male mice. They reported that cell proliferation follows an inverted U-shape in response to voluntary running in these adult mice. Cell proliferation rate was found to peak at 3 days of voluntary running, and then return to basal levels after 32 days of running. This finding is echoed by Snyder, Glover, Sanzone, Kamhi, and Cameron (2009), reporting that voluntary running for 12 days increased cell proliferation in male C57BL/6 mice. Of note, voluntary running displays a transient influence on proliferation with increases in the number of proliferating cells after 12 days, but not 19 days of running. Conversely, running leads to a consecutive increase in the number of newborn neurons with increasing days running (Kronenberg et al., 2006) and voluntary running for 5 and 12 days has also been shown to increase neurogenesis (Farioli-Vecchioli et al., 2014; Fischer, Walker, Overall, Brandt, & Kempermann, 2014). Snyder et al. (2009) also found that 5 and 12 days of running significantly increased cell survival rate at both time points. Three weeks of voluntary running increases cell proliferation in 21-day-old mice (Kitamura, Mishina, & Sugiyama, 2003). Interestingly, prolonged running in 9-month-old mice for several months can restore the running-induced increase in hippocampal cell proliferation, which is absent in 19-day running in Synder's study (Snyder et al., 2009), suggesting that a long-term sustained running is able to reverse age-associated decline in cell proliferation.

Running distance is the sole parameter of voluntary running that can be measured as a physical assessment. Holmes et al. (2004) studied the effects of circadian rhythm and the dose-dependent effects of voluntary running activity on adult hippocampal cell proliferation and neurogenesis in C57BL/6 mice. Mice were allowed access to the running wheel for 0, 1, or 3 h at three different time points including inactive phase (0700), onset of active phase (1300), and active phase (1900), respectively. Following 8 days of running, only mice allowed to run in their active period for 3 h displayed a significant increase in cell proliferation and neurogenesis compared to their sedentary counterparts. This study indicates that the neurogenic effect of physical running is not purely dependent on total amount of activity and that the duration or circadian phase can significantly influence this activity-induced increase in cell proliferation and neurogenesis.

Kitamura et al. (2003) reported a significant increase in cell proliferation in C57/Bl6 mice that were housed six to a cage equipped with a running wheel for 21 days. In mice with an approximately twofold decrease in running distance, determined by removing the running wheels from the home cage every other day, they showed an increase in cell proliferation together with an increase in hippocampal brain-derived neurotrophic factor (BDNF) levels comparable to those of the mice with continuous running for 21 days, indicating that such decrease in running distance is still effective in inducing a significant increase in cell proliferation in their mice.

Adding to the aforementioned strain difference, species differences in the response to running-induced neurogenesis have also been found. The inverted U-shape response of cell proliferation in response to running in adult C57BL/6J mice was absent in adult male Sprague–Dawley (SD) rats. In study using adult SD rats, an acute effect of voluntary running on cell proliferation was observed at the 3-day time point, which is similar to that reported in mice (Patten et al., 2013). The other time points including 7, 14, and 28 days were also able to promote cell proliferation in these rats. Acute running for 3, 5, or 7 days did not increase neuronal differentiation, whereas chronic running for at least 14 days enhanced differentiation (Patten et al., 2013; Yau et al., 2011, 2012). Taken together, voluntary running for at least 7 days is likely to be the minimal duration of voluntary running to observe a significant increase in both cell proliferation and neuronal differentiation in SD rats.

Lee, Inoue, et al. (2013) investigated whether resistance wheel running could have an effect on promoting neurogenesis comparable to that observed in wheel running without resistant load. Resistant voluntary wheel running was introduced by increasing work levels with a minimum resistance for the first week and then progressively increasing the resistance to 35% of body weight for the 4-week running period. They reported that both types of running increased cell proliferation and neurogenesis. Their data showed that despite a shorter running distance (approximately 470 m/day which is equivalent to half of that with the free-load running wheel), rats with a sevenfold increase in work levels showed an increase in cell proliferation and neurogenesis comparable to the rats with a nonresistant running wheel. All animals that exercised, regardless of resistance level, had improved spatial memory and increased BDNF levels in comparison to their sedentary counterparts. Selectively bred low capacity runner rats display a similar increase in hippocampal mRNA BDNF levels despite their lower running activity (sevenfold lower than the high capacity runners) (Groves-Chapman et al., 2011). This study demonstrated that hippocampal mRNA BDNF levels can be increased with a minimum of 500 m/day of voluntary wheel running. It is unknown if there is a comparable increase in neurogenesis between high and low capacity runners, because neurogenesis is not yet examined in these runners.

Genetic Influence

Different strains of mice display a large variation in basal levels of precursor cell proliferation (Kempermann, Chesler, Lu, Williams, & Gage, 2006; Kempermann & Gage, 2002b; Kempermann, Kuhn, & Gage, 1997). This intrinsic variation in levels of adult neurogenesis may lead to differential sensitivity in response to physical exercise; therefore, genetic variation in experimental animals should be taken into account for investigating the exercise-elicited beneficial effects on adult neurogenesis. A positive association between increased neurogenesis and enhanced hippocampal-dependent learning and memory has been repeatedly found in C57BL/6J mice (Marlatt, Potter, Lucassen, & van Praag, 2012; Mustroph et al., 2012; van Praag et al., 2005); however, such associations were absent in the mice that were selectively bred for increased voluntary wheel running. Although these mice displayed large increases in neurogenesis, they did not show improvements in spatial learning in the Morris water maze (Rhodes et al., 2003). Furthermore, with equal duration and amount of voluntary running activity to those of C57BL/6 mice that showed increased cell proliferation, DBA/2 mice showed no significant increase in cell proliferation (Overall et al., 2013), suggesting that the strain of animals can significantly influence neurogenesis levels following running.

With prior characterization of strain differences in basal levels of proliferation by the Kempermann group, Clark et al. studied the effects of genetic influence on physical running-induced adult hippocampal neurogenesis in 12 mouse strains. They reported that increases in hippocampal neurogenesis can be largely influenced by genetic background whereby some strains show a relatively higher increase in neurogenesis with a comparable amount of running activities to others (Clark et al., 2011). In fact, there is a significant effect of strain on running activity with varied daily running distance ranging from approximately 2 to 9 km/day among 12 strains. Notably, running can increase adult hippocampal neurogenesis in all strains but the magnitude of the increase is strain dependent. The least responsive strain is the C57BL/6J mouse line with only a 1.6-fold increase in neurogenesis, whereas other strains showed a more than twofold increase in neurogenesis, with four- to fivefold increases in the most responsive strains including BALV/cByJ, CAST/EiJ, AKR/J, B6129SF1/J, and SM/J in response to voluntary running. Of note, there was a significant correlation between individual running distance and density of newborn neurons in all strains, indicating that running distance is still the main factor for inducing hippocampal neurogenesis. The authors argued that the low levels of increases observed in C57BL/6J mice may be due to the relatively high levels of basal neurogenesis and lowest level of running activity. On the other hand, it is worth emphasizing that home cage physical activity, which is measured in the home cage without a running wheel, was not correlated with hippocampal neurogenesis regardless of strains (Clark et al., 2011), suggesting that a certain threshold amount of physical activity is required for observable and quantifiable changes in neurogenesis.

Pioneer work by van Praag, Christie, et al. (1999) has shown that C57BL/6J mice subjected to voluntary wheel running showed a significant increase in hippocampal neurogenesis in association with improved spatial learning and memory. With the discovery of low responders (129S1, D2, and B6 strains) and high responders (B6D2F1 and B6129F1 strains) to exercise-induced neurogenesis, Merritt and Rhodes (2015) examined if there is a positive correlation between the levels of neurogenesis and the degree of cognitive improvement in water maze task. They have shown a main effect of strain in determining the average running activity across 30 days of access to a running wheel and increases in neurogenesis levels following running. However, a significant relationship between the number of newborn neurons and levels of running activity is only observed in some strains and appears to be limited to 129S1 and B6129F1 strains. In contrast, B6, D2, and B6D2F1 strains displayed a null or negative relationship, suggesting that increases in neurogenesis in response to running may depend on running distance only in some specific mouse strains. Enhancements in hippocampal neurogenesis were associated with improvement in performance in the plus water maze in all strains used in their study. However, the level of neurogenesis was not necessarily proportional to the improvements in learning, since the 129S1 strain showed the lowest increase in neurogenesis following running, but displayed an improvement in spatial learning similar to that of the B6 strain which had the highest increase in neurogenesis among the strains. It is well known that in addition to neurogenesis, physical exercise triggers a variety of changes in the adult brains that may enhance synaptic plasticity, for example, dendritic enrichment (Eadie et al., 2005), increasing production of neurotrophic factors including insulin-like growth factor-1 (IGF-1) (Carro, Nunez, Busiguina, & Torres-Aleman, 2000), vascular endothelial growth factor (VEGF) (Fabel et al., 2003), and BDNF (Neeper, Gomez-Pinilla, Choi, & Cotman, 1996) (see Section Cellular Mechanism for further discussion). Variations in these factors among strains may also contribute to differential results in learning behavior in response to physical exercise. Therefore, learning and neurogenesis may be dissociated in some mouse strains.

In selectively bred mice that tend to have higher running activity (10 km/day, approximately threefold higher than normal control mice), it has been shown that the greatest increase in physical exercise-induced neurogenesis (fivefold increase) is observed when compared to nonrunning counterparts. Normal control runners show a fourfold increase in neurogenesis following a 40-day running paradigm as compared to their nonrunning counterparts. A strong correlation between the amount of neurogenesis and the running distance was replicated in the normal control mice, but such correlation was absent in their selectively bred mice with higher running activity (Rhodes et al., 2003), suggesting that a ceiling effect of running-induced neurogenesis may occur in these mice. On the other hand, running improves learning in control mice, but not in selectively bred runners, supporting the above-noted findings that increases in neurogenesis may not be necessarily associated with improvement in spatial learning.

FORCED EXERCISE

Forced Treadmill Running

Although treadmill training may trigger a stress response concurrent with elevated glucocorticoid hormones which is suppressive to neurogenesis (Gould, Tanapat, McEwen, Flugge, & Fuchs, 1998), this training regime allows researchers to investigate the effects of exercise intensity or duration on neurogenesis and changes in its related molecules that may correspond to the impact of exercise intensity.

Li et al. (2013) conducted an experiment in mice with three regimes of treadmill running which divided the animals into four groups (1) controls nonrunner; (2) regular runners with daily running on treadmill at the same time, speed, and duration; (3) irregular duration runners that ran at the same time and speed, but different duration; and (4) irregular time-of-day runners that ran at the same duration and speed, but different time of day. Both acute (8 days) treadmill running and long-term (29 days) treadmill running were able to increase cell survival in the DG of the mice regardless of time of day or duration of physical training. However, the mice with the regular exercise regime showed the highest increase in survival rate of newborn neurons among the runner groups. The regular runners also displayed the highest increase in neuronal differentiation when compared to the other groups. This study not only suggests that a regular pattern of physical exercise yields a greater influence on hippocampal neurogenesis in association with significant improvement in hippocampal learning and memory, but also indicates that forced treadmill running can be as effective as voluntary running in inducing adult hippocampal neurogenesis, even with an acute exercise training for 7 days (Uda et al., 2006).

Unlike voluntary wheel running, a suitable intensity or duration of treadmill running may be critical for observing a positive effect on hippocampal neurogenesis owing to the stress response that may be triggered by forced treadmill training. Lou, Liu, Chang, and Chen (2008) investigated if different intensities of treadmill running would have differential effects on neurogenesis and lead to changes in neurogenesis-related molecules in parallel to the training intensity in juvenile SD rats. Low-intensity, intermediate-intensity, and high-intensity exercise was chosen based on the speeds of treadmill running. The study concluded that low-intensity treadmill running is optimal for increasing hippocampal neurogenesis in parallel with the greatest increase in exercise-related molecules such as BDNF, VEGF, and NMDA receptor subunit 1, whereas rats with intermediate- and high-intensity exercise showed no such effects (Lou et al., 2008).

Animals subjected to voluntary running usually run a much longer distance and duration than those subjected to forced running. To compare the effects of forced and voluntary running on behavioral and neural changes, Leasure and Jones (2008) carefully matched running activity from these two methods following 8 weeks of running in female Long-Evans rats. Both voluntary and forced running can significantly increase hippocampal neurogenesis when compared to sedentary controls. Interestingly, compared to voluntary runners, a significantly higher number of survival neurons were found in rats subjected to forced treadmill exercise. Stress is apparently not a factor that contributes to the differential effects on neurogenesis in this study, because the author reported that both forced and voluntary runners display a comparable level of fecal corticosterone levels, which may be attributed to adaptation to the chronic running paradigm (Fediuc, Campbell, & Riddell, 2006). Notably, differences in speed and duration in this study may be a critical factor underlying the differential effects of these two exercise regimes. The author reported that voluntary runners ran for less time at a higher speed, whereas forced runners ran for a longer period of time at a lower speed. These data suggest that an appropriate combination of speed and duration of forced physical training may be important for optimal enhancements in neurogenesis.

Forced Swimming

Pioneer work by van Praag, Kempermann, et al. (1999) compared cell proliferation and survival in mice under different housing conditions including (1) control group; (2) voluntary running with three to four mice housed in a cage with a running wheel; (3) water maze learners with two trials per day over 30 days; (4) swimmers placed in a water maze twice per day for 30 days (each trial consists of approximately between 12 and 40 s per day); (5) environment enriched group with social interaction, running wheel, toys, and tunnels. In this study, only runner groups showed a significant increase in cell proliferation while there was no significant increase in cell survival in mice with running wheels or environment enrichment. Neither spatial learning training nor swimming in the water maze affected proliferation or survival of newborn cells (van Praag, Kempermann, et al., 1999). In contrast, another study using more trials per day in the water maze showed a remarkable increase in neurogenesis (Gould et al., 1999).

Collectively taken, the data suggest that the intensity of water maze swimming was too small in the van Praag study and therefore was not adequate to stimulate adult hippocampal neurogenesis. It may also be due to a forced swimming-induced stress response, because stress can significantly suppress adult neurogenesis (Gould et al., 1998) and this may override the positive influence of limited physical exercise on neurogenesis.

THE EFFECT OF AGE ON EXERCISE-INDUCED HIPPOCAMPAL NEUROGENESIS

Drastic reductions in adult neurogenesis can be observed in aged adult rats (Drapeau et al., 2003; Heine, Maslam, Joels, & Lucassen, 2004; Kuhn, Dickinson-Anson, & Gage, 1996; McDonald & Wojtowicz, 2005; Rao, Hattiangady, & Shetty, 2006), mice (Kronenberg et al., 2006), and marmosets (Leuner, Kozorovitskiy, Gross, & Gould, 2007). Substantial declines in neurogenesis with aging also occur in human hippocampus across the life span (Knoth et al., 2010), raising the possibility that declines in neurogenesis may contribute to cognitive decline in aged individuals. This decline throughout the life span can be due to a decreasing pool of proliferating precursor cells residing in the SGZ (Kuhn et al., 1996).

The majority of studies examining hippocampal neurogenesis and exercise utilize young adult rats between postnatal days (PND) 60–120. However, there are studies that have examined animals outside of this time frame, including juvenile animals (adolescents classified as between PND 25 and PND 50) and aged animals (classified as over PND 150). In the section, we discuss the effects of age on exercise-induced neurogenesis, with a focus on studies using rats and mice.

Juvenile Rats and Mice

In humans there is evidence to suggest that aerobic exercise in childhood may be beneficial for the brain later in life (Dik, Deeg, Visser, & Jonker, 2003). Some of these benefits may result from boosts in neurogenesis. Kitamura et al. (2003) have shown that 3 weeks of exercise starting in C57BL/6 mice at PND 21 can positively increase cell proliferation by examining changes in BrdU-positive cells in the DG. Furthermore, these authors determined that exercise-induced increases in proliferation were reliant on NMDA receptors, because when mice lacking the NMDA ε1 subunit were allowed access to a running wheel for 3 weeks, no changes in BrdU-positive cells were observed (Kitamura et al., 2003). Interestingly, the increase in proliferation observed in the control animals was only apparent directly following the exercise. In animals that exercised for 3 weeks and then remained sedentary for 3 weeks following, proliferation levels were comparable to animals who did not exercise at all, showing that the effects of exercise on cell proliferation in these animals were transient (Kitamura et al., 2003). It is unknown, however, whether neurogenesis (as assessed by immature or mature neuronal markers such as doublecortin or NeuN) would still be upregulated at this time.

Similar effects of exercise on cell proliferation have been observed in adolescent rats. Helfer, Goodlett, Greenough, and Klintsova (2009) determined that 12 days of running from PND 30 to PND 42 can significantly enhance the number of BrdU-positive cells in the DG, indicating increases in cell proliferation. A second cohort of animals was given access to a running wheel from PND 30 to PND 42 and then put back into standard housing conditions for an additional month before sacrifice. Cell survival and neurogenesis (as measured by BrdU-positive/NeuN cells) were also increased when compared to animals that did not exercise (Helfer et al., 2009).

Aged Rats and Mice

Neurogenesis experiences an age-related decline which begins as early as 1.5–3 months in mice (Gil-Mohapel et al., 2013), and this decrease has been linked to reductions in cognitive function which are common around this time point (Erickson & Barnes, 2003; Gil-Mohapel et al., 2013). Proliferation appears to be the most affected by age (Kuhn et al., 1996); however, neuronal differentiation and cell survival also decline with age (Gil-Mohapel et al., 2013). Interestingly, despite these age-related reductions in neurogenesis, the neurons that are born in the aged brain still appear to be functionally equivalent to those in the young brain (Morgenstern, Lombardi, & Schinder, 2008; Toni et al., 2008).

The ability of exercise to enhance neurogenesis is maintained in older animals, and exercise from 3 to 9 months of age (PND 90–270) can significantly reduce the age-dependent reduction in cell proliferation, which leads to an increase in the number of mature cells in the DG (Kronenberg et al., 2006). In fact, even short periods of running (10 days), initiated late in life (17 months), can increase cell proliferation in mice (Kannangara et al., 2011). Whether these increases in proliferation lead to overall changes in neurogenesis was not examined in this study.

Further studies have indicated that running can also boost neurogenesis and prevent age-related declines even when wheel running does not begin until middle-age (Marlatt et al., 2012; Wu et al., 2008) or senescence (van Praag et al., 2005). These increases in neurogenesis were also accompanied by significant improvements in hippocampal-dependent learning and memory (Marlatt et al., 2012; van Praag et al., 2005), indicating that exercise has a beneficial effect on age-related cognitive decline through boosting neurogenesis. Interestingly, van Praag et al. (2005), who examined the effects of exercise in 19-month-old mice, found that neurogenesis was boosted to approximately

50% of the young control animals following running, but a study conducted by Creer, Romberg, Saksida, van Praag, and Bussey (2010) did not see any benefits of running on neurogenesis when running commenced at 22 months. These discrepancies may be due to the length of time that animals had access to the running wheel (45 days vs 8 days), indicating that longer periods of exercise may be required to enhance neurogenesis in the aged brain.

The beneficial effects of exercise on hippocampal neurogenesis may also be important in neurodegenerative diseases such as Alzheimer's disease. Rodriguez et al. (2008) showed that neurogenesis was significantly reduced in the 3xTG model of Alzheimer's disease from 9 months of age, but access to a running wheel improved levels of neurogenesis in this aggressive model of Alzheimer's disease (Rodriguez et al., 2011). Unfortunately, when it comes to Huntington's disease, the results are not as promising. While declines in neurogenesis have been documented, even in the presymptomatic and early-symptomatic stages of the disease (Simpson et al., 2011), exercise does not appear to reduce or rescue these deficits (Kohl et al., 2007; Potter, Yuan, Ottenritter, Mughal, & van Praag, 2010).

The results from the above studies indicate that exercise, implemented at any age, can be beneficial for hippocampal neurogenesis. It appears that the younger the animals are when exercise begins the greater the enhancements, although the period of running also seems to play a role.

SEX DIFFERENCES IN EXERCISE-INDUCED ENHANCEMENTS IN HIPPOCAMPAL NEUROGENESIS

The male and female brains are significantly different, both structurally and functionally, and can differ in their response to interventions such as exercise. It is therefore important to examine whether exercise differentially affects hippocampal neurogenesis between the sexes. It is first important to acknowledge that "female" hormones such as estrogen and progesterone and "male" hormones such as testosterone have been linked to changes in cell proliferation and cell survival in the adult hippocampus (Galea, Spritzer, Barker, & Pawluski, 2006). For example, studies have shown that the amount of cell proliferation in the DG is significantly greater in females than in males, but this effect depends on the endocrine state of the female, and there are no differences in cell survival between the sexes (Galea & McEwen, 1999; Tanapat, Hastings, Reeves, & Gould, 1999).

The effects of exercise on cell proliferation and neurogenesis are well documented in male (Mustroph et al., 2012; Patten et al., 2013) and female (Boehme et al., 2011; Brown et al., 2003; Kannangara et al., 2011; Kobilo et al., 2011; Marlatt et al., 2012; van Praag, Kempermann, et al., 1999) rats and mice. In studies that have used both males and females, voluntary exercise has been shown to increase cell proliferation and neurogenesis in wild-type mice, with no differences noted between the sexes (Clark et al., 2008). In a study that utilized forced running on a treadmill, both male and female mice had similar increases in cell proliferation (measured by BrdU after 4 days of running), and in cell survival (measured by BrdU after 4 weeks of running, 3 times per week) (Ma, Hamadeh, Christie, Foster, & Tarnopolsky, 2012). Interestingly, Ma et al. (2012) also examined the effects of forced running on neurogenesis in a mouse model of amyotrophic lateral sclerosis. These animals had higher basal levels of cell proliferation than wild types, an effect that was more pronounced in males, and exercise actually led to a trend toward a decrease in cell proliferation ($p=0.056$) in these genetically modified animals (Ma et al., 2012). These results indicate that under specific disease conditions, exercise might not always be beneficial, and sex-specific differences may be revealed (Ma et al., 2012).

All of the studies noted above differ significantly in the methods of exercise employed (forced vs voluntary, running wheel vs treadmill), the length of the exercise period, the animal strain used (rats vs mice, strain of rat/mouse used), the age of animals when tested, and the markers used to assess proliferation and neurogenesis. However, the commonality among all these studies is that exercise can significantly enhance the levels of new neurons in the DG of both male and female animals and in the majority of cases, exercise does not affect males and females differently.

CELLULAR MECHANISMS

The majority of studies examining voluntary exercise note that cell proliferation is greatly increased, and that it is probably this increase in proliferation that leads to a net increase in neurogenesis (Olson, Eadie, Ernst, & Christie, 2006). Several neurotrophic factors such as BDNF, IGF-1, and VEGF have been suggested to play a role in neurogenic enhancements in the DG. Other evidence suggests roles for the glutamatergic system and increased gene expression. Overall, the effects of exercise are multifaceted and it is likely that many of these mechanisms occur concurrently, leading to an increase in hippocampal neurogenesis coupled with enhanced hippocampal function and learning and memory capacity. There are many ways through which voluntary exercise can increase cell proliferation including changes in blood vasculature and increases in neurotrophins, which are discussed in more detail below.

Neurotrophins

BDNF is increased in the hippocampus following exercise (Cotman & Berchtold, 2002; Boehme et al., 2011; Farmer et al., 2004; Neeper, Gomez-Pinilla, Choi, & Cotman, 1995; Vaynman & Gomez-Pinilla, 2005).

This increase occurs after only 2–7 days of running (Neeper et al., 1995) and continues for as long as the animals have access to a running wheel (Berchtold, Chinn, Chou, Kesslak, & Cotman, 2005), and for at least 2 weeks after cessation of running (Berchtold, Castello, & Cotman, 2010). An increase in BDNF that occurs due to voluntary exercise may not only promote neurogenesis but also enhance functional plasticity (eg, long-term potentiation) and improve learning and memory performance (Bekinschtein, Oomen, Saksida, & Bussey, 2011; Farmer et al., 2004; van Praag, Christie, et al., 1999). A direct link between BDNF and neurogenesis has been evidenced by remarkable reduction in adult neurogenesis in the mouse with knocking down BDNF expression in the DG using RNA interference and lentiviral injections (Taliaz, Stall, Dar, & Zangen, 2010). Other studies have found that blocking BDNF centrally using a postnatal BDNF knockout mouse model ($BDNF^{2L/2LCk\text{-}Cre}$) does not affect cell survival (and actually increases proliferation) but that the neurons that do survive following blockade of BDNF fail to mature fully (ie, they lack calbindin expression), have reduced dendritic differentiation, and do not migrate properly (Chan, Cordeira, Calderon, Iyer, & Rios, 2008). This suggests that BDNF may not be a fundamental requirement for neurogenesis to occur, but it does play an important role in ensuring complete neuronal differentiation and maturation. Additional studies have further cemented a role of BDNF in neurogenesis by overexpressing BDNF in the brain using osmotic pumps (Scharfman et al., 2005). One month after a 2-week BDNF infusion ended, animals that received BDNF showed a significant increase in neuronal differentiation, indicating that BDNF is able to increase mature neuronal populations (Scharfman et al., 2005).

Hippocampal gene expression after acute (3 days) or chronic (7 or 28 days) exercise was examined by Molteni, Ying, and Gomez-Pinilla (2002) using microarray technology. Exercise upregulated genes involved in synaptic trafficking, signal transduction, and transcription regulators as well as genes associated with the glutamatergic system (Molteni et al., 2002). However, the most pronounced increase, seen at all time points, was that of BDNF, which has interactions with most other upregulated genes. This suggests that BDNF may be responsible for many of the effects seen with exercise, in particular synaptic plasticity and neurogenesis (Molteni et al., 2002). BDNF can act both pre- and postsynaptically through the TrkB receptor which is also upregulated in response to exercise (Molteni et al., 2002). TrkB receptor activation can affect both synaptic vesicle formation and release as well as MAPK and CaMKII pathways, which can increase the expression of downstream genes such as CREB, another gene upregulated in response to exercise (Molteni et al., 2002).

Other neurotrophic factors such as nerve growth factor and FGF-2 are also upregulated following exercise (Gomez-Pinilla, Dao, & So, 1997; Neeper et al., 1996). In particular, after 3 days of exercise, levels of these neurotrophins were significantly increased, but were not significantly different from controls after 7 or 28 days of exercise (Molteni et al., 2002). This suggests that they might play a role in the early effects of exercise on the brain, but that this effect is not long term.

Another growth factor upregulated by exercise is IGF-1 (Trejo, Carro, & Torres-Aleman, 2001). In particular, IGF-1 is taken up by the hippocampus from the bloodstream and if this process is blocked using subcutaneous infusions of blocking IGF-1 antiserum, the increase in neurogenesis seen with exercise is inhibited (Trejo et al., 2001). Systemic IGF-1 injections in sedentary rats can also mimic the effects of exercise, leading to enhancements in neurogenesis (Carro et al., 2000). When IGF-1 is taken up by neurons there is increased firing and sensitivity of the neuron and this may stimulate BDNF and c-fos expression (Carro et al., 2000) as well as increase neurogenesis in surrounding areas.

Cerebral Blood Flow and Angiogenesis

Exercise increases cerebral blood flow to the brain and this may increase neuronal activity and potentially enhance neurogenesis. Angiogenesis and vascular function are increased in response to exercise in many areas of the brain, which may improve normal neuronal function as well as offering protection if insult occurs (Christie et al., 2008). Magnetic resonance imaging (MRI), used in mice and humans, has also indicated that there may be a correlation among exercise, blood flow in the DG, and neurogenesis; however, in this particular study, histological examination did not find vasculature changes in response to exercise in the mice (Pereira et al., 2007).

Increased blood flow may also increase exposure to growth factors (Neeper et al., 1995; Spier et al., 2004), which may influence neurogenesis. VEGF, a neurotrophin that stimulates angiogenesis, is upregulated by exercise (Fabel et al., 2003) and also appears to play a role in the enhancement of neurogenesis (Cao et al., 2004; Jin et al., 2002). New neurons in the DG often cluster around local microvasculature (Fabel et al., 2003; Palmer, Willhoite, & Gage, 2000). Furthermore, if VEGF is blocked, exercise-induced increases in neurogenesis do not occur (Fabel et al., 2003).

Adipocyte-Secreted Hormone: Adiponectin

Although neurotrophic factors such as BDNF, IGF-1, and VEGF have been suggested as the key mediators of exercise-induced hippocampal neurogenesis (Cotman & Berchtold, 2002), recent discoveries have found that adiponectin, an adipokine secreted by adipose tissue, serves as an additional key peripheral factor that mediates physical exercise-induced hippocampal cell proliferation (Yau, Li, et al., 2014). An in vitro study has demonstrated that adiponectin stimulates proliferation, but shows no effect on neuronal or glial differentiation of adult hippocampal neural progenitor cells (Zhang, Guo, Zhang, & Lu, 2011). Intracerebroventricular injection of recombinant adiponectin (Liu et al., 2012) or adenovirus expressing recombinant adiponectin (Yau, Li, et al., 2014) mimics the antidepressant effect of physical exercise. In addition, hippocampal adiponectin levels can be significantly increased following 14-day voluntary running and is concurrent with increases in hippocampal cell proliferation in mice (Yau, Li, et al., 2014). Furthermore, increasing adiponectin levels in the brain by injecting adenovirus expressing adiponectin into the ventricle mimics physical exercise-induced hippocampal neurogenesis and decreased depressive-like behaviors, whereas adiponectin knockout completely diminishes both the antidepressant and the neurogenic effects of physical exercise (Yau, Li, et al., 2014).

Changes in the Glutamatergic System and Other Considerations

Exercise can upregulate genes related to the glutamatergic system while downregulating genes involved in the GABAergic system (Molteni et al., 2002). Exercise can also increase the availability of the NR2A and NR2B subunits of the glutamatergic NMDA receptor in the hippocampus (Molteni et al., 2002). If the NR2A (ε-1) subunit is knocked out, the increases in neurogenesis and BDNF levels that occur with exercise do not occur (Kitamura et al., 2003).

Other than the above-noted factors, serotonin is another important factor for running-induced neurogenesis since lack of serotonin abolished running-induced proliferation, though the basal hippocampal neurogenesis is normal (Klempin et al., 2013). In addition to enhancing neurogenesis, exercise may also impact the neuronal phenotype in the DG. New neurons formed in the SGZ migrate through the granule cell layer and increase dendritic arborization as they mature (Altman & Bayer, 1990; Green & Juraska, 1985). Redila and Christie (2006) showed that voluntary exercise increases dendritic complexity in the DG. Further, exercise also increases dendritic spine size and quantity which may play a role in increased synaptic strength (Stranahan, Khalil, & Gould, 2007).

MEASURING HIPPOCAMPAL NEUROGENESIS IN THE HUMAN BRAIN

Evidence showing the occurrence of adult neurogenesis in human brain was first shown with immunostaining of fixed postmortem tissues (Eriksson et al., 1998). This result was further confirmed by later experiments with isolation of human neural progenitor cells from tissue biopsies (Johansson et al., 1999; Kukekov et al., 1999; Roy et al., 2000). However, knowledge about the functional significance of adult neurogenesis in the human brain is still very limited. Refinement of currently available in vivo imaging techniques and neurogenesis-related peripheral biomarkers will open an era for investigating the role of adult neurogenesis in exercise-elicited cognitive gain in the human brain.

In Vivo Imaging

Physical exercise improves neurogenesis in association with enhanced angiogenesis (van Praag et al., 2005; Van der Borght et al., 2009). Using MRI techniques, Pereira et al. demonstrated that the increase in cerebral blood volume in the hippocampus was positively correlated with cognitive improvement in individuals subjected to a 12-week physical training. The result corresponds to their finding in exercised mice showing a positive correlation between hippocampal-specific increase in cerebral blood volume and BrdU-positive cells in the DG (Pereira et al., 2007). Proton nuclear magnetic resonance spectroscopy (^{1}H NMR) detects the levels of a small metabolite of neural progenitor cells. *N*-Acetylaspartate is another possible way to detect neural progenitor cells in living tissue (Manganas et al., 2007). However, further studies will need to validate the feasibility and reliability of these two methods to examine levels of neurogenesis in the live human brain. With recent findings showing the functional significance of hippocampal neurogenesis in pattern separation of learning and memory formation, a pattern separation task (Dery et al., 2013) would be an alternate method for studying alterations in human hippocampal neurogenesis.

Neurogenesis Biomarkers in the Human Brain

BDNF, IGF-1, and VEGF are believed to be the primary mediators of adult neurogenesis as aforementioned in Section 6. The effects of exercise on peripheral levels of these factors are still controversial. Therefore, the relationship between exercise-induced changes in peripheral and central levels of these neurotrophic factors needs to be fully validated before drawing conclusions that these neurotrophic factors are potential neurogenesis biomarkers. With the emerging technique of in vivo imaging, future investigations will need to focus on the relationship between changes in peripheral levels of neurotrophins and adult neurogenesis in human subjects.

Brain BDNF is found to be the major contributor to the increase in plasma BDNF in response to exercise (Rasmussen et al., 2009). However, animal studies have demonstrated dissociation between changes in plasma BDNF levels and hippocampal neurogenesis in rats with voluntary exercise (Yau et al., 2012). A dissociation between central and peripheral BDNF levels has also been found in the clinical setting, reporting an increase in the brain levels of BDNF (in blood samples from the internal jugular vein), but no change in peripheral BDNF levels following 3 months of endurance training in healthy subjects (Seifert et al., 2010). Although transient increases in peripheral levels of BDNF have been reported following acute exercise (Knaepen, Goekint, Heyman, & Meeusen, 2010), the timing of blood collection after exercise may contribute to these discrepancies. Elevated BDNF levels return to baseline quickly within 1 h postexercise and then reach to a level lower than baseline (Knaepen et al., 2010) or drop below baseline 2 and 3 h after acute exercise (Castellano & White, 2008; Yarrow, White, McCoy, & Borst, 2010). Furthermore, extended physical training may affect the resting serum levels of BDNF and VEGF, because their levels are significantly decreased in resting serum levels in adolescent athletes (Lee, Wong, et al., 2013).

Several clinical studies have reported a positive correlation between serum levels of IGF-1 and cognitive functions (Aleman et al., 1999, 2000; Arwert, Deijen, & Drent, 2005). However, the relationship between IGF-1 and sustained physical exercise is equivocal. Although acute exercise training increases peripheral IGF-1 levels in both middle-aged men with two trials of 60-min cycling exercise (Manetta et al., 2003) and road cyclist athletes (Zebrowska, Gasior, & Langfort, 2009), the relationship between IGF-1 levels and cognitive function is not reported in these studies. Controversial results have been reported with sustained exercise training with no effect (Cearlock & Nuzzo, 2001) or negative effects (Karatay, Yildirim, Melikoglu, Akcay, & Senel, 2007) on IGF-1 levels in healthy subjects. Decreased IGF-1 levels were also found in athletes (Lee et al., 2014) and individuals with 6-week low-intensity cycling (Nishida et al., 2010).

Acute exercise can increase levels of VEGF in skeletal muscle (Gustafsson et al., 2002; Hoffner, Nielsen, Langberg, & Hellsten, 2003), with its expression reaching peak levels immediately after exercise training, and gradually returning to basal levels (Breen et al., 1996). VEGF mRNA expression in human muscle was elevated 30 min after cessation of exercise (Gustafsson et al., 2002). However, plasma VEGF protein levels were decreased in the femoral vein following 3 h of two-legged kicking ergometry (Hiscock, Fischer, Pilegaard, & Pedersen, 2003). Similarly, plasma arterial VEGF was lower following a 10-day exercise training (Gustafsson et al., 2002). Differential responses of peripheral VEGF to acute exercise training in well-trained and sedentary subjects may suggest that duration of physical training may affect levels of VEGF (Kraus, Stallings, Yeager, & Gavin, 2004).

EXERCISE IMPROVES COGNITIVE FUNCTION IN HUMANS

Erickson et al. (2011) reported a significant increase in hippocampal volume in association with improved spatial memory in seniors who engaged in a 1-year program of aerobic exercise training when compared to sedentary controls. In a longitudinal study, there is a positive correlation between weekly walking distance and increases in neocortical and hippocampal volumes measured by MRI 9 years later (Erickson et al., 2009). Aerobic fitness in children is positively correlated with neurocognitive function (Hillman et al., 2009) in addition to a better performance in executive functions (Buck, Hillman, & Castelli, 2008). Children with cerebral palsy who engaged in an exercise training program (8 weeks with 1-h exercise, three times a week) showed a significant improvement in cognitive function and quality of life. These beneficial effects of exercise can exist 4 months after the cessation of exercise training (Verschuren et al., 2007). Physical activity is also known to attenuate cognitive decline in association with neurodegenerative disorders, such as Parkinson's and Alzheimer's disease (Yau, Gil-Mohapel, et al., 2014). Individuals aged 60–79 years with 6 months of aerobic exercise (three times a week for an hour), but not nonaerobic exercise, showed an increase in brain regions of the frontal lobe that are involved in memory processing (Colcombe et al., 2004). Regular moderate- to high-resistance training is also beneficial for cognitive function (Cassilhas et al., 2007; Liu-Ambrose et al., 2010; Perrig-Chiello, Perrig, Ehrsam, Staehelin, & Krings, 1998). Because of the difficulties in assessing the effects of resistance training in animal studies and inadequate studies on examining the effect of resistance training in humans, the relationship between cognitive improvement and resistance training is still inconclusive to date.

The effect of exercise can be significantly influenced by exercise duration, intensity, and type as evidenced by the aforementioned animal and human studies. However, it remains unclear whether there is a threshold level of exercise needed to improve cognitive function in humans. It would be of great interest to know how much and what types of aerobic physical exercise are needed to improve cognitive function. Colcombe and Kramer (2003) concluded that cardiovascular fitness training enhances cognitive performance regardless of task type in longitudinal fitness training. They also concluded that combining aerobic exercise and strength training in elderly and with exercise longer than 30 min per session produces the greatest cognitive benefits in the elderly. Their conclusion is in fact echoed by the findings in animals showing that both chronic voluntary and forced exercise training can improve spatial learning and memory (Ang et al., 2006; van Praag, Christie, et al., 1999).

Clinical trials with resistance training regimes show that 6 months of moderate or high resistance exercise training in seniors results in a comparable cognitive improvement (Cassilhas et al., 2007). In contrast, another clinical study reported that exercise duration is critical as cognitive improvement was only observed at the 12-month but not the 6-month time point (Liu-Ambrose et al., 2010). Contradictory results between these two studies may be due to differences in exercise intensity or age of the subjects. Of note, the later study reported that cognitive improvement in individuals with 12-month resistance training once a week was comparable to that with the same training twice a week (Liu-Ambrose et al., 2010).

Limited studies have specifically addressed the duration, intensity, and type of exercise for optimal benefits on the human brain. Some studies suggest that long-term moderate intensity aerobic exercise is sufficient to improve cognitive function and enhance hippocampal volume (Erickson et al., 2011; Lautenschlager et al., 2008). Overtraining could be detrimental in causing symptoms such as depression and fatigue (McKenzie, Lama, Potts, Sheel, & Coutts, 1999; Morgan, Brown, Raglin, O'Connor, & Ellickson, 1987). It is likely that long-term aerobic exercise with a moderate intensity will potentially lead to physical fitness and brain health. Individuals may not display equal enhancements in cognition from physical training. Routine exercise regimes may vary in individuals depending on personal capabilities and physical fitness to reach the moderate aerobic exercise intensity that is thought to be the appropriate amount of physical exercise across individuals (Ahlskog, Geda, Graff-Radford, & Petersen, 2011).

CONCLUSIONS

Convergent evidence from animal and human studies has suggested that regular long-term physical exercise can benefit cognitive function regardless of age. The adult brain is plastic across the life span and can adapt to environmental challenges through plasticity in structural and functional modification, including adult hippocampal neurogenesis. Adult neurogenesis, which may play a critical role in learning and memory and may be important for maintaining hippocampal plasticity throughout the life span, can be significantly impacted by the duration, intensity, and type of physical training, as well as by gender and age. Although there is not enough evidence for prescribing a reliable exercise program for improving brain health in humans, maintaining a physically active lifestyle is definitely helpful for preventing age-related cognitive decline.

References

Ahlskog, J. E., Geda, Y. E., Graff-Radford, N. R., & Petersen, R. C. (2011). Physical exercise as a preventive or disease-modifying treatment of dementia and brain aging. *Mayo Clinic Proceedings, 86*, 876–884.

Aleman, A., de Vries, W. R., de Haan, E. H., Verhaar, H. J., Samson, M. M., & Koppeschaar, H. P. (2000). Age-sensitive cognitive function, growth hormone and insulin-like growth factor 1 plasma levels in healthy older men. *Neuropsychobiology, 41*, 73–78.

Aleman, A., Verhaar, H. J., De Haan, E. H., De Vries, W. R., Samson, M. M., Drent, M. L., et al. (1999). Insulin-like growth factor-I and cognitive function in healthy older men. *The Journal of Clinical Endocrinology and Metabolism, 84*, 471–475.

Altman, J., & Bayer, S. A. (1990). Migration and distribution of two populations of hippocampal granule cell precursors during the perinatal and postnatal periods. *The Journal of Comparative Neurology, 301*, 365–381.

Ang, E. T., Dawe, G. S., Wong, P. T., Moochhala, S., & Ng, Y. K. (2006). Alterations in spatial learning and memory after forced exercise. *Brain Research, 1113*, 186–193.

Arida, R. M., Scorza, C. A., da Silva, A. V., Scorza, F. A., & Cavalheiro, E. A. (2004). Differential effects of spontaneous versus forced exercise in rats on the staining of parvalbumin-positive neurons in the hippocampal formation. *Neuroscience Letters, 364*, 135–138.

Arwert, L. I., Deijen, J. B., & Drent, M. L. (2005). The relation between insulin-like growth factor I levels and cognition in healthy elderly: a meta-analysis. *Growth Hormone and IGF Research: Official Journal of the Growth Hormone Research Society and the International IGF Research Society, 15*, 416–422.

Bekinschtein, P., Oomen, C. A., Saksida, L. M., & Bussey, T. J. (2011). Effects of environmental enrichment and voluntary exercise on neurogenesis, learning and memory, and pattern separation: BDNF as a critical variable? *Seminars in Cell and Developmental Biology, 22*, 536–542.

Berchtold, N. C., Castello, N., & Cotman, C. W. (2010). Exercise and time-dependent benefits to learning and memory. *Neuroscience, 167*, 588–597.

Berchtold, N. C., Chinn, G., Chou, M., Kesslak, J. P., & Cotman, C. W. (2005). Exercise primes a molecular memory for brain-derived neurotrophic factor protein induction in the rat hippocampus. *Neuroscience, 133*, 853–861.

Bjornebekk, A., Mathe, A. A., & Brene, S. (2005). The antidepressant effect of running is associated with increased hippocampal cell proliferation. *The International Journal of Neuropsychopharmacology/Official Scientific Journal of the Collegium Internationale Neuropsychopharmacologicum, 8*, 357–368.

Boehme, F., Gil-Mohapel, J., Cox, A., Patten, A., Giles, E., Brocardo, P. S., et al. (2011). Voluntary exercise induces adult hippocampal neurogenesis and BDNF expression in a rodent model of fetal alcohol spectrum disorders. *European Journal of Neuroscience, 33*, 1799–1811.

Breen, E. C., Johnson, E. C., Wagner, H., Tseng, H. M., Sung, L. A., & Wagner, P. D. (1996). Angiogenic growth factor mRNA responses in muscle to a single bout of exercise. *Journal of Applied Physiology, 81*, 355–361.

Brown, J., Cooper-Kuhn, C. M., Kempermann, G., Van Praag, H., Winkler, J., Gage, F. H., et al. (2003). Enriched environment and physical activity stimulate hippocampal but not olfactory bulb neurogenesis. *European Journal of Neuroscience, 17*, 2042–2046.

Buck, S. M., Hillman, C. H., & Castelli, D. M. (2008). The relation of aerobic fitness to stroop task performance in preadolescent children. *Medicine and Science in Sports and Exercise, 40*, 166–172.

Cameron, H. A., Woolley, C. S., McEwen, B. S., & Gould, E. (1993). Differentiation of newly born neurons and glia in the dentate gyrus of the adult rat. *Neuroscience, 56*, 337–344.

Cao, L., Jiao, X., Zuzga, D. S., Liu, Y., Fong, D. M., Young, D., et al. (2004). VEGF links hippocampal activity with neurogenesis, learning and memory. *Nature Genetics, 36*, 827–835.

Carro, E., Nunez, A., Busiguina, S., & Torres-Aleman, I. (2000). Circulating insulin-like growth factor I mediates effects of exercise on the brain. *Journal of Neuroscience, 20*, 2926–2933.

Cassilhas, R. C., Viana, V. A., Grassmann, V., Santos, R. T., Santos, R. F., Tufik, S., et al. (2007). The impact of resistance exercise on the cognitive function of the elderly. *Medicine and Science in Sports and Exercise, 39*, 1401–1407.

Castellano, V., & White, L. J. (2008). Serum brain-derived neurotrophic factor response to aerobic exercise in multiple sclerosis. *Journal of the Neurological Sciences, 269*, 85–91.

Cearlock, D. M., & Nuzzo, N. A. (2001). Effects of sustained moderate exercise on cholesterol, growth hormone and cortisol blood levels in three age groups of women. *Clinical Laboratory Science: Journal of the American Society for Medical Technology, 14*, 108–111.

Chan, J. P., Cordeira, J., Calderon, G. A., Iyer, L. K., & Rios, M. (2008). Depletion of central BDNF in mice impedes terminal differentiation of new granule neurons in the adult hippocampus. *Molecular and Cellular Neuroscience, 39*, 372–383.

Christie, B. R., Eadie, B. D., Kannangara, T. S., Robillard, J. M., Shin, J., & Titterness, A. K. (2008). Exercising our brains: how physical activity impacts synaptic plasticity in the dentate gyrus. *NeuroMolecular Medicine, 10*, 47–58.

Clark, P. J., Brzezinska, W. J., Thomas, M. W., Ryzhenko, N. A., Toshkov, S. A., & Rhodes, J. S. (2008). Intact neurogenesis is required for benefits of exercise on spatial memory but not motor performance or contextual fear conditioning in C57BL/6J mice. *Neuroscience, 155*, 1048–1058.

Clark, P. J., Kohman, R. A., Miller, D. S., Bhattacharya, T. K., Brzezinska, W. J., & Rhodes, J. S. (2011). Genetic influences on exercise-induced adult hippocampal neurogenesis across 12 divergent mouse strains. *Genes, Brain, and Behavior, 10*, 345–353.

Colcombe, S., & Kramer, A. F. (2003). Fitness effects on the cognitive function of older adults: a meta-analytic study. *Psychological Science, 14*, 125–130.

Colcombe, S. J., Kramer, A. F., Erickson, K. I., Scalf, P., McAuley, E., Cohen, N. J., et al. (2004). Cardiovascular fitness, cortical plasticity, and aging. *Proceedings of the National Academy of Sciences of the United States of America, 101*, 3316–3321.

Cotman, C. W., & Berchtold, N. C. (2002). Exercise: a behavioral intervention to enhance brain health and plasticity. *Trends in Neuroscience, 25*, 295–301.

Creer, D. J., Romberg, C., Saksida, L. M., van Praag, H., & Bussey, T. J. (2010). Running enhances spatial pattern separation in mice. *Proceedings of the National Academy of Sciences of the United States of America, 107*, 2367–2372.

Deng, W., Aimone, J. B., & Gage, F. H. (2010). New neurons and new memories: how does adult hippocampal neurogenesis affect learning and memory? *Nature Reviews Neuroscience, 11*, 339–350.

Dery, N., Pilgrim, M., Gibala, M., Gillen, J., Wojtowicz, J. M., Macqueen, G., et al. (2013). Adult hippocampal neurogenesis reduces memory interference in humans: opposing effects of aerobic exercise and depression. *Frontiers in Neuroscience, 7*, 66.

Dik, M., Deeg, D. J., Visser, M., & Jonker, C. (2003). Early life physical activity and cognition at old age. *Journal of Clinical and Experimental Neuropsychology, 25*, 643–653.

Drapeau, E., Mayo, W., Aurousseau, C., Le Moal, M., Piazza, P. V., & Abrous, D. N. (2003). Spatial memory performances of aged rats in the water maze predict levels of hippocampal neurogenesis. *Proceedings of the National Academy of Sciences of the United States of America, 100*, 14385–14390.

Eadie, B. D., Redila, V. A., & Christie, B. R. (2005). Voluntary exercise alters the cytoarchitecture of the adult dentate gyrus by increasing cellular proliferation, dendritic complexity, and spine density. *The Journal of Comparative Neurology, 486*, 39–47.

Erickson, C. A., & Barnes, C. A. (2003). The neurobiology of memory changes in normal aging. *Experimental Gerontology, 38*, 61–69.

Erickson, K. I., Prakash, R. S., Voss, M. W., Chaddock, L., Hu, L., Morris, K. S., et al. (2009). Aerobic fitness is associated with hippocampal volume in elderly humans. *Hippocampus, 19*, 1030–1039.

Erickson, K. I., Voss, M. W., Prakash, R. S., Basak, C., Szabo, A., Chaddock, L., et al. (2011). Exercise training increases size of hippocampus and improves memory. *Proceedings of the National Academy of Sciences of the United States of America, 108*, 3017–3022.

Eriksson, P. S., Perfilieva, E., Bjork-Eriksson, T., Alborn, A. M., Nordborg, C., Peterson, D. A., et al. (1998). Neurogenesis in the adult human hippocampus. *Nature Medicine, 4*, 1313–1317.

Fabel, K., Fabel, K., Tam, B., Kaufer, D., Baiker, A., Simmons, N., et al. (2003). VEGF is necessary for exercise-induced adult hippocampal neurogenesis. *European Journal of Neuroscience, 18*, 2803–2812.

Farioli-Vecchioli, S., Mattera, A., Micheli, L., Ceccarelli, M., Leonardi, L., Saraulli, D., et al. (2014). Running rescues defective adult neurogenesis by shortening the length of the cell cycle of neural stem and progenitor cells. *Stem Cells, 32*, 1968–1982.

Farmer, J., Zhao, X., van Praag, H., Wodtke, K., Gage, F. H., & Christie, B. R. (2004). Effects of voluntary exercise on synaptic plasticity and gene expression in the dentate gyrus of adult male Sprague-Dawley rats in vivo. *Neuroscience, 124*, 71–79.

Fediuc, S., Campbell, J. E., & Riddell, M. C. (2006). Effect of voluntary wheel running on circadian corticosterone release and on HPA axis responsiveness to restraint stress in Sprague-Dawley rats. *Journal of Applied Physiology, 100*, 1867–1875.

Fischer, T. J., Walker, T. L., Overall, R. W., Brandt, M. D., & Kempermann, G. (2014). Acute effects of wheel running on adult hippocampal precursor cells in mice are not caused by changes in cell cycle length or S phase length. *Frontiers in Neuroscience, 8*, 314.

Fordyce, D. E., & Wehner, J. M. (1993). Physical activity enhances spatial learning performance with an associated alteration in hippocampal protein kinase C activity in C57BL/6 and DBA/2 mice. *Brain Research, 619*, 111–119.

Fuss, J., Abdallah, B. N. -M., Vogt, M. A., Touma, C., Pacifici, P. G., Palme, R., et al. (2009). Voluntary exercise induces anxiety-like behavior in adult C57BL/6J mice correlating with hippocampal neurogenesis. *Hippocampus, 20*, 364–376.

Galea, L. A., & McEwen, B. S. (1999). Sex and seasonal differences in the rate of cell proliferation in the dentate gyrus of adult wild meadow voles. *Neuroscience, 89*, 955–964.

Galea, L. A., Spritzer, M. D., Barker, J. M., & Pawluski, J. L. (2006). Gonadal hormone modulation of hippocampal neurogenesis in the adult. *Hippocampus, 16*, 225–232.

Garrett, L., Lie, D. C., Martin Hrabé de Angelis, M. H., Wurst, W., & Hölter, S. M. (2012). *BMC Neuroscience, 13*, 61–71.

Gheusi, G., & Lledo, P. M. (2007). Control of early events in olfactory processing by adult neurogenesis. *Chemical Senses, 32*, 397–409.

Gil-Mohapel, J., Brocardo, P. S., Choquette, W., Gothard, R., Simpson, J. M., & Christie, B. R. (2013). Hippocampal neurogenesis levels predict WATERMAZE search strategies in the aging brain. *PLoS One, 8*, e75125.

Gomez-Pinilla, F., Dao, L., & So, V. (1997). Physical exercise induces FGF-2 and its mRNA in the hippocampus. *Brain Research, 764*, 1–8.

Gould, E., Beylin, A., Tanapat, P., Reeves, A., & Shors, T. J. (1999). Learning enhances adult neurogenesis in the hippocampal formation. *Nature Neuroscience, 2*, 260–265.

Gould, E., Tanapat, P., McEwen, B. S., Flugge, G., & Fuchs, E. (1998). Proliferation of granule cell precursors in the dentate gyrus of adult monkeys is diminished by stress. *Proceedings of the National Academy of Sciences of the United States of America, 95*, 3168–3171.

Green, E. J., & Juraska, J. M. (1985). The dendritic morphology of hippocampal dentate granule cells varies with their position in the granule cell layer: a quantitative Golgi study. *Experimental Brain Research, 59*, 582–586.

Groves-Chapman, J. L., Murray, P. S., Stevens, K. L., Monroe, D. C., Koch, L. G., Britton, S. L., et al. (2011). Changes in mRNA levels for brain-derived neurotrophic factor after wheel running in rats selectively bred for high- and low-aerobic capacity. *Brain Research, 1425*, 90–97.

Gustafsson, T., Knutsson, A., Puntschart, A., Kaijser, L., Nordqvist, A. C., Sundberg, C. J., et al. (2002). Increased expression of vascular endothelial growth factor in human skeletal muscle in response to short-term one-legged exercise training. *Pflugers Archiv: European Journal of Physiology, 444*, 752–759.

Heine, V. M., Maslam, S., Joels, M., & Lucassen, P. J. (2004). Prominent decline of newborn cell proliferation, differentiation, and apoptosis in the aging dentate gyrus, in absence of an age-related hypothalamus-pituitary-adrenal axis activation. *Neurobiology Aging, 25*, 361–375.

Helfer, J. L., Goodlett, C. R., Greenough, W. T., & Klintsova, A. Y. (2009). The effects of exercise on adolescent hippocampal neurogenesis in a rat model of binge alcohol exposure during the brain growth spurt. *Brain Research, 1294*, 1–11.

Hillman, C. H., Pontifex, M. B., Raine, L. B., Castelli, D. M., Hall, E. E., & Kramer, A. F. (2009). The effect of acute treadmill walking on cognitive control and academic achievement in preadolescent children. *Neuroscience, 159*, 1044–1054.

Hiscock, N., Fischer, C. P., Pilegaard, H., & Pedersen, B. K. (2003). Vascular endothelial growth factor mRNA expression and arteriovenous balance in response to prolonged, submaximal exercise in humans. *American Journal of Physiology Heart and Circulatory Physiology, 285*, H1759–H1763.

Hoffner, L., Nielsen, J. J., Langberg, H., & Hellsten, Y. (2003). Exercise but not prostanoids enhance levels of vascular endothelial growth factor and other proliferative agents in human skeletal muscle interstitium. *The Journal of Physiology, 550*, 217–225.

Holmes, M. M., Galea, L. A., Mistlberger, R. E., & Kempermann, G. (2004). Adult hippocampal neurogenesis and voluntary running activity: circadian and dose-dependent effects. *Journal of Neuroscience Research, 76*, 216–222.

Jin, K., Zhu, Y., Sun, Y., Mao, X. O., Xie, L., & Greenberg, D. A. (2002). Vascular endothelial growth factor (VEGF) stimulates neurogenesis in vitro and in vivo. *Proceedings of the National Academy of Sciences of the United States of America, 99*, 11946–11950.

Johansson, C. B., Momma, S., Clarke, D. L., Risling, M., Lendahl, U., & Frisen, J. (1999). Identification of a neural stem cell in the adult mammalian central nervous system. *Cell, 96*, 25–34.

Kannangara, T. S., Lucero, M. J., Gil-Mohapel, J., Drapala, R. J., Simpson, J. M., Christie, B. R., et al. (2011). Running reduces stress and enhances cell genesis in aged mice. *Neurobiology Aging, 32*, 2279–2286.

Karatay, S., Yildirim, K., Melikoglu, M. A., Akcay, F., & Senel, K. (2007). Effects of dynamic exercise on circulating IGF-1 and IGFBP-3 levels in patients with rheumatoid arthritis or ankylosing spondylitis. *Clinical Rheumatology, 26*, 1635–1639.

Kempermann, G., Chesler, E. J., Lu, L., Williams, R. W., & Gage, F. H. (2006). Natural variation and genetic covariance in adult hippocampal neurogenesis. *Proceedings of the National Academy of Sciences of the United States of America, 103*, 780–785.

Kempermann, G., & Gage, F. H. (2002a). Genetic determinants of adult hippocampal neurogenesis correlate with acquisition, but not probe trial performance, in the water maze task. *European Journal of Neuroscience, 16*, 129–136.

Kempermann, G., & Gage, F. H. (2002b). Genetic influence on phenotypic differentiation in adult hippocampal neurogenesis. *Brain Research Developmental Brain Research, 134*, 1–12.

Kempermann, G., Kuhn, H. G., & Gage, F. H. (1997). Genetic influence on neurogenesis in the dentate gyrus of adult mice. *Proceedings of the National Academy of Sciences of the United States of America, 94*, 10409–10414.

Kitamura, T., Mishina, M., & Sugiyama, H. (2003). Enhancement of neurogenesis by running wheel exercises is suppressed in mice lacking NMDA receptor epsilon 1 subunit. *Neuroscience Research, 47*, 55–63.

Klaus, F., & Amerin, I. (2012). Running in laboratory and wild rodents: Differences in context sensitivity and plasticity of hippocampal neurogenesis. *Behavioural Brain Research, 227* 3630–370.

Klempin, F., Beis, D., Mosienko, V., Kempermann, G., Bader, M., & Alenina, N. (2013). Serotonin is required for exercise-induced adult hippocampal neurogenesis. *Journal of Neuroscience, 33*, 8270–8275.

Knaepen, K., Goekint, M., Heyman, E. M., & Meeusen, R. (2010). Neuroplasticity–exercise-induced response of peripheral brain-derived neurotrophic factor: a systematic review of experimental studies in human subjects. *Sports Medicine, 40*, 765–801.

Knoth, R., Singec, I., Ditter, M., Pantazis, G., Capetian, P., Meyer, R. P., et al. (2010). Murine features of neurogenesis in the human hippocampus across the lifespan from 0 to 100 years. *PLoS One, 5*, e8809.

Kobilo, T., Liu, Q. R., Gandhi, K., Mughal, M., Shaham, Y., & van Praag, H. (2011). Running is the neurogenic and neurotrophic stimulus in environmental enrichment. *Learning and Memory (Cold Spring Harbor, NY), 18*, 605–609.

Kohl, Z., Kandasamy, M., Winner, B., Aigner, R., Gross, C., Couillard-Despres, S., et al. (2007). Physical activity fails to rescue hippocampal neurogenesis deficits in the R6/2 mouse model of Huntington's disease. *Brain Research, 1155*, 24–33.

Kraus, R. M., Stallings, H. W., 3rd, Yeager, R. C., & Gavin, T. P. (2004). Circulating plasma VEGF response to exercise in sedentary and endurance-trained men. *Journal of Applied Physiology, 96*, 1445–1450.

Kronenberg, G., Bick-Sander, A., Bunk, E., Wolf, C., Ehninger, D., & Kempermann, G. (2006). Physical exercise prevents age-related decline in precursor cell activity in the mouse dentate gyrus. *Neurobiology Aging, 27*, 1505–1513.

Kuhn, H. G., Dickinson-Anson, H., & Gage, F. H. (1996). Neurogenesis in the dentate gyrus of the adult rat: age-related decrease of neuronal progenitor proliferation. *Journal of Neuroscience, 16*, 2027–2033.

Kukekov, V. G., Laywell, E. D., Suslov, O., Davies, K., Scheffler, B., Thomas, L. B., et al. (1999). Multipotent stem/progenitor cells with similar properties arise from two neurogenic regions of adult human brain. *Experimental Neurology, 156*, 333–344.

Lautenschlager, N. T., Cox, K. L., Flicker, L., Foster, J. K., van Bockxmeer, F. M., Xiao, J., et al. (2008). Effect of physical activity on cognitive function in older adults at risk for Alzheimer disease: a randomized trial. *Journal of American Medical Association, 300*, 1027–1037.

Leasure, J. L., & Jones, M. (2008). Forced and voluntary exercise differentially affect brain and behavior. *Neuroscience, 156*, 456–465.

Lee, M. C., Inoue, K., Okamoto, M., Liu, Y. F., Matsui, T., Yook, J. S., et al. (2013). Voluntary resistance running induces increased hippocampal neurogenesis in rats comparable to load-free running. *Neuroscience Letters, 537*, 6–10.

Lee, T. M., Wong, M. L., Lau, B. W., Lee, J. C., Yau, S. Y., & So, K. F. (2014). Aerobic exercise interacts with neurotrophic factors to predict cognitive functioning in adolescents. *Psychoneuroendocrinology, 39*, 214–224.

Leuner, B., Kozorovitskiy, Y., Gross, C. G., & Gould, E. (2007). Diminished adult neurogenesis in the marmoset brain precedes old age. *Proceedings of the National Academy of Sciences of the United States of America, 104*, 17169–17173.

Li, H., Liang, A., Guan, F., Fan, R., Chi, L., & Yang, B. (2013). Regular treadmill running improves spatial learning and memory performance in young mice through increased hippocampal neurogenesis and decreased stress. *Brain Research, 1531*, 1–8.

Liu-Ambrose, T., Nagamatsu, L. S., Graf, P., Beattie, B. L., Ashe, M. C., & Handy, T. C. (2010). Resistance training and executive functions: a 12-month randomized controlled trial. *Archives of Internal Medicine, 170*, 170–178.

Liu, J., Guo, M., Zhang, D., Cheng, S. Y., Liu, M., Ding, J., et al. (2012). Adiponectin is critical in determining susceptibility to depressive behaviors and has antidepressant-like activity. *Proceedings of the National Academy of Sciences of the United States of America, 109*, 12248–12253.

Liu, H. L., Zhao, G., Cai, K., Zhao, H. H., & Shi, L. D. (2011). Treadmill exercise prevents decline in spatial learning and memory in APP/PS1 transgenic mice through improvement of hippocampal long-term potentiation. *Behavioural Brain Research, 218*, 308–314.

Lou, S. J., Liu, J. Y., Chang, H., & Chen, P. J. (2008). Hippocampal neurogenesis and gene expression depend on exercise intensity in juvenile rats. *Brain Research, 1210*, 48–55.

Ma, X., Hamadeh, M. J., Christie, B. R., Foster, J. A., & Tarnopolsky, M. A. (2012). Impact of treadmill running and sex on hippocampal neurogenesis in the mouse model of amyotrophic lateral sclerosis. *PLoS One, 7*, e36048.

Manetta, J., Brun, J. F., Maimoun, L., Fedou, C., Prefaut, C., & Mercier, J. (2003). The effects of intensive training on insulin-like growth factor I (IGF-I) and IGF binding proteins 1 and 3 in competitive cyclists: relationships with glucose disposal. *Journal of Sports Sciences, 21*, 147–154.

Manganas, L. N., Zhang, X., Li, Y., Hazel, R. D., Smith, S. D., Wagshul, M. E., et al. (2007). Magnetic resonance spectroscopy identifies neural progenitor cells in the live human brain. *Science, 318*, 980–985.

Marlatt, M. W., Potter, M. C., Lucassen, P. J., & van Praag, H. (2012). Running throughout middle-age improves memory function, hippocampal neurogenesis and BDNF levels in female C57Bl/6J mice. *Developmental Neurobiology, 72*, 943–952.

McDonald, H. Y., & Wojtowicz, J. M. (2005). Dynamics of neurogenesis in the dentate gyrus of adult rats. *Neuroscience Letters, 385*, 70–75.

McKenzie, D. C., Lama, I. L., Potts, J. E., Sheel, A. W., & Coutts, K. D. (1999). The effect of repeat exercise on pulmonary diffusing capacity and EIH in trained athletes. *Medicine and Science in Sports and Exercise, 31*, 99–104.

Merritt, J. R., & Rhodes, J. S. (2015). Mouse genetic differences in voluntary wheel running, adult hippocampal neurogenesis and learning on the multi-strain-adapted plus water maze. *Behavioural Brain Research, 280*, 62–71.

Mistlberger, R. E., & Holmes, M. M. (2000). Behavioral feedback regulation of circadian rhythm phase angle in light-dark entrained mice. *American Journal of Physiology Regulatory, Integrative and Comparative Physiology, 279*, R813–R821.

Molteni, R., Ying, Z., & Gomez-Pinilla, F. (2002). Differential effects of acute and chronic exercise on plasticity-related genes in the rat hippocampus revealed by microarray. *European Journal of Neuroscience, 16*, 1107–1116.

Morgan, W. P., Brown, D. R., Raglin, J. S., O'Connor, P. J., & Ellickson, K. A. (1987). Psychological monitoring of overtraining and staleness. *British Journal of Sports Medicine, 21*, 107–114.

Morgenstern, N. A., Lombardi, G., & Schinder, A. F. (2008). Newborn granule cells in the ageing dentate gyrus. *The Journal of Physiology, 586*, 3751–3757.

Mustroph, M. L., Chen, S., Desai, S. C., Cay, E. B., DeYoung, E. K., & Rhodes, J. S. (2012). Aerobic exercise is the critical variable in an enriched environment that increases hippocampal neurogenesis and water maze learning in male C57BL/6J mice. *Neuroscience, 219*, 62–71.

Naylor, A. S., Bull, C., Nilsson, M. K. L., Zhu, C., Björk-Eriksson, T., Eriksson, P. S., et al. (2008). *Proceedings of the National Academy of Sciences of the United States of America, 105*, 14632–14637.

Neeper, S. A., Gomez-Pinilla, F., Choi, J., & Cotman, C. (1995). Exercise and brain neurotrophins. *Nature, 373*, 109.

Neeper, S. A., Gomez-Pinilla, F., Choi, J., & Cotman, C. W. (1996). Physical activity increases mRNA for brain-derived neurotrophic factor and nerve growth factor in rat brain. *Brain Research, 726*, 49–56.

Nishida, Y., Matsubara, T., Tobina, T., Shindo, M., Tokuyama, K., Tanaka, K., et al. (2010). Effect of low-intensity aerobic exercise on insulin-like growth factor-I and insulin-like growth factor-binding proteins in healthy men. *International Journal of Endocrinology, 2010*.

Olson, A. K., Eadie, B. D., Ernst, C., & Christie, B. R. (2006). Environmental enrichment and voluntary exercise massively increase neurogenesis in the adult hippocampus via dissociable pathways. *Hippocampus, 16*, 250–260.

Overall, R. W., Walker, T. L., Leiter, O., Lenke, S., Ruhwald, S., & Kempermann, G. (2013). Delayed and transient increase of adult hippocampal neurogenesis by physical exercise in DBA/2 mice. *PLoS One, 8*, e83797.

Palmer, T. D., Willhoite, A. R., & Gage, F. H. (2000). Vascular niche for adult hippocampal neurogenesis. *The Journal of Comparative Neurology, 425*, 479–494.

Patten, A. R., Sickmann, H., Hryciw, B. N., Kucharsky, T., Parton, R., Kernick, A., et al. (2013). Long-term exercise is needed to enhance synaptic plasticity in the hippocampus. *Learning and Memory (Cold Spring Harbor, NY), 20*, 642–647.

Pereira, A. C., Huddleston, D. E., Brickman, A. M., Sosunov, A. A., Hen, R., McKhann, G. M., et al. (2007). An in vivo correlate of exercise-induced neurogenesis in the adult dentate gyrus. *Proceedings of the National Academy of Sciences of the United States of America, 104*, 5638–5643.

Perrig-Chiello, P., Perrig, W. J., Ehrsam, R., Staehelin, H. B., & Krings, F. (1998). The effects of resistance training on well-being and memory in elderly volunteers. *Age and Ageing, 27*, 469–475.

Potter, M. C., Yuan, C., Ottenritter, C., Mughal, M., & van Praag, H. (2010). Exercise is not beneficial and may accelerate symptom onset in a mouse model of Huntington's disease. *PLoS Currents, 2*, RRN1201.

Redila, V. A., & Christie, B. R. (2006). Exercise-induced changes in dendritic structure and complexity in the adult hippocampal dentate gyrus. *Neuroscience, 137*, 1299–1307.

Rei, A., Takashi, M., Shim, J. H., Keitaro, Y., & Masugi, N. (2011). Involvement of prograulin in the enhancement of hippocampal neurogenesis by voluntary exercise. *NeuroReport, 22*, 881–886.

Borght, Van der, Ferrari, F., Klauke, K., Roman, V., Havekes, R., Sgoifo, A., van der Zee, E. A., et al. (2006). Hippocampal cell proliferation across the day: Increase by running wheel activity, but no effect of sleep and wakefulness. *Behavior Brain Research, 167*, 36–41.

van Praag, H., Christie, B. R., Sejnowski, T. J., & Gage, F. H. (1999). Running enhances neurogenesis, learning, and long-term potentiation in mice. *Proceedings of the National Academy of Sciences of the United States of America, 96*, 13427–13431.

van Praag, H., Kempermann, G., & Gage, F. H. (1999). Running increases cell proliferation and neurogenesis in the adult mouse dentate gyrus. *Nature Neuroscience, 2*, 266–270.

van Praag, H., Schinder, A. F., Christie, B. R., Toni, N., Palmer, T. D., & Gage, F. H. (2002). Functional neurogenesis in the adult hippocampus. *Nature, 415*, 1030–1034.

van Praag, H., Shubert, T., Zhao, C., & Gage, F. H. (2005). Exercise enhances learning and hippocampal neurogenesis in aged mice. *Journal of Neuroscience, 25*, 8680–8685.

Rao, M. S., Hattiangady, B., & Shetty, A. K. (2006). The window and mechanisms of major age-related decline in the production of new neurons within the dentate gyrus of the hippocampus. *Aging Cell, 5*, 545–558.

Rasmussen, P., Brassard, P., Adser, H., Pedersen, M. V., Leick, L., Hart, E., et al. (2009). Evidence for a release of brain-derived neurotrophic factor from the brain during exercise. *Experimental Physiology, 94*, 1062–1069.

Rhodes, J. S., van Praag, H., Jeffrey, S., Girard, I., Mitchell, G. S., Garland, T., Jr., et al. (2003). Exercise increases hippocampal neurogenesis to high levels but does not improve spatial learning in mice bred for increased voluntary wheel running. *Behavioral Neuroscience, 117*, 1006–1016.

Rodriguez, J. J., Jones, V. C., Tabuchi, M., Allan, S. M., Knight, E. M., LaFerla, F. M., et al. (2008). Impaired adult neurogenesis in the dentate gyrus of a triple transgenic mouse model of Alzheimer's disease. *PLoS One, 3*, e2935.

Rodriguez, J. J., Noristani, H. N., Olabarria, M., Fletcher, J., Somerville, T. D., Yeh, C. Y., et al. (2011). Voluntary running and environmental enrichment restores impaired hippocampal neurogenesis in a triple transgenic mouse model of Alzheimer's disease. *Current Alzheimer Research, 8*, 707–717.

Roy, N. S., Wang, S., Jiang, L., Kang, J., Benraiss, A., Harrison-Restelli, C., et al. (2000). In vitro neurogenesis by progenitor cells isolated from the adult human hippocampus. *Nature Medicine, 6*, 271–277.

Santarelli, L., Saxe, M., Gross, C., Surget, A., Battaglia, F., Dulawa, S., et al. (2003). Requirement of hippocampal neurogenesis for the behavioral effects of antidepressants. *Science, 301*, 805–809.

Scharfman, H., Goodman, J., Macleod, A., Phani, S., Antonelli, C., & Croll, S. (2005). Increased neurogenesis and the ectopic granule cells after intrahippocampal BDNF infusion in adult rats. *Experimental Neurology, 192*, 348–356.

Seifert, T., Brassard, P., Wissenberg, M., Rasmussen, P., Nordby, P., Stallknecht, B., et al. (2010). Endurance training enhances BDNF release from the human brain. *American Journal of Physiology Regulatory, Integrative and Comparative Physiology, 298*, R372–R377.

Simpson, J. M., Gil-Mohapel, J., Pouladi, M. A., Ghilan, M., Xie, Y., Hayden, M. R., et al. (2011). Altered adult hippocampal neurogenesis in the YAC128 transgenic mouse model of Huntington disease. *Neurobiology of Disease, 41*, 249–260.

Snyder, J. S., Glover, L. R., Sanzone, K. M., Kamhi, J. F., & Cameron, H. A. (2009). The effects of exercise and stress on the survival and maturation of adult-generated granule cells. *Hippocampus, 19*, 898–906.

Snyder, J. S., Soumier, A., Brewer, M., Pickel, J., & Cameron, H. A. (2011). Adult hippocampal neurogenesis buffers stress responses and depressive behaviour. *Nature, 476*, 458–461.

Spier, S. A., Delp, M. D., Meininger, C. J., Donato, A. J., Ramsey, M. W., & Muller-Delp, J. M. (2004). Effects of ageing and exercise training on endothelium-dependent vasodilatation and structure of rat skeletal muscle arterioles. *The Journal of Physiology, 556*, 947–958.

Stranahan, A. M., Khalil, D., & Gould, E. (2006). Social isolation delays the positive effects of running on adult neurogenesis. *Nature Neuroscience, 9*, 526–533.

Stranahan, A. M., Khalil, D., & Gould, E. (2007). Running induces widespread structural alterations in the hippocampus and entorhinal cortex. *Hippocampus, 17*, 1017–1022.

Surget, A., Saxe, M., Leman, S., Ibarguen-Vargas, Y., Chalon, S., Griebel, G., et al. (2008). Drug-dependent requirement of hippocampal neurogenesis in a model of depression and of antidepressant reversal. *Biological Psychiatry, 64*, 293–301.

Taliaz, D., Stall, N., Dar, D. E., & Zangen, A. (2010). Knockdown of brain-derived neurotrophic factor in specific brain sites precipitates behaviors associated with depression and reduces neurogenesis. *Molecular Psychiatry, 15*, 80–92.

Tanapat, P., Hastings, N. B., Reeves, A. J., & Gould, E. (1999). Estrogen stimulates a transient increase in the number of new neurons in the dentate gyrus of the adult female rat. *Journal of Neuroscience, 19*, 5792–5801.

Titterness, A. K., Wiebe, E., Kwasnica, A., Keyes, G., & Christie, B. R. (2011). Voluntary exercise does not enhance long-term potentiation in the adolescent female dentate gyrus. *Neuroscience, 183*, 25–31.

Toni, N., Laplagne, D. A., Zhao, C., Lombardi, G., Ribak, C. E., Gage, F. H., et al. (2008). Neurons born in the adult dentate gyrus form functional synapses with target cells. *Nature Neuroscience, 11*, 901–907.

Trejo, J. L., Carro, E., & Torres-Aleman, I. (2001). Circulating insulin-like growth factor I mediates exercise-induced increases in the number of new neurons in the adult hippocampus. *Journal of Neuroscience, 21*, 1628–1634.

Uda, M., Ishido, M., Kami, K., & Masuhara, M. (2006). Effects of chronic treadmill running on neurogenesis in the dentate gyrus of the hippocampus of adult rat. *Brain Research, 1104*, 64–72.

Van der Borght, K., Kobor-Nyakas, D. E., Klauke, K., Eggen, B. J., Nyakas, C., Van der Zee, E. A., et al. (2009). Physical exercise leads to rapid adaptations in hippocampal vasculature: temporal dynamics and relationship to cell proliferation and neurogenesis. *Hippocampus, 19*, 928–936.

Vaynman, S., & Gomez-Pinilla, F. (2005). License to run: exercise impacts functional plasticity in the intact and injured central nervous system by using neurotrophins. *Neurorehabilitation and Neural Repair, 19*, 283–295.

Verschuren, O., Ketelaar, M., Gorter, J. W., Helders, P. J., Uiterwaal, C. S., & Takken, T. (2007). Exercise training program in children and adolescents with cerebral palsy: a randomized controlled trial. *Archives of Pediatrics and Adolescent Medicine, 161*, 1075–1081.

Wu, C. W., Chang, Y. T., Yu, L., Chen, H. I., Jen, C. J., Wu, S. Y., et al. (2008). Exercise enhances the proliferation of neural stem cells and neurite growth and survival of neuronal progenitor cells in dentate gyrus of middle-aged mice. *Journal of Applied Physiology, 105*, 1585–1594.

Wyss, J. M., Chambless, B. D., Kadish, I., & van Groen, T. (2000). Age-related decline in water maze learning and memory in rats: strain differences. *Neurobiology Aging, 21*, 671–681.

Yarrow, J. F., White, L. J., McCoy, S. C., & Borst, S. E. (2010). Training augments resistance exercise induced elevation of circulating brain derived neurotrophic factor (BDNF). *Neuroscience Letters, 479*, 161–165.

Yau, S. Y., Gil-Mohapel, J., Christie, B. R., & So, K. F. (2014). Physical exercise-induced adult neurogenesis: a good strategy to prevent cognitive decline in neurodegenerative diseases? *BioMed Research International, 2014*, 403120.

Yau, S. Y., Lau, B. W., Tong, J. B., Wong, R., Ching, Y. P., Qiu, G., et al. (2011). Hippocampal neurogenesis and dendritic plasticity support running-improved spatial learning and depression-like behaviour in stressed rats. *PLoS One, 6*, e24263.

Yau, S. Y., Lau, B. W., Zhang, E. D., Lee, J. C., Li, A., Lee, T. M., et al. (2012). Effects of voluntary running on plasma levels of neurotrophins, hippocampal cell proliferation and learning and memory in stressed rats. *Neuroscience, 222*, 289–301.

Yau, S. Y., Li, A., Hoo, R. L., Ching, Y. P., Christie, B. R., Lee, T. M., et al. (2014). Physical exercise-induced hippocampal neurogenesis and antidepressant effects are mediated by the adipocyte hormone adiponectin. *Proceedings of the National Academy of Sciences of the United States of America, 111*, 15810–15815.

Zebrowska, A., Gasior, Z., & Langfort, J. (2009). Serum IGF-I and hormonal responses to incremental exercise in athletes with and without left ventricular hypertrophy. *Journal of Sports Science and Medicine, 8*, 67–76.

Zhang, D., Guo, M., Zhang, W., & Lu, X. Y. (2011). Adiponectin stimulates proliferation of adult hippocampal neural stem/progenitor cells through activation of p38 mitogen-activated protein kinase (p38MAPK)/glycogen synthase kinase 3beta (GSK-3beta)/beta-catenin signaling cascade. *The Journal of Biological Chemistry, 286*, 44913–44920.

CHAPTER

6

Dietary and Nutritional Regulation

T. Murphy, S. Thuret
King's College London, London, United Kingdom

INTRODUCTION

In the previous chapters of this section (Part II: Neurogenesis in Health and Wellbeing), the strong, enduring, and positive effects elicited by environmental exposures such as learning and physical activity on adult hippocampal neurogenesis (AHN) have been described. A great body of literature centered on animal studies supports the notion that diet also modulates brain structure and function, exerting its influence throughout the life span of an organism. In this chapter we focus on positive inducers of AHN in the form of (1) dietary paradigms such as calorie restriction (CR: a consistent reduction of total daily food intake) and intermittent fasting (IF, every-other-day feeding) as well as (2) nutritional supplementation with polyphenols and polyunsaturated fatty acids (PUFAs) (Fig. 6.1).

We discuss the effects of these factors in physiological contexts and also outline how diet and nutrition can be used to reverse negative changes and preserve the integrity of AHN under conditions of psychopathology, aging, and disease with supporting evidence from epidemiological and intervention human studies.

Diet is an important lifestyle factor that influences health, disease, and longevity. Food consumption directly influences the availability of nutrients as well as the levels of growth factors and hormones in the circulatory system, with subsequent impacts on tissue homeostasis throughout the body (Rafalski, Mancini, & Brunet, 2012). A fundamental role of stem cells is to change their activities throughout the life span in response to changing environmental, growth, and regeneration demands, processes intimately linked to nutritional changes (Signer & Morrison, 2013).

Adult Neurogenesis in the Hippocampus
http://dx.doi.org/10.1016/B978-0-12-801977-1.00006-4

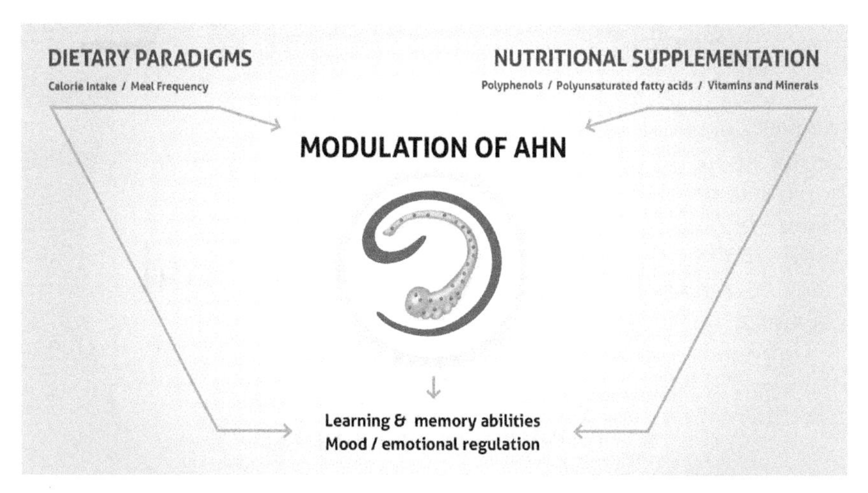

FIGURE 6.1 **Dietary paradigms and nutritional supplementation have been demonstrated to positively and negatively modulate adult hippocampal neurogenesis (AHN) in animal studies.** This in turn impacts on AHN-mediated learning and memory performance, as well as depressive and anxiety phenotypes. Many animal studies have also demonstrated that these dietary interventions impact on behavior without assessing changes in AHN as a potential underlying cellular mechanism. Schematic representation of a sagittal view of the hippocampal dentate gyrus which harbors neural stem cells (in green). Immature granule neurons derived from these stem cells will integrate into the hippocampal circuit, receiving information from the entorhinal cortex (not shown) and projecting this onto CA3 neurons (red shape), mediating the hippocampus to fulfill its behavioral functions. Illustration by Improve Presentation (improvepresentation.com).

It is increasingly recognized that mammalian stem cells are able to respond to extrinsic signals derived from systemic factors to alter tissue composition and growth (Mihaylova, Sabatini, & Yilmaz, 2014). This is due largely to the design of the specialized niche, or microenvironment, where stem cells reside. A stem cell niche is a three-dimensional entity composed of many different cell types that are intimately associated with the surrounding vasculature. This close proximity enables stem cells to respond to extrinsic cues. These may include paracrine "local" or "short-range" signals from cells that compose the niche or delivery of "long-range" cues from the circulating or systemic environment (Villeda & Wyss-Coray, 2013), including those linked to diet and nutrition. Notably, the dentate gyrus is richly concentrated around blood vessels, facilitating potential communication between the neural precursor cells (NPCs) of this niche and the systemic environment (Palmer, Willhoite, & Gage, 2000).

Whereas stem cells residing in the muscle or liver tissue would typically become activated in response to changing growth demands, NPCs are engaged by a requirement of the hippocampus to process spatial cues in complex and novel environments (Kempermann, 2012). More recently, it is also appreciated that AHN is a critical cellular process in mood control and mediating the response to particular types of antidepressants in animal studies (Eisch & Petrik, 2012).

So how do diet and nutrition impact on the levels of AHN in the context of their functional role in cognitive and emotional processing? To address this question, we first discuss the profound effects of CR and IF on AHN. In particular, it is thought that there is a selective advantage in keeping cognitively sharp during the "lean" times (Mattson, 2012). Increased AHN during times of food restriction is possibly an adaptive mechanism to provide a cognitive flexibility to adapt to diverse environmental conditions to maximize the chances of feeding or finding shelter (Kempermann, 2012).

Throughout, we will give an overview of the influences of different dietary paradigms and nutritional supplementation on both proliferation and differentiation of NPCs in the dentate gyrus.

DIETARY REGULATION OF ADULT HIPPOCAMPAL NEUROGENESIS

Table 6.1 summarizes the promising animal studies that have positioned CR and IF as robust dietary paradigms that positively modulate AHN across a number of different physiological and disease contexts. Table 6.2 describes the many animal studies that have evaluated the impacts of these paradigms on cognition and the associated molecular and cellular mechanisms, without specifically assessing changes in AHN; notable studies and key observations are discussed in more detail in this section.

Calorie Restriction

Calorie restriction (CR) typically involves limiting calorie intake by 20–40% compared to baseline unrestricted or ab libitum (AL) consumption, in the absence of micro- or macronutrient deficiency. Lifelong CR extends the average and maximum life span of model organisms by a corresponding 20–40%. CR also entails significant benefits for "health span," a term that describes the years of our life lived free of pathology and disease. For example, CR improves insulin sensitivity and autonomic function and reduces inflammation, as well as delays the onset of age-related processes and diseases in many model organisms including worms, flies, and rodents (Anderson & Weindruch, 2012).

A long-term study in nonhuman primates showed that a 30% CR did not improve survival outcomes, but did reduce the incidence of cancer and type 2 diabetes, when initiated in young and older age animals (Mattison et al., 2012). It must be noted that four of the CR group in this study lived longer than 40 years (the equivalent of 120 years for humans) whereas only one monkey lived to this age in the control group. In a similar study, additional positive benefits of a 30% CR diet included reduced brain atrophy, age-associated

TABLE 6.1 Animal Studies of Dietary Regulation of Adult Hippocampal Neurogenesis

Model	CR or IF	Duration	Impact on AHN	References
C57BL/6 mice	40% CR	From 16 weeks of age for remainder of life	Increased PRO but no effect on survival of NPCs; increased survival of glial cells.	Bondolphi et al. (2004)
Wistar strain albino rats	ADF to bring about 30–40% CR	From 3 to 4 months of age for 12 weeks	Increased expression of the immature neuronal marker PSA–NCAM and BDNF and NT-3.	Kumar, Parkash, Kataria, and Kaur (2009)
Male Sprague–Dawley rats	ADF to bring about 30% CR	From 3 months of age for 3 months	Increased survival of NPCs and BDNF expression.	Lee et al. (2000)
Male C57BL/6 mice	ADF to bring about 30% CR	From 8 weeks of age for 3 months	Increased survival of NPCs at 4 weeks.	Lee et al. (2002)
Male C57Bl6J	IF mice were fed for 8 h out of every 24-h period	From the age of 10 weeks for 3 months before surgery	Enhances basal and suppresses stroke-induced increases in proliferation; reduces cell death; no effect on differentiation.	Manzanero et al. (2014)

ADF, alternate day fasting; *BDNF*, brain-derived neurotrophic factor; *CR*, calorie restriction; *IF*, intermittent fasting; *NPC*, neural precursor cell; *PRO*, proliferation; *PSA–NCAM*, polysialic acid neural cell adhesion molecule.

pathologies, and all-course mortality when restriction is initiated in young adult monkeys (7–14 years) (Colman et al., 2009, 2014). In humans, individuals belonging to the CR society are voluntarily undergoing reduced calorie intake with early reports of the same positive adaptations that occur in CR rodents and monkeys. These include decreased metabolic, hormonal, and inflammatory risk factors for type 2 diabetes, cardiovascular disease, and some types of cancer (Anderson & Weindruch, 2012).

CR is thought to activate adaptive cellular stress response pathways which enable a shift of energy resources away from anabolism and reproduction toward somatic maintenance (Mair & Dillin, 2008). In this way, cells become resilient against external stressors. Further, it has been proposed that the organism-wide benefits of CR are partly derived by its effects on stem cell number, integrity, and function across multiple tissues (Rafalski et al., 2012). Indeed, CR reverses or slows down age-related declines in the activity of stem

TABLE 6.2 Animal Studies Evaluating the Impact of Dietary Regulation on Learning and Memory

Model	CR or IF	Duration	Effect on depressive behavior	Effect on learning and memory	Cellular and molecular mechanisms	References
Male F344xBN hybrid rats	40% CR	Lifelong from age of 17 weeks		Reduced age-related decline in spatial learning (MWM)	Stabilization of key synaptic proteins (synatophysin, NMDA receptor subunits)	Adams et al. (2008)
Male swiss mice	ADF	6–8 months from age of 8 weeks		Enhances learning and consolidation processes (rotarod, operant conditioning)	Increased expression of NR2B subunit of the NMDA receptor	Fontan-Lozano et al. (2007)
Mouse (3 x TgAD)	40% CR or alternate day IF	For 7 or 14 months		Enhance locomotion and exploratory behavior; CR and IF diets ameliorate age-related memory deficits	Significantly lower levels of amyloid peptides in the hippocampus of mice in the CR group compared to the AL and IF groups. IF preserves performance even in the presence of pathology	Halagappa et al. (2007)
	Acute IF (+antidepressant imipramine)	9 h	Reduced total immobility time (FST); no alteration in locomotor activity (OFT) Additive effects of imipramine and 9 h fasting on FST		Increased the phosphorylation of CREB	Li et al. (2014)
Adult C57BL/6 mice	40% CR	10 days	Reduced total immobility time (FST); increased social interaction		Dependent on hypothalamic neuropeptide orexin	Lutter et al. (2008)

Continued

TABLE 6.2 Animal Studies Evaluating the Impact of Dietary Regulation on Learning and Memory—cont'd

Model	CR or IF	Duration	Effect on depressive behavior	Effect on learning and memory	Cellular and molecular mechanisms	References
Wistar strain male albino rats	Alternate day IF	3 months from the age of 3 and 24 months		Improved age-associated impairment in motor coordination (rotarod) and memory function (MWM)	Attenuation of age-related increases in protein carbonylation and oxidative damage. Increased mitochondrial complex IV activity and levels of synaptic proteins	Singh et al. (2012)
Transgenic PS1 and PS2 double KO mouse model of AD	30% CR	4 months		Improved novel object recognition and contextual fear conditioning memory	Reduced ventricle enlargement, caspase-3 activation, astrogliosis, and tau phosphorylation. Upregulation of AHN and synaptic plasticity genes	Wu et al. (2008)

Cognitive test used is indicated in parentheses.
ADF, alternate day fasting; *AD*, Alzheimer's disease; *CR*, calorie restriction; *FST*, forced swim test; *IF*, intermittent fasting; *MWM*, Morris water maze; *NPC*, neural precursor cell; *PRO*, proliferation.

cells throughout the body, such as in the muscle (Bondolfi, Ermini, Long, Ingram, & Jucker, 2004; Cerletti, Jang, Finley, Haigis, & Wagers, 2012).

CR increased survival of NPC at 4 weeks in young adult mice (Lee, Duan, Long, Ingram, & Mattson, 2000). These changes in AHN were accompanied by changes in markers of brain plasticity related to AHN, such as increases in both neurotrophin expression in the dentate gyrus and brain-derived neurotrophic factor (BDNF) in the CA1 and CA3 subregions of the hippocampus (Lee et al., 2000). The generation of neurotrophic factors such as BDNF is an important adaptive and neuroprotective response to CR; this contributes to the enhancement of AHN by CR (Rothman, Griffioen, Wan, & Mattson, 2012). Maintenance on a CR regimen increases markers of synaptic plasticity and is linked to greater electrical and synaptic activity throughout neuronal circuits when compared to satiated and resting states (Stranahan & Mattson, 2012). These examples illustrate an important concept in dietary and nutritional regulation of AHN, namely that complementary processes related to brain plasticity are also engaged by the broad spectrum impacts of these interventions.

To our knowledge no study has evaluated AHN in response to CR in an animal model of depression but an interesting report demonstrated that 10 days of CR resulted in antidepressant-like responses, an effect dependent on the hypothalamic neuropeptide orexin (Lutter et al., 2008).

AHN markedly declines with age and aging has been reported to be a major contributor to the reduced proliferation of NPCs (Artegiani & Calegari, 2012). CR reverses age-related declines in proliferation but not the survival of NPCs (Bondolfi et al., 2004). In middle-aged mice, CR from 3 to 11 months of age enhanced the survival of newborn glial cells in the DG (Bondolfi et al., 2004; Gillette-Guyonnet & Vellas, 2008). The preserving effect of CR to counter age-related decline in spatial and working memory is well established (Mattson, 2012). Mechanistically, CR stabilizes the levels of glutamate receptors and synaptic proteins required for excitatory transmission and thought to underlie hippocampal-dependent spatial learning and memory (Adams et al., 2008). In addition, it is not only lifelong CR that exerts positive effects in combating age-related processes but also starting CR in midlife or short exposures even at a late age also entail beneficial effects on cognitive performance (Redman & Ravussin, 2011). CR is thought to achieve these benefits through preservation of synaptic proteins, reductions in oxidative stress, and reductions in fatty acid metabolism (Stranahan & Mattson, 2012).

As discussed in Part III, alterations in the levels of hippocampal neurogenesis are increasingly regarded as an integral aspect of Alzheimer's disease (AD). Notably, the putative role of diet and nutrition in both contributing to and mitigating cognitive impairment in AD has emerged as a topic of increasing scientific and public interest. A number of studies have demonstrated that CR reduces AD pathology in animal models of the disease (Mouton, Chachich, Quigley, Spangler, & Ingram, 2009; Qin et al., 2008; Wang et al., 2005). An interesting study reported that 30% CR for 4 months in a transgenic mouse model reduced a number of pathological features, including ventricle enlargement, hippocampal atrophy, cell death, reactive gliosis, and tau phosphorylation and also demonstrated upregulation of genes associated with neurogenesis and synaptic plasticity in the CR hippocampus, along with downregulation of genes associated with the expression of inflammatory markers (Wu et al., 2008). The combined reduction of disease pathology and enhancement of AHN by CR is likely to underlie the amelioration of memory deficits as assessed in this study.

Intermittent Fasting: Effects of Expanding Time Between Meals

Intermediate or intermittent fasting (IF) refers to alternate periods of AL intake with complete or partial restriction of calories. Similar to CR, IF does not mean severe nutrient deprivation/starvation and all IF regimens are on a background of adequate vitamin and mineral intake (Longo & Mattson, 2014). The most widely used IF paradigm in animal studies has been alternate day fasting, which has been adapted for humans in the form of the increasingly popular "5, 2" diet, which involves 2 days of complete or partial CR during a weekly period (Mattson et al., 2014). Notably, IF does not necessarily reduce overall caloric consumption or lower body weight, since subjects may compensate for reduced intake during the restriction period by overeating on the AL phase, although in many studies implementation of IF results in a 20–30% reduction in caloric consumption over time (Longo & Mattson, 2014).

A recent study put forward the complementary concept of time-restricted feeding, where subjects can only consume during a regular daily window of 8–9 h, where the benefits are proportional to fasting duration (Chaix, Zarrinpar, Miu, & Panda, 2014). Moreover, this study also showed that benefits are maintained when the time-restricted feeding is interrupted by AL periods, such as during weekends, a finding of particular relevance to humans. This study also reports that periods of time-restricted feeding "protect" against, or rather reduce, the negative impacts of an obesogenic diet (Chaix et al., 2014).

IF regimens have been demonstrated to induce a multitude of positive impacts on metabolic profiles and age-related diseases in animal studies; these include the attenuation or prevention of diabetes-like phenotypes and cardiovascular disease, as well as increasing maximal life span (Longo & Mattson, 2014). At the molecular level, IF is thought to engage adaptive cellular stress response pathways and appears to engage many of the pathways described above for CR (Mattson et al., 2014).

Many studies have adopted an alternate day feeding paradigm to bring about reduced calorie intake; as such, it remains a topic of open discussion whether it is CR or the period of fasting itself which is driving the beneficial effects. For example, a 30% CR achieved through alternate day fasting increases the survival of newly born cells in the DG, resulting in enhanced neurogenesis and gliogenesis (Lee, Seroogy, & Mattson, 2002). Long-term IF has also been reported to increase neuronal excitability and hippocampal expression of the NMDA receptor subunit NR2B, resulting in enhanced learning and consolidation processes (Fontan-Lozano et al., 2007). Similar to CR, IF engages multiple processes to elicit its positive effects on brain plasticity.

A recent study evaluated the impacts of IF on AHN after transient cerebral artery occlusion and found that IF enhances basal and suppresses stroke-induced increases in proliferation in the SGZ and SVZ. IF reduced cell death but had no effect on the differentiation of neurons or glia. IF also reduced the sensorimotor impairment and infarct size following ischemia and reperfusion, suggesting that IF protects against neurological damage in ischemic stroke brain, possibly through reductions in the circulating levels of leptin (Manzanero et al., 2014).

To our knowledge, no animal study investigating IF as a treatment for mood/anxiety disorders has evaluated AHN as a candidate cellular mediator. Nonetheless, a recent study has shown that as a little as a 9-h fasting in mice, but not 3 or 18h, leads to significant antidepressant effects and increases the phosphorylation of CREB (Li et al., 2014). A 9-h fasting coadministered with the antidepressant imipramine led to additive antidepressant effects and further increased the phosphorylation of CREB in the hippocampus and frontal cortex (Li et al., 2014).

Similar to studies of mood/anxiety, no study investigating IF in the context of aging has evaluated the levels of AHN in animals. Some studies have shown that IF preserves synaptic plasticity with increasing age (Djordjevic, Djordjevic, Adzic, & Radojcic, 2012). For example, short-term late-onset exposure to an IF regimen has been shown to improve age-related declines in cognitive and motor functions, changes that are associated with increased expression of synaptic plasticity markers as well as reductions in oxidative damage-enhanced mitochondrial complex IV activity both in the brain and in peripheral organs (Singh et al., 2012). These findings are supported by the similar effects of CR in mitigating age-related declines in markers of synaptic plasticity (Adams et al., 2008).

In a study using a triple transgenic mice model of AD, long-term maintenance of these mice (14months) on either a 40% CR or IF diet, prior to the onset of the disease phenotype, rescued cognitive deficits (Halagappa et al., 2007). The beneficial effects of CR on cognition were associated with significantly reduced levels of β-amyloid protein and phospho-tau. Intriguingly, IF appeared to protect neurons from injury despite the fact that there are no reductions in either Aβ or tau, suggesting that this dietary approach may protect neurons downstream of Aβ and tau aggregation. Notably, extensive β-amyloid deposition has been well documented in elderly persons in the absence of cognitive deficits (Driscoll & Troncoso, 2011). Together, these data suggest that stimulation of adaptive stress responses following IF is a possible mechanism by which cognition is preserved despite marked signs of disease pathology (Stranahan & Mattson, 2012).

NUTRITIONAL SUPPLEMENTATION

Table 6.3 summarizes the many animal studies that have investigated the impacts of nutritional supplementation on AHN across a number of different physiological and disease contexts. Table 6.4 describes the many animal studies that have evaluated the impacts of these paradigms on depressive behavior and cognition, along with associated molecular and cellular mechanisms, without specifically assessing changes in AHN; notable studies and key observations are discussed in more detail in this section.

Polyphenols

Polyphenols are present in a wide variety of fruits, teas, and consumed plants, including cocoa (Dias et al., 2012). They contain high levels of antioxidants and this is thought to be one of the mechanisms underlying their beneficial effects. Curcuminoids and flavonoids represent two of the main subtypes of polyphenols demonstrated to act on cellular and molecular processes to elicit positive impacts (Murphy, Dias, & Thuret, 2014).

Curcumin, the most ubiquitous curcuminoid and active ingredient in the spice turmeric, has been consumed for medicinal purposes for thousands of years (Gupta, Patchva, Koh, & Aggarwal, 2012). A distinctive feature of curcumin, like dietary intervention by CR and IF, is the ability to modulate a multitude of signaling molecules and its varied properties, including being antibacterial, anti-inflammatory, chemotherapeutic, and neuroprotective (Gupta et al., 2012).

Resveratrol is another naturally occurring polyphenol found in the skin of red grapes, nuts, and several other plants that has been extensively researched owing to its neuroprotective effects on the brain (Gomez-Pinilla & Nguyen, 2012). Resveratrol is a mimetic of CR, extending the maximal life span in yeast, worms, flies, and fish.

Due to their well-documented effects on neural plasticity as well as on cognition, mood/anxiety, and aging, this section will focus on findings from studies investigating curcumin and resveratrol.

In rodents, treatment with polyphenols converges on CREB, a pivotal transcription factor linked to the expression of BDNF and other hippocampal plasticity markers glutamatergic receptors subunits such

TABLE 6.3 Animal Studies Investigating the Impacts of Nutritional Regulation on Adult Hippocampal Neurogenesis

Model	Nutritional supplementation	Duration	Impact on AHN	References
Male aged (19 months) C57B6/J mice	n-3 PUFA mixture, or olive oil or no dietary supplement	8 weeks	Increased proliferation and differentiation, increased dendritic arborization of the newborn neurons and a reduction in astrocytosis and cell death. Increased DG volume.	Cutuli et al. (2014)
Aged male Sprague–Dawley rats	Curcumin-fortified diets (480 mg/kg)	6 and 12 weeks from age of 15 months	Increased proliferation with 12 week treatment only + upregulation of genes involved in neurotransmission, neuronal development, signal transduction, and metabolism	Dong et al. (2012)
Mature adult C57BL/6 mice	Low, medium, and high n-3:n-6 dietary ratio of PUFAs	Given from the age of 3–7 months	High group increased proliferation marker KI67 and differentiation marker DCX. No change in the number of glial cells or cell death. Reduced hippocampal tissue expression of TNF-α.	Grundy et al. (2014)
Male 129/SvXC57BL/6 hybrid mice	Resveratrol; red wine containing 20 mg/L of resveratrol; red wine containing 3.1 mg/L of resveratrol	3 weeks oral administration of resveratrol	Resveratrol and red wine containing 20 mg/L increased total proliferation and enhancement of angiogenesis and neurogenesis.	Harada et al. (2011)
Transgenic fat-1 mice (in vivo) NPCs derived from mouse ES cells (in vitro)	Fat-1 mice rich in endogenous n-3 PUFA DHA, 5 μM DHA (in vitro)	10–12 weeks (in vivo) 7 days Differentiation assay (in vitro)	Increased brain DHA enhances proliferation and density of dendritic spines (in vivo). Promotes differentiation and neurite outgrowth as well as increasing proliferation of cells undergoing differentiation into neuronal lineages (in vitro).	He et al. (2009)

Continued

TABLE 6.3 Animal Studies Investigating the Impacts of Nutritional Regulation on Adult Hippocampal Neurogenesis—cont'd

Model	Nutritional supplementation	Duration	Impact on AHN	References
ApoE4-carrier and ApoE knockout mice	Multinutrient diet Fortasyn (FC), containing DHA, EPA	7 or 9 months from the age of 2 months	No change in the levels of synaptophysin and neurogenesis; MRS revealed decreased levels of glutamate in both the apoE knockout and the wild-type mice increase in CBV in a region of midbrain in the apoE KO and wild-type mice fed.	Jansen et al. (2013)
Male transgenic mouse models of AD (APP–PS1)	Diet enriched with DHA, EPA, and UMP (DEU diet) or diet enriched with DHA, EPA, and other minerals	7 or 9 months from the age of 2 months	FC but not the DEU diet had a significantly higher amount of doublecortin-positive cells; both diets had no effect on reversing declines in the levels of *N*-acetylaspartylglutamate.	Jansen et al. (2013)
In vitro; NPCs obtained from 15.5-day-old rat embryos	DHA In vitro: 10 μM In vivo: 300 mg/kg/day	4 and 7 days (in vitro) 7 weeks (in vivo)	Increased number of neurons and their morphological maturity, increased proliferation and reduced cell death (in vitro). Increased number of BrdU+/NeuN + neurons in the DG (in vivo).	Kawakita et al. (2006)
Pregnant Wistar rats	Resveratrol (10 mg/kg body weight, oral)	Supplementation throughout pregnancy in groups subjected to restraint stressors during early or late gestational period	Improved the expression of DCX +ve neurons and BDNF expression.	Madhyastha et al. (2013)

C17.2 NPCs and cortical NPCs derived from C57BL/6 mice and C57BL/6 mice	In vitro: resveratrol (0.1–50 μm); in vivo: 1 or 10 mg/kg, intraperitoneal	In vitro: 12, 24, or 48 h; in vivo: 14 days	Impaired NPC proliferation by activating AMPK in vitro; decreases proliferation in the DG without activating glial cells or causing neuronal damage but reduces the survival of newly generated hippocampal cells and impairs neurogenesis.	Park et al. (2012)
Neurospheres derived from hippocampi of male C57Bl/6 mice	0.1–5 μM resveratrol	5–6 days Differentiation assay	Dose-dependent decrease in the number of hippocampal neurospheres that produce neurons.	Saharan et al. (2013)
Adult male C57BL/6	Flavanol epicatechin (4 mg per day in water)	10 weeks	No difference in the number of surviving newborn DG cells and BrdU/NeuN-positive cells. Increased hippocampal BDNF.	Stringer et al. (2015)
Swiss male mice	PUFA-enriched diet (2.5 g/kg, oral dose)	6 weeks from the age of 4 weeks	Increased number of dividing cells as measured by BrdU in the DG. Also increase in synatophysin, BDNF, and hippocampal volume.	Venna et al. (2009)

AD, Alzheimer's disease; *BDNF*, brain-derived neurotrophic factor; *DCX*, doublecortin; *DHA*, docosahexaenoic acid; *DG*, dentate gyrus; *EPA*, eicosapentaenoic acid; *MRS*, magnetic resonance spectroscopy; *NPC*, neural precursor cell; *PRO*, proliferation; *PUFA*, polyunsaturated fatty acid.

TABLE 6.4 Animal Studies Investigating the Impact of Nutritional Regulation on Depressive Behavior and Cognition, Along With Associated Molecular Mechanisms

Model	Nutritional supplementation	Duration	Effect on depressive behavior	Effect on learning and memory	Cellular and molecular mechanisms	References
Rats	Blueberry diet (2% w/w)	7 weeks from the age of 2 months		Improves spatial memory performance and the rate of leaning (MWM)	Activation of ERK1/2, increases in total CREB, and elevated levels of BDNF. BDNF elevated in the DG.	Rendeiro et al. (2012)
C57BL/6J mice	Intraventricular injection of resveratrol (5 μg/μL)	1 week from the age of 8–9 months		Improved long-term memory (FC)	Reduced expression of miR-134 and miR-124, leading to upregulation of CREB and BDNF.	Zhao et al. (2013)
Male Wistar rats	Pure flavanols, pure anthocyanins and blueberry diet (2% w/w)	6 weeks from the age of 18 months		All treatments improved spatial memory to a similar extent	Increased in hippocampal BDNF particularly enhanced by anthocyanins.	Rendeiro et al. (2013)
Male Wistar rats	Resveratrol (80 mg/kg, ip)	Chronic administration of resveratrol for 5 weeks during unpredictable chronic mild stress		Improved cognitive performance as measured by MWM and NOR	Decrease in serum corticosterone and increases in BDNF (prefrontal cortex and hippocampus).	Li et al. (2014)
Mice	Resveratrol treatment (20, 40, and 80 mg/kg, ip)	21 days	Reduced immobility time (FST and TST), comparable to the effects of the antidepressant fluoxetine.		Reduced serum corticosterone levels; increased BDNF protein and ERK phosphorylation levels in the prefrontal cortex and hippocampus, comparable to the effects of the antidepressant fluoxetine.	Wang et al. (2013)

Aged Wistar rats	Curcumin (75 mg/kg by gavage)	8 days		Improved learning and spatial memory (MWM)	Increased expression of NCAM PSA (necessary for consolidation of hippocampus-based learning) in the dentate infragranular zone	Conboy et al. (2009)
Male adult Wistar rats	Free curcumin (50 mg/kg/day) or curcumin-loaded lipid-core nanocapsules (2.5 mg/kg/day), intraperitoneally	10 days		Improved learning and memory behavior (Y maze and NOR)	Reduced synaptotoxicity, neuroinflammation, tau hyperphosphorylation, and cell signaling. Signaling disturbances triggered by Aβ. Increased BDNF. Encapsulated curcumin elicited similar neuroprotective response at a 20-fold lower concentration.	Hoppe et al. (2013)
Young adult male Wistar rats	Two dietary doses of flavonoids (Dose I: 8.7 mg/day and Dose II: 17.4 mg/day)	3 weeks		Enhanced spatial memory acquisition and consolidation (MWM)	Increased PSA–NCAM, NR2B subunit of NMDA, BDNF and mTOR signaling in the hippocampus.	Rendeiro et al. (2014)
Rat model of the metabolic syndrome	Administration of TAK-085 (highly purified and concentrated PUFAs containing EPA and DHA, 300 mg/kg)	13 weeks		Reduced the number of reference memory-related errors but no effect on working memory errors (radial maze)	Decreased the lipid peroxide levels and reactive oxygen species and increased the levels of BDNF in the cerebral cortex and hippocampus.	Hashimoto et al. (2013)
Pregnant rats and their offspring	N-3 deficient diet	15 weeks	Increased anxiety-like behavior in the male offspring (OFT, elevated +)		Reduced levels of BDNF. Disrupted insulin signaling. Reduction in the anxiolytic-related neuropeptide Y-1 receptor and increase in the anxiogenic-related glucocorticoid receptor in the cognitive-related frontal cortex, hypothalamus, and hippocampus.	Bhatia et al. (2011)

Continued

TABLE 6.4 Animal Studies Investigating the Impact of Nutritional Regulation on Depressive Behavior and Cognition, Along With Associated Molecular Mechanisms—cont'd

Model	Nutritional supplementation	Duration	Effect on depressive behavior	Effect on learning and memory	Cellular and molecular mechanisms	References
Young (3–4 months) and aged (20–22 months) rats	DPA, a metabolite of EPA (200 mg/kg/day)	56 days		Improved spatial learning (MWM)	Downregulate microglial activation and decreases sphingomyelinase and caspase 3 activation.	Kelly et al. (2011)
Young (3-month-old) and aged (22-month-old) C57Bl6/J mice	Daily diet enriched in EPA (10%) and DHA (7%)	Switch from control to EPA/DHA-enriched diet for 2 months		Improved spatial learning (Y maze) and memory (NOR)	Reduced expression of proinflammatory cytokines and astrocytic morphology.	Labrousse et al. (2012)
Two-year-old rats	Fish oil (11% DHA)	1 month		No improvement in spatial learning (MWM)	No significant changes in gene expression were observed in the aged mice.	Barceló-Coblijn et al. (2003)
C57BL/6J mice homozygous for either the apoE3 (3/3) or the apoE4 (4/4) allele	Fish oil diet enriched in DHA or cholesterol	4 months		Improved object recognition memory (NOR)	Reversed phospholipid disruption and reduced intraneuronal amyloid-β levels in apoEε4 mice.	Kariv-Inbal et al. (2012)

Behavioral test used is indicated in parentheses. *BDNF*, brain derived neurotrophic factor; *DG*, dentate gyrus; *DHA*, docosahexaenoic acid; *EPA*, eicosapentaenoic acid; *FC*, fear conditioning; *FST*, forced swim test; *MRS*, magnetic resonance spectroscopy; *MWM*, Morris water maze; *NOR*, novel object recognition; *NPC*, neural precursor cell; *OFT*, open field test; *PRO*, proliferation; *PUFA*, polyunsaturated fatty acid.

as NMDA–NR2B and polysialated neural cell adhesion molecule (PSA–NCAM: required for the establishment of durable memories) to improve spatial and long-term memory (Rendeiro et al., 2012, 2013; Zhao et al., 2013). Similar to dietary interventions such as CR and IF, the benefits of polyphenol treatment appear to be brought about by the additive effects on multiple molecular and cellular processes related to enhancing brain plasticity.

Through activation of SIRT1, resveratrol has been consistently demonstrated, alongside novel regulators such as microRNAs miR-134 and miR-124, to upregulate CREB levels and subsequently promotion of BDNF synthesis in the hippocampus (Zhao et al., 2013). SIRT-1 activation, however, is dramatically reduced during neuronal differentiation in the adult hippocampus, while being expressed in proliferating cells (Saharan, Jhaveri, & Bartlett, 2013), suggesting other targets through which resveratrol triggers elevated neuronal plasticity. Further to this, resveratrol supplementation, resulting in SIRT1 activation, prevents NPCs differentiating into neurons (Saharan et al., 2013).

It has been shown that orally administered resveratrol acts on the hippocampus via indirect synergistic loops (Harada, Zhao, Kurihara, Nakagata, & Okajima, 2011). This involves resveratrol stimulating mice sensory neurons of the gastrointestinal tract to release peptides linked to brain plasticity. Abolishment of calcitonin gene-related peptide, a potent vasodilator released by these neurons, reverses the beneficial effects of resveratrol on hippocampal angiogenesis, neurogenesis, and cognition (Harada et al., 2011). Further, this study also reported that resveratrol was undetectable in the hippocampus of the mice under investigation, adding supporting evidence for an indirect mechanism of action, possibly mediated by sensory neurons (Harada et al., 2011).

It must be noted that not all studies have reported a beneficial effect of polyphenols on brain plasticity; for example, resveratrol has been shown to reduce the proliferation of NPCs derived from young mice in a dose-dependent manner as well as reduce the survival of newly generated hippocampal cells, impairing neurogenesis as a consequence (Park, Kong, Yu, Mattson, & Lee, 2012). In this study, resveratrol activated AMPK signaling and downregulated the levels of phosphorylated CREB and BDNF in the hippocampus, which were accompanied by deficits in spatial learning and memory.

Growing evidence suggests that the regular consumption of dietary polyphenols can lead to positive effects on mental health-related behaviors, a process probably involving brain plasticity (Dias et al., 2012). Chronic administration of resveratrol was able to prevent a wide range of detrimental effects on cognition induced by the unpredictable chronic mild stress model of depression, likely mediated through the reduction in serum corticosterone levels, increased levels of BDNF, phosphorylated extracellular signal-regulated kinase (pERK), and pCREB in the prefrontal cortex and hippocampus (Liu et al., 2014). Similar biological measures were found in nonstressed mice administered with resveratrol, which also presented with a reduced depressive phenotype comparable to those of fluoxetine-treated animals (Wang et al., 2013). Resveratrol has also been proposed to buffer prenatal stress in the adult offspring, likely through protective mechanisms against oxidative stress (Madhyastha, Sekhar, & Rao, 2013) and through enhancement of hippocampal BDNF and postnatal neurogenesis levels during early and late gestational stressors (Madhyastha, Sahu, & Rao, 2014).

A recent report showed that daily consumption of the flavanol epicatechin reduced anxiety through a nonneurogenic mechanism, with this behavioral change being proposed to be driven by elevating hippocampal and cortical tyrosine hydroxylase with complementary reductions in cortical monoamine oxidase levels (Stringer, Guerrieri, Vivar, & van Praag, 2015). Epicatechin also increased hippocampal BDNF levels. In further support of these effects not being mediated by neurogenesis, epicatechin did not elicit any improvements on a test of pattern separation (Stringer et al., 2015).

Many different polyphenols have been reported to retard age-related declines in CNS function, cognition, and behavior. For example, blueberries, which are rich in both the anthocyanin and the flavanol subset of polyphenols, are effective in persevering spatial working memory in aged animals (Spencer, 2010), a process likely mediated by the marked increases in both NSC proliferation and IFG-1 signaling in the hippocampi of aged rats supplemented with these berries (Casadesus et al., 2004). Even short-term (8 days) curcumin supplementation to aged rats enhances the expression of neuronal plasticity markers in the DG and markedly improved both spatial learning and memory (Conboy et al., 2009). In a study that assessed the effects of a 6- and 12-week curcumin-supplemented diet on hippocampal cellular proliferation, cognitive function, and transcriptional responses in aged rats, only the 12-week supplemented diet improved spatial memory, suggesting that prolonged curcumin consumption is required to prevent or slow down the decline of cognitive function with aging (Dong et al.2012), contrary to the benefits of short-term exposure described above (Conboy et al., 2009). Notably, only the 12-week intervention enhanced AHN, suggesting that the generation of new neuronal cells may require cumulative exposure to the active metabolites of resveratrol (Dong et al., 2012).

Furthermore, results from an exon array expression study of the hippocampal and cortical rat tissue revealed differentially expressed genes related to brain development, cognition, and neurogenesis, whereby expression was dependent on the length of curcumin treatment (Dong et al., 2012). For example, expression of the NeuroD1 gene, integral for

AHN and survival of neuronal progenitors, was increased following a 6-week treatment, whereas a 12-week curcumin treatment markedly upregulated genes implicated in synaptic transmission and memory formation in the hippocampus.

Polyphenols derived from multiple dietary sources are able to counteract cognitive deterioration and reduce neuropathology in different animal models of AD (Choi, Lee, Hong, & Lee, 2012). Recent animal studies also support a role for these compounds in promoting brain plasticity in the context of AD. For example, a 12-week curcumin treatment of aged rats also results in differential expression of genes thought to participate in both the AD neurodegenerative process and the synaptic plasticity such as the Cav1 gene implicated in alterations of cholesterol distribution in the AD brain, spatial memory formation, and age-related working memory decline (Dong et al., 2012).

A detailed study measured the accumulation of polyphenol metabolites in the mouse brain of an AD-transgenic model following human-relevant oral dosage with a Cabernet Sauvignon red wine (abundant in resveratrol) and reported that one of the targeted metabolites, quercetin-3-*O*-glucuronide, reduced neuronal generation of β-amyloid peptides and prevented the formation of the toxic oligomeric species of this peptide (Ho, Ferruzzi, et al., 2013). Moreover, treatment with this specific metabolite markedly reversed AD-type deficits in hippocampal basal synaptic transmission and LTP, via a likely mechanism involving the activation of signaling pathways that result in phosphorylation of CREB (Ho, Ferruzzi, et al., 2013). Similarly, rats treated intraperitoneally with free or nanoencapsulated curcumin for 10 days did not show β-amyloid-induced cognitive impairment or typical decreases in both hippocampal synaptophysin and BDNF levels (Hoppe et al., 2013).

It is of great interest to see how AHN and learning, memory, and mood are altered in these experimental paradigms, particularly in response to exposure to these targeted and specific resveratrol metabolites.

Polyunsaturated Fatty Acids

Omega-3 or PUFAs are classified as essential, meaning that their levels depend on dietary intake (Davidson, 2013). Long-chain PUFAs such as docosahexaenoic acid (DHA) and eicosapentaenoic acid (EPA) are fundamental to CNS function. Notably PUFAs are important structural components of neural cell membranes and modulate cholesterol-induced reductions in membrane fluidity by displacing cholesterol from the plasma membrane and improving neurotransmission and signaling (Su, 2010). PUFAs are also important regulators of gene expression related to neuronal processes (Schuchardt, Huss, Stauss-Grabo, & Hahn, 2010). Although fish is the major source of the EPA and DHA consumed (71%), these fatty acids are also contained in other foods, such as meat (20%), eggs (6%), and plant foods such as leek and cereal-based products (3%) (Meyer et al., 2003).

Sufficiently high intake of PUFAs induces pleiotropic effects and is increasingly regarded as efficacious antiinflammatory agents for chronic inflammatory conditions such as rheumatoid arthritis (Calder, 2006).

PUFA supplementation has also been associated with enhancement of AHN, thought to be a key mechanism underlying associated improvements in cognition with this intervention (He, Qu, Cui, Wang, & Kang, 2009; Kawakita, Hashimoto, & Shido, 2006). Further, dietary supplementation with n-3 PUFAs restored neurogenic markers in the DG of mouse models of chronic autoimmune conditions such as systemic lupus erythematosus and Sjögren's syndrome, characterized by lower levels of AHN among other biological deficits (Crupi et al., 2012).

Perinatal supplementation with PUFAs also mitigated impairments in working, short-term memory and altered fear responses observed in rats as a consequence of chemical-induced neurotoxicity, an effect likely mediated through decreased apoptosis and increased proliferation of NSCs (Lei et al., 2013). Perinatal supplementation with the DHA precursor α-linolenic acid resulted in increased levels of AHN in the offspring, but only when the dam had been exposed to the enriched diet also during pregnancy (Niculescu, Lupu, & Craciunescu, 2011), highlighting the importance of appropriate timings when delivering dietary interventions for maximized results.

In a test of the long-term efficacy of PUFAs, a higher ratio of an n-3:n-6 PUFA diet increased hippocampal levels of PUFA with increases in spatial memory and also anxiety (Grundy, Toben, Jaehne, Corrigan, & Baune, 2014). At the molecular and cellular levels, this diet also reduced TNF-α levels and increased markers of proliferation and differentiation in the DG without changing the numbers of glial cells or cell death.

The metabolic syndrome is also characterized by impaired cognitive ability, a recent study reported that PUFA formulation containing EPA and DHA reduced the number of memory-related errors committed by rats in a model of the syndrome (Hashimoto et al., 2013). This formulation both reduced the number of reactive oxygen species and increased the levels of BDNF in the cerebral cortex and hippocampus of these rats.

Chronic supplementation with a PUFA-enriched diet induced antidepressant behavioral changes that showed an additive effect when combined with a low dose of the antidepressant imipramine (Venna et al., 2009). These behavioral changes were associated with an increase in hippocampal volume and elevated expression of plasticity markers such as synaptophysin and BDNF as well as an increase in the number of proliferating cells. Similarly, supplementation with a PUFA-enriched diet containing DHA and EPA improved performance in an avoidance conditioning paradigm and reduced plasma corticosterone levels, suggesting that PUFA supplementation improves learning and has a strong antistress effect (Perez et al., 2013). Rats exposed to a PUFA-deficient diet during the gestational period and lactation presented with decreased brain levels of DHA and plasticity markers, such as BDNF and reduced activation of CREB in adulthood (Bhatia et al., 2011). At the behavioral level, these animals exhibited increased anxiety-related phenotypes.

Supplementation of docosapentaenoic acid (DPA), an intermediate molecule in the metabolism of DHA from EPA, preserves hippocampal function in the aged brain as evidenced by enhanced LTP and improved spatial learning when compared to control-fed counterparts (Kelly et al., 2011). Similarly, EPA has also been reported to preserve LTP in aged rats (Lynch et al., 2007). Like CR, IF, and supplementation of polyphenols, PUFAs also elicit a multitude of molecular and cellular effects. Switching aged mice for 2 months from a control to an EPA/DHA-enriched diet reduced the expression of proinflammatory cytokine and astrocytic morphology, changes thought to underlie improved spatial memory deficits in these aged mice (Labrousse et al., 2012).

With regard to AHN, supplementation of diet with an n-3 PUFA mixture, but not olive oil, to aged mice resulted in enhanced neurogenesis, accompanied by increased dendritic arborization of the newborn neurons and a reduction in astrocytosis and cell death (Cutuli et al., 2014).These n-3 PUFA-treated mice demonstrated improved object recognition and spatial memory as well as aversive response retention (Cutuli et al., 2014). Other studies, however, have reported little or no protective effects of DHA/EPA-enriched diets on aged-associated cognitive decline (Barcelo-Coblijn et al., 2003). Such discrepancies in the literature can be partly explained by differences in length and/or composition of PUFA supplementation.

Supplementing transgenic mouse models of the disease with a diet enriched mainly in DHA significantly lowers the formation of amyloid plaques (Amtul, Uhrig, Rozmahel, & Beyreuther, 2011). With regard to brain plasticity, EPA supplementation has been shown to prevent the inhibition of LTP induced by β-amyloid peptides (Lynch et al., 2007).

Apolipoprotein ε4 (apoEε4) is a major genetic risk factor for vascular dementia and sporadic AD. Interestingly, when compared to the apoEε2 and apoEε3 isoforms, apoEε4 functions inefficiently as a cholesterol transporter, potentially impacting synaptogenesis and synaptic maintenance through altered release of cholesterol (Pfrieger, 2003). In support of this notion, maintaining mice with compromised apoEε4 function on a fish-oil diet improves behavioral and cognitive performance (Kariv-Inbal et al., 2012).

It is increasingly recognized that the beneficial effects of diet will be gleaned by multidietary approaches, rather than from the supplementation of a single nutrient to patients at risk or diagnosed with AD (von Arnim, Gola, & Biesalski, 2010). At the behavioral level, a multinutrient diet containing DHA, EPA, phospholipids, uridine monophosphate (UMP), choline, B vitamins, and antioxidants exerted anxiolytic effects on both apoE KO and wild-type mice; an important effect given that anxiety and restlessness are significantly correlated with impairments in activities of daily living in AD (Jansen et al., 2014). The multinutrient diet improved learning and spatial memory, but only in the apoE KO mice and was independent of any change in both synaptophysin-labeled presynaptic boutons and in the number of neuroblasts in the DG.

In a similar study, the anxiolytic effect of the PUFA-diet enriched with phospholipids, choline, folic acid, vitamins B6, B12, C, and E, and selenium in a transgenic mouse model of AD was also observed (Jansen et al., 2013). Notably, n-3 PUFA supplementation on its own did not exert these effects, suggesting that it is the combination of additional nutrients that underlies the reductions in anxiety-like behavior. Both the PUFA only and PUFA plus additional nutrients diet had no effect in attenuating spatial learning, but mice in the latter group did show significant increases in doublecortin-positive cells, suggesting that the multinutrient diet restored neurogenesis in this model of AD (Jansen et al., 2013).

HUMAN STUDIES

While cognitive behavioral testing in human subjects and animals is a necessary and appropriate means of assessing performance following dietary intervention, there is a pressing need to relate cognition, mood/emotionality, and behavior to in vivo structural and dynamic quantitative assessments of AHN to diet-induced effects (Spencer, 2010). AHN appears to be altered following dietary and nutritional intervention but investigation of both the dynamics and the functions of AHN in humans has remained challenging owing to the absence of accepted macroscopic neuroimaging readouts (Ho, Hooker, Sahay, Holt, & Roffman, 2013). Given that AHN constitutes, in quantitative terms, a small number of cells, neuroimaging modalities do not yet have the spatial and temporal sensitivity to profile these cells in vivo.

Calorie Restriction

Despite the difficulties in imaging AHN, a study with healthy elderly subjects (mean age, 60.5 years; mean BMI, 28) reported that CR improves verbal memory by 20% three months after the intervention when compared to intake of polyunsaturated fatty acids and the control AL group (Witte, Fobker, Gellner, Knecht, & Floel, 2009). This is the first, and to date, only evidence that a daily 30% CR intervention positively impacts on cognitive performance in aged cohorts in a trial setting.

Epidemiological studies report that high caloric diets are associated with an increased risk of developing AD (Gillette-Guyonnet & Vellas, 2008), but in some instances only among individuals carrying the ApoE4 allele, a major genetic risk factor for AD (Luchsinger, Tang, Shea, & Mayeux, 2002). Similarly, a recent population-based case–control study reports that a high calorific intake is associated with a nearly twofold increase in the risk of having mild cognitive impairment, regarded as being an important early clinical indicator that an individual has an increased chance of developing AD (Geda et al., 2013). Conversely, epidemiological studies reveal that those who habitually consume fewer calories have a reduced risk of developing AD (Gustafson, Rothenberg, Blennow, Steen, & Skoog, 2003).

It must be noted that the long-term effects of CR (over 1 year) in elderly persons remain to be elucidated and as such caution must be followed before engaging in a CR regimen. While there are encouraging signs from studies in animal models regarding the benefit of CR and/or IF to mitigate or delay the onset of AD, it must be noted that malnutrition is a common problem among the elderly; as such, dietary recommendations will have to take into consideration this background (Pasinetti & Eberstein, 2008).

Intermittent Fasting

Studies of CR among individuals voluntarily undergoing this paradigm and evidence from the cardiology field have provided proof of principle that dietary interventions based on reducing energy intake are not only safely tolerated by humans but also elicit beneficial effects on a myriad of general health markers including insulin sensitivity, markers of oxidative stress, hypertension, and inflammation (Cava & Fontana, 2013). Together, these results suggest that IF is a feasible intervention in humans, however, it is unlikely that such a regimen could be satisfactorily tolerated over a prolonged period owing to a likely increase in aversive subjective states on the "off" day (Heilbronn, Smith, Martin, Anton, & Ravussin, 2005). A practical suggestion is the addition of a small meal on the fasting day which improves the tolerability and adherence to IF.

In a recent qualitative review, 92 studies were identified that have investigated the potential of IF as an intervention for mood disorders (Fond, Macgregor, Leboyer, & Michalsen, 2013). For example, a rather stringent study from the 1970s reported that depression remitted by 86% among patients subjected to 10 days of complete fasting. Recently, studies have shown that under conditions of IF or CR healthy volunteers or those with conditions such as irritable bowel syndrome showed improved energy levels as well as reduced depressive and anxiety symptoms (Fond et al., 2013).

Fasting has historically been undertaken for religious reasons and studies of undertaking IF for religious regions have shown that fasting individuals have reduced scores on clinical measures of anxiety, depression, and tension (Fond et al., 2013). Further, for patients with bipolar affective disorder, a reduction of both manic and depressive symptoms is observed following fasting. It must be noted that there have been human studies which have not shown a benefit of fasting on depressive or anxiety symptoms (Fond et al., 2013).

Polyphenols

Brain imaging studies reveal that efficient cerebral blood flow (CBF) is integral to optimal brain function as evidenced by marked reductions in CBF among individuals with impoverished cognitive performance and patients diagnosed with dementia (Nagahama et al., 2003; Ruitenberg et al., 2005). While not regarded as a definitive readout of AHN, the observation that 1–2h of infusion of subtypes of polyphenol markedly and rapidly increases CBF measures in human subjects is still of great interest (Fisher, Sorond, & Hollenberg, 2006). Further to this, intervention with a flavonol-rich drink derived largely from cocoa increased blood flow, as measured by the blood oxygen level-dependent signal on fMRI, in certain regions of the brain as well as modifying the BOLD response during a task switching test (Francis, Head, Morris, & Macdonald, 2006). While these studies primarily focus on the flavanol subset of polyphenols (eg, cocoa), it is highly plausible that other polyphenols, particularly those rich in other flavonols such as grapes and blackcurrants, may also impact positively CBF and therefore AHN (Spencer, 2010).

A number of epidemiological studies have reported that increased consumption of different polyphenols improves various aspects of cognitive function in the aging population (Bhullar & Rupasinghe, 2013). The recent report of a 13-year-long clinical study builds on these findings by investigating the relationship between specific classes of polyphenols and cognitive performance in a cohort of 2574 middle-aged adults (age range: 35–60 at baseline) (Kesse-Guyot et al., 2012). Assessment of participants dietary records revealed that increased intake of catechin, theaflavins, flavonols, and hydroxybenzoic acids was particularly associated with improved episodic memory and in some cases with preserved verbal memory. It must be noted that unexpected negative associations were also observed; for example, increased intake of the subclasses catechins; proanthocyanidins and flavanols were linked to poorer performance in tests of executive function.

There is a lack of prospective clinical studies and trials investigating the therapeutic potential of polyphenols in AD. However, results from a randomized control trial where participants were required to twice daily consume a 120mg dose of Egb761 for 5 years were recently reported (Vellas et al., 2012). Disappointingly, long-term use of the polyphenol extract Egb761 did not reduce the risk of progression to AD when compared to placebo. The authors do point out that the number of dementia-related events in their study was much lower than expected, resulting in reduced statistical power to differences between the groups. Indeed, overrecruitment of healthy volunteers is a problematic issue in prevention trials, particularly in dementia research where baseline control individuals appear healthier and better educated than the general elderly population (Dangour, Allen, Richards, Whitehouse, & Uauy, 2010).

Polyunsaturated Fatty Acids

Frequent fish consumption has been associated with decreased risk of depression and suicidal ideation (Tanskanen et al., 2001), higher self-reported mental health status (Silvers & Scott, 2002), and decreased prevalence of postpartum depression (Hibbeln, 2002). Moreover, low concentrations of PUFAs in the blood of depressed patients also support a potential role of altered levels of PUFAs in the development of depression (Adams, Lawson, Sanigorski, & Sinclair, 1996; Peet, Murphy, Shay, & Horrobin, 1998). Conversely, in patients diagnosed with unipolar depression, EPA treatment decreased symptoms of depression to a level similar to that of fluoxetine (Jazayeri et al., 2008). Notably, the effects of combined EPA and fluoxetine treatment were superior when compared to either intervention alone.

Evidence from observational studies largely suggests that PUFA-enriched diets protect individuals from or slow down cognitive decline in the elderly population (Luchtman & Song, 2013; Sydenham, Dangour, & Lim, 2012). Results from a 5-year population-based prospective study of cognitive decline in relation to fish, EPA, and DHA consumption in a cohort of nondemented elderly men (n=210, age range: 70–89) from the Netherlands revealed that among individuals consuming fish, cognitive decline was significantly lower in a dose-dependent manner when compared to their counterparts who consumed no fish (van Gelder, Tijhuis, Kalmijn, & Kromhout, 2007). Similarly, a 5-week crossover trail evaluating the impact of fish oil PUFA intake on cognitive performance revealed that PUFA supplementation in a Swedish cohort of healthy middle-aged to elderly subjects (n=40, age range: 51–72 years) significantly improved working memory (Nilsson, Radeborg, Salo, & Bjorck, 2012).

A recent neuroimaging study explored whether EPA and DPA intake impacted not only cognitive performance at 75 years but also brain volume in a healthy cohort of Swedish participants (n=252) with a baseline age of 70 and a low dietary intake of these fatty acids at the start of study; self-reported increases in EPA and DHA intake at 70 years positively associated with both cognitive performance and global gray matter volume (Titova et al., 2013).

A Cochrane review by reports that the pooling of data from three randomized control trials of high methodological quality failed to show any benefit of n-3 PUFA supplementation on cognitive function in a sample of healthy elderly persons (age >60 years), where n-3 PUFA supplementation occurred for at least a 6 month duration (Sydenham et al., 2012). Intervention studies based on supplementation with n-3 PUFA protecting against AD have also proved disappointing, with many studies failing to show any protective effect against the disease (Dangour, Allen, Elbourne, et al., 2010). It is possible that a protective effect of PUFA intake on cognitive decline may only be apparent with advancing age and marked cognitive decline, or with longer exposures that begin earlier in adulthood.

Further, these studies often test patients already with mild to moderate dementia, too late a time point in the pathophysiological course of AD. These negative results may reflect intrinsic limitations in the design of these trials such as the preponderance to recruit healthier elderly participants who are more motivated to adhere to instructions to eat healthier and undertake exercise more, thus confounding results by not being a truly representative sample of the target population (Dangour, Allen, Elbourne, et al., 2010).

As described above, many other dietary components/regimens possess diverse properties to promote brain plasticity. Considering the complementary roles of polyphenols and PUFA, it is not surprising that the specific multinutrient diets containing a combination of these factors have been developed to mitigate risk factors associated with AD (Scheltens et al., 2010). Indeed, two randomized, double-blind controlled clinical trials reported that daily consumption of such multinutrient component diets improved memory performance in patients with mild AD (Scheltens et al., 2010).

TRANSLATIONAL CONSIDERATIONS: FROM ANIMALS TO HUMANS

As we have described above, diet influences multiple aspects of brain plasticity, including neurodevelopment, neurotrophins, neurogenesis, synaptogenesis, and ultimate activity at the brain network level. Together these processes underlie and influence cognitive and mood/emotional processing, thus positioning diet as a key modulator of brain structure and function.

Importantly, diet-induced changes in AHN and others forms of brain plasticity potentially represent low-cost and effective means for protecting against the debilitating effects of psychiatric and neurological disorders. This viewpoint is strongly supported by many of the cellular and molecular animal studies discussed which together indicate that in addition to being a noninvasive intervention diet entails a "broad spectrum of action" (Gomez-Pinilla, 2008), an advantageous feature for the treatment of neurological disorders characterized by diffuse pathology and deficits across multiple cellular domains. This is further supported by the promising results from epidemiological studies that have shown that these dietary factors entail improvements in emotional and cognitive domains in humans.

Diet and Exercise

Feeding and exercising are interconnected determinants of energy balance that have influenced the evolution of the modern brain over thousands of years. Given that the brain places the largest demand on oxygen consumption, it is not surprising that energy metabolism has a profound influence on brain function (Gomez-Pinilla & Nguyen, 2012). In particular, both food consumption and physical activity stimulate mitochondrial activity and thus energy provision to the brain, which in turn modulate the signaling pathways linked to neuronal function and brain plasticity (Gomez-Pinilla & Nguyen, 2012). This interconnectivity is also exemplified by the role of BDNF; in addition to its role in enhancing brain plasticity, BDNF has also been implicated in modulating brain energy metabolism, as evidenced by studies that demonstrate that perturbed BDNF signaling can manifest in metabolic disorders such as obesity (Rothman et al., 2012).

The combination of exercise and CR is particularly noteworthy. In this paradigm, typically, 12.5% of the energy restriction comes from adherence to a restricted diet and the other 12.5% comes from increased energy expenditure via exercise. The principal advantage of combining CR with exercise is that individuals may find it easier to comply with energy restriction if this is split between CR and exercise. Some studies, however, have reported that CR plus exercise does not elicit positive changes to health other than those elicited by CR (Deruisseau et al., 2006; Holloszy, 1997; Seo et al., 2006).

Animal versus Human Brain Plasticity

Perhaps more important than the technical challenges related to sensitively capturing dynamic processes such as neurogenesis in the human brain is to acknowledge possible fundamental neurobiological and functional differences when extrapolating from findings in animals to humans. While the levels of AHN appear comparable in middle-aged rodents and humans (Spalding et al., 2013), it is now appreciated that age-related declines in rodents are potentially more pronounced than in humans (Spalding et al., 2013). Such an observation has implications for dietary regimens known to specifically increase neurogenesis in older animals, as there is a possibility that these same regimens will not elicit the same magnitude of increases because the baseline levels of neurogenesis are potentially higher in elderly humans.

Supplementation and Blood–Brain Barrier Permeability

It must be emphasized that the effects of nutritional supplementation are dependent on duration of the intervention, baseline levels, and the mode of administration (Gomez-Pinilla & Tyagi, 2013). DHA, for example, elicits a multitude of actions at the cellular and molecular level, making it difficult to ascribe recommended levels of intake (Gomez-Pinilla & Tyagi, 2013). Indeed, a number of studies have demonstrated that a varied range of PUFA intake can have beneficial effects on mental health (Sinn, Milte, & Howe, 2010).

Owing to the effects of gut enzymes on the metabolism of polyphenols, their metabolites exhibit a wide range of bioavailability. Another consideration is the ability of the metabolites of polyphenols to cross the blood–brain barrier (BBB). On this note, while polyphenols have been claimed to exert neuroprotective effects, many polyphenols have also been demonstrated to be incapable of crossing the BBB, suggesting that benefits may arise via indirect or other pathways (Gomez-Pinilla & Nguyen, 2012). The profiling of the bioavailability and pharmacokinetics of specific metabolites of polyphenols that penetrate the BBB and act on their targets is an important area of research to further develop.

CONCLUSIONS

The studies above position dietary and nutritional regulation as a potent environmental regulator of AHN in health and well-being. The notion that alterations in AHN represent a cellular response to rapidly changing dietary and nutritional signals in the systemic milieu is now supported by a wealth of robust studies in animals and appears to be encouragingly borne out by studies in humans, although these require more definitive proof in the intervention/trial setting. As such, the modulation of AHN by diet has the potential to be an efficacious, noninvasive, and cost-effective means to counterbalance disturbances to brain plasticity in the context of aging, mood disorders, and neurological disorders.

References

Adams, P. B., Lawson, S., Sanigorski, A., & Sinclair, A. J. (1996). Arachidonic acid to eicosapentaenoic acid ratio in blood correlates positively with clinical symptoms of depression. *Lipids*, *31*(Suppl.), S157–S161.

Adams, M. M., Shi, L., Linville, M. C., Forbes, M. E., Long, A. B., Bennett, C., et al. (2008). Caloric restriction and age affect synaptic proteins in hippocampal CA3 and spatial learning ability. *Experimental Neurology*, *211*, 141–149.

Amtul, Z., Uhrig, M., Rozmahel, R. F., & Beyreuther, K. (2011). Structural insight into the differential effects of omega-3 and omega-6 fatty acids on the production of Abeta peptides and amyloid plaques. *The Journal of Biological Chemistry*, *286*, 6100–6107.

Anderson, R. M., & Weindruch, R. (2012). The caloric restriction paradigm: implications for healthy human aging. *American Journal of Human Biology*, *24*, 101–106.

von Arnim, C. A., Gola, U., & Biesalski, H. K. (2010). More than the sum of its parts? Nutrition in Alzheimer's disease. *Nutrition (Burbank, Los Angeles County, Calif.)*, *26*, 694–700.

Artegiani, B., & Calegari, F. (2012). Age-related cognitive decline: can neural stem cells help us? *Aging*, *4*, 176–186.

Barcelo-Coblijn, G., Hogyes, E., Kitajka, K., Puskas, L. G., Zvara, A., Hackler, L., Jr., et al. (2003). Modification by docosahexaenoic acid of age-induced alterations in gene expression and molecular composition of rat brain phospholipids. *Proceedings of the National Academy of Sciences of the United States of America*, *100*, 11321–11326.

Bhatia, H. S., Agrawal, R., Sharma, S., Huo, Y. X., Ying, Z., & Gomez-Pinilla, F. (2011). Omega-3 fatty acid deficiency during brain maturation reduces neuronal and behavioral plasticity in adulthood. *PLoS One*, *6*, e28451.

Bhullar, K. S., & Rupasinghe, H. P. (2013). Polyphenols: multipotent therapeutic agents in neurodegenerative diseases. *Oxidative Medicine and Cellular Longevity*, *2013*, 891748.

Bondolfi, L., Ermini, F., Long, J. M., Ingram, D. K., & Jucker, M. (2004). Impact of age and caloric restriction on neurogenesis in the dentate gyrus of C57BL/6 mice. *Neurobiology of Aging*, *25*, 333–340.

Calder, P. C. (2006). n-3 polyunsaturated fatty acids, inflammation, and inflammatory diseases. *The American Journal of Clinical Nutrition*, *83*, 1505S–1519S.

Casadesus, G., Shukitt-Hale, B., Stellwagen, H. M., Zhu, X., Lee, H. G., Smith, M. A., et al. (2004). Modulation of hippocampal plasticity and cognitive behavior by short-term blueberry supplementation in aged rats. *Nutritional Neuroscience*, *7*, 309–316.

Cava, E., & Fontana, L. (2013). Will calorie restriction work in humans? *Aging*, *5*, 507–514.

Cerletti, M., Jang, Y. C., Finley, L. W., Haigis, M. C., & Wagers, A. J. (2012). Short-term calorie restriction enhances skeletal muscle stem cell function. *Cell Stem Cell*, *10*, 515–519.

Chaix, A., Zarrinpar, A., Miu, P., & Panda, S. (2014). Time-restricted feeding is a preventative and therapeutic intervention against diverse nutritional challenges. *Cell Metabolism*, *20*, 991–1005.

Choi, D. Y., Lee, Y. J., Hong, J. T., & Lee, H. J. (2012). Antioxidant properties of natural polyphenols and their therapeutic potentials for Alzheimer's disease. *Brain Research Bulletin*, *87*, 144–153.

Colman, R. J., Anderson, R. M., Johnson, S. C., Kastman, E. K., Kosmatka, K. J., Beasley, T. M., et al. (2009). Caloric restriction delays disease onset and mortality in rhesus monkeys. *Science (New York, N.Y.)*, *325*, 201–204.

Colman, R. J., Beasley, T. M., Kemnitz, J. W., Johnson, S. C., Weindruch, R., & Anderson, R. M. (2014). Caloric restriction reduces age-related and all-cause mortality in rhesus monkeys. *Nature Communications*, *5*, 3557.

Conboy, L., Foley, A. G., O'Boyle, N. M., Lawlor, M., Gallagher, H. C., Murphy, K. J., et al. (2009). Curcumin-induced degradation of PKC delta is associated with enhanced dentate NCAM PSA expression and spatial learning in adult and aged Wistar rats. *Biochemical Pharmacology*, *77*, 1254–1265.

Crupi, R., Cambiaghi, M., Deckelbaum, R., Hansen, I., Mindes, J., Spina, E., et al. (2012). n-3 fatty acids prevent impairment of neurogenesis and synaptic plasticity in B-cell activating factor (BAFF) transgenic mice. *Preventive Medicine*, *54*(Suppl.), S103–S108.

Cutuli, D., De Bartolo, P., Caporali, P., Laricchiuta, D., Foti, F., Ronci, M., et al. (2014). n-3 polyunsaturated fatty acids supplementation enhances hippocampal functionality in aged mice. *Frontiers in Aging Neuroscience*, *6*, 220.

Dangour, A. D., Allen, E., Elbourne, D., Fasey, N., Fletcher, A. E., Hardy, P., et al. (2010). Effect of 2-y n-3 long-chain polyunsaturated fatty acid supplementation on cognitive function in older people: a randomized, double-blind, controlled trial. *The American Journal of Clinical Nutrition*, *91*, 1725–1732.

Dangour, A. D., Allen, E., Richards, M., Whitehouse, P., & Uauy, R. (2010). Design considerations in long-term intervention studies for the prevention of cognitive decline or dementia. *Nutrition Reviews, 68*(Suppl. 1), S16–S21.

Davidson, M. H. (2013). Omega-3 fatty acids: new insights into the pharmacology and biology of docosahexaenoic acid, docosapentaenoic acid, and eicosapentaenoic acid. *Current Opinion in Lipidology, 24*, 467–474.

Deruisseau, K. C., Kavazis, A. N., Judge, S., Murlasits, Z., Deering, M. A., Quindry, J. C., et al. (2006). Moderate caloric restriction increases diaphragmatic antioxidant enzyme mRNA, but not when combined with lifelong exercise. *Antioxidants & Redox Signaling, 8*, 539–547.

Dias, G. P., Cavegn, N., Nix, A., do Nascimento Bevilaqua, M. C., Stangl, D., Zainuddin, M. S., et al. (2012). The role of dietary polyphenols on adult hippocampal neurogenesis: molecular mechanisms and behavioural effects on depression and anxiety. *Oxidative Medicine and Cellular Longevity, 2012*, 541971.

Djordjevic, J., Djordjevic, A., Adzic, M., & Radojcic, M. B. (2012). Effects of chronic social isolation on Wistar rat behavior and brain plasticity markers. *Neuropsychobiology, 66*, 112–119.

Dong, S., Zeng, Q., Mitchell, E. S., Xiu, J., Duan, Y., Li, C., et al. (2012). Curcumin enhances neurogenesis and cognition in aged rats: implications for transcriptional interactions related to growth and synaptic plasticity. *PLoS One, 7*, e31211.

Driscoll, I., & Troncoso, J. (2011). Brain resilience and plasticity in the face of Alzheimer pathology. *Current Alzheimer Research, 8*, 329.

Eisch, A. J., & Petrik, D. (2012). Depression and hippocampal neurogenesis: a road to remission? *Science (New York, N.Y.), 338*, 72–75.

Fisher, N. D., Sorond, F. A., & Hollenberg, N. K. (2006). Cocoa flavanols and brain perfusion. *Journal of Cardiovascular Pharmacology, 47*(Suppl. 2), S210–S214.

Fond, G., Macgregor, A., Leboyer, M., & Michalsen, A. (2013). Fasting in mood disorders: neurobiology and effectiveness. A review of the literature. *Psychiatry Research, 209*, 253–258.

Fontan-Lozano, A., Saez-Cassanelli, J. L., Inda, M. C., de los Santos-Arteaga, M., Sierra-Dominguez, S. A., Lopez-Lluch, G., et al. (2007). Caloric restriction increases learning consolidation and facilitates synaptic plasticity through mechanisms dependent on NR2B subunits of the NMDA receptor. *The Journal of Neuroscience, 27*, 10185–10195.

Francis, S. T., Head, K., Morris, P. G., & Macdonald, I. A. (2006). The effect of flavanol-rich cocoa on the fMRI response to a cognitive task in healthy young people. *Journal of Cardiovascular Pharmacology, 47*(Suppl. 2), S215–S220.

Geda, Y. E., Ragossnig, M., Roberts, L. A., Roberts, R. O., Pankratz, V. S., Christianson, T. J., et al. (2013). Caloric intake, aging, and mild cognitive impairment: a population-based study. *Journal of Alzheimer's Disease, 34*, 501–507.

van Gelder, B. M., Tijhuis, M., Kalmijn, S., & Kromhout, D. (2007). Fish consumption, n-3 fatty acids, and subsequent 5-y cognitive decline in elderly men: the Zutphen Elderly Study. *The American Journal of Clinical Nutrition, 85*, 1142–1147.

Gillette-Guyonnet, S., & Vellas, B. (2008). Caloric restriction and brain function. *Current Opinion in Clinical Nutrition and Metabolic Care, 11*, 686–692.

Gomez-Pinilla, F., & Nguyen, T. T. (2012). Natural mood foods: the actions of polyphenols against psychiatric and cognitive disorders. *Nutritional Neuroscience, 15*, 127–133.

Gomez-Pinilla, F., & Tyagi, E. (2013). Diet and cognition: interplay between cell metabolism and neuronal plasticity. *Current Opinion in Clinical Nutrition and Metabolic Care, 16*, 726–733.

Gomez-Pinilla, F. (2008). Brain foods: the effects of nutrients on brain function. *Nature reviews. Neuroscience, 9*, 568–578.

Grundy, T., Toben, C., Jaehne, E. J., Corrigan, F., & Baune, B. T. (2014). Long-term omega-3 supplementation modulates behavior, hippocampal fatty acid concentration, neuronal progenitor proliferation and central TNF-alpha expression in 7 month old unchallenged mice. *Frontiers in Cellular Neuroscience, 8*, 399.

Gupta, S. C., Patchva, S., Koh, W., & Aggarwal, B. B. (2012). Discovery of curcumin, a component of golden spice, and its miraculous biological activities. *Clinical and Experimental Pharmacology & Physiology, 39*, 283–299.

Gustafson, D., Rothenberg, E., Blennow, K., Steen, B., & Skoog, I. (2003). An 18-year follow-up of overweight and risk of Alzheimer disease. *Archives of Internal Medicine, 163*, 1524–1528.

Halagappa, V. K., Guo, Z., Pearson, M., Matsuoka, Y., Cutler, R. G., Laferla, F. M., et al. (2007). Intermittent fasting and caloric restriction ameliorate age-related behavioral deficits in the triple-transgenic mouse model of Alzheimer's disease. *Neurobiology of Disease, 26*, 212–220.

Harada, N., Zhao, J., Kurihara, H., Nakagata, N., & Okajima, K. (2011). Resveratrol improves cognitive function in mice by increasing production of insulin-like growth factor-I in the hippocampus. *The Journal of Nutritional Biochemistry, 22*, 1150–1159.

Hashimoto, M., Inoue, T., Katakura, M., Tanabe, Y., Hossain, S., Tsuchikura, S., et al. (2013). Prescription n-3 fatty acids, but not eicosapentaenoic acid alone, improve reference memory-related learning ability by increasing brain-derived neurotrophic factor levels in SHR.Cg-Lepr(cp)/NDmcr rats, a metabolic syndrome model. *Neurochemical Research, 38*, 2124–2135.

He, C., Qu, X., Cui, L., Wang, J., & Kang, J. X. (2009). Improved spatial learning performance of fat-1 mice is associated with enhanced neurogenesis and neuritogenesis by docosahexaenoic acid. *Proceedings of the National Academy of Sciences of the United States of America, 106*, 11370–11375.

Heilbronn, L. K., Smith, S. R., Martin, C. K., Anton, S. D., & Ravussin, E. (2005). Alternate-day fasting in nonobese subjects: effects on body weight, body composition, and energy metabolism. *The American Journal of Clinical Nutrition, 81*, 69–73.

Hibbeln, J. R. (2002). Seafood consumption, the DHA content of mothers' milk and prevalence rates of postpartum depression: a cross-national, ecological analysis. *Journal of Affective Disorders, 69*, 15–29.

Ho, L., Ferruzzi, M. G., Janle, E. M., Wang, J., Gong, B., Chen, T. Y., et al. (2013). Identification of brain-targeted bioactive dietary quercetin-3-O-glucuronide as a novel intervention for Alzheimer's disease. *FASEB Journal, 27*, 769–781.

Ho, N. F., Hooker, J. M., Sahay, A., Holt, D. J., & Roffman, J. L. (2013). In vivo imaging of adult human hippocampal neurogenesis: progress, pitfalls and promise. *Molecular Psychiatry, 18*, 404–416.

Holloszy, J. O. (1997). Mortality rate and longevity of food-restricted exercising male rats: a reevaluation. *Journal of Applied Physiology (Bethesda, MD: 1985), 82*, 399–403.

Hoppe, J. B., Coradini, K., Frozza, R. L., Oliveira, C. M., Meneghetti, A. B., Bernardi, A., et al. (2013). Free and nanoencapsulated curcumin suppress beta-amyloid-induced cognitive impairments in rats: involvement of BDNF and Akt/GSK-3beta signaling pathway. *Neurobiology of Learning and Memory, 106*, 134–144.

Jansen, D., Zerbi, V., Arnoldussen, I. A., Wiesmann, M., Rijpma, A., Fang, X. T., et al. (2013). Effects of specific multi-nutrient enriched diets on cerebral metabolism, cognition and neuropathology in AbetaPPswe-PS1dE9 mice. *PLoS One, 8*, e75393.

Jansen, D., Zerbi, V., Janssen, C. I., van Rooij, D., Zinnhardt, B., Dederen, P. J., et al. (2014). Impact of a multi-nutrient diet on cognition, brain metabolism, hemodynamics, and plasticity in apoE4 carrier and apoE knockout mice. *Brain Structure & Function, 219*, 1841–1868.

Jazayeri, S., Tehrani-Doost, M., Keshavarz, S. A., Hosseini, M., Djazayery, A., Amini, H., et al. (2008). Comparison of therapeutic effects of omega-3 fatty acid eicosapentaenoic acid and fluoxetine, separately and in combination, in major depressive disorder. *The Australian and New Zealand Journal of Psychiatry, 42*, 192–198.

Kariv-Inbal, Z., Yacobson, S., Berkecz, R., Peter, M., Janaky, T., Lutjohann, D., et al. (2012). The isoform-specific pathological effects of apoE4 in vivo are prevented by a fish oil (DHA) diet and are modified by cholesterol. *Journal of Alzheimer's Disease, 28*, 667–683.

Kawakita, E., Hashimoto, M., & Shido, O. (2006). Docosahexaenoic acid promotes neurogenesis in vitro and in vivo. *Neuroscience, 139*, 991–997.

Kelly, L., Grehan, B., Chiesa, A. D., O'Mara, S. M., Downer, E., Sahyoun, G., et al. (2011). The polyunsaturated fatty acids, EPA and DPA exert a protective effect in the hippocampus of the aged rat. *Neurobiology of Aging, 32*(2318), e2311–e2315.

Kempermann, G. (2012). New neurons for 'survival of the fittest'. Nature reviews. *Neuroscience, 13*, 727–736.

Kesse-Guyot, E., Fezeu, L., Andreeva, V. A., Touvier, M., Scalbert, A., Hercberg, S., et al. (2012). Total and specific polyphenol intakes in midlife are associated with cognitive function measured 13 years later. *The Journal of Nutrition, 142*, 76–83.

Kumar, S., Parkash, J., Kataria, H., & Kaur, G. (2009). Interactive effect of excitotoxic injury and dietary restriction on neurogenesis and neurotrophic factors in adult male rat brain. *Neuroscience Research, 65*, 367–374.

Labrousse, V. F., Nadjar, A., Joffre, C., Costes, L., Aubert, A., Gregoire, S., et al. (2012). Short-term long chain omega3 diet protects from neuroinflammatory processes and memory impairment in aged mice. *PLoS One, 7*, e36861.

Lee, J., Duan, W., Long, J. M., Ingram, D. K., & Mattson, M. P. (2000). Dietary restriction increases the number of newly generated neural cells, and induces BDNF expression, in the dentate gyrus of rats. *Journal of Molecular Neuroscience, 15*, 99–108.

Lee, J., Seroogy, K. B., & Mattson, M. P. (2002). Dietary restriction enhances neurotrophin expression and neurogenesis in the hippocampus of adult mice. *Journal of Neurochemistry, 80*, 539–547.

Lei, X., Zhang, W., Liu, T., Xiao, H., Liang, W., Xia, W., et al. (2013). Perinatal supplementation with omega-3 polyunsaturated fatty acids improves sevoflurane-induced neurodegeneration and memory impairment in neonatal rats. *PLoS One, 8*, e70645.

Li, B., Zhao, J., Lv, J., Tang, F., Liu, L., Sun, Z., et al. (2014). Additive antidepressant-like effects of fasting with imipramine via modulation of 5-HT2 receptors in the mice. *Progress in Neuro-Psychopharmacology & Biological Psychiatry, 48*, 199–206.

Liu, D., Zhang, Q., Gu, J., Wang, X., Xie, K., Xian, X., et al. (2014). Resveratrol prevents impaired cognition induced by chronic unpredictable mild stress in rats. *Progress in Neuro-Psychopharmacology & Biological Psychiatry, 49*, 21–29.

Longo, V. D., & Mattson, M. P. (2014). Fasting: molecular mechanisms and clinical applications. *Cell Metabolism, 19*, 181–192.

Luchsinger, J. A., Tang, M. X., Shea, S., & Mayeux, R. (2002). Caloric intake and the risk of Alzheimer disease. *Archives of Neurology, 59*, 1258–1263.

Luchtman, D. W., & Song, C. (2013). Cognitive enhancement by omega-3 fatty acids from childhood to old age: findings from animal and clinical studies. *Neuropharmacology, 64*, 550–565.

Lutter, M., Sakata, I., Osborne-Lawrence, S., Rovinsky, S. A., Anderson, J. G., Jung, S., et al. (2008). The orexigenic hormone ghrelin defends against depressive symptoms of chronic stress. *Nature Neuroscience, 11*, 752–753.

Lynch, A. M., Loane, D. J., Minogue, A. M., Clarke, R. M., Kilroy, D., Nally, R. E., et al. (2007). Eicosapentaenoic acid confers neuroprotection in the amyloid-beta challenged aged hippocampus. *Neurobiology of Aging, 28*, 845–855.

Madhyastha, S., Sekhar, S., & Rao, G. (2013). Resveratrol improves postnatal hippocampal neurogenesis and brain derived neurotrophic factor in prenatally stressed rats. *International Journal of Developmental Neuroscience, 31*, 580–585.

Madhyastha, S., Sahu, S. S., & Rao, G. (2014). Resveratrol for prenatal-stress-induced oxidative damage in growing brain and its consequences on survival of neurons. *Journal of Basic and Clinical Physiology and Pharmacology, 25*, 63–72.

Mair, W., & Dillin, A. (2008). Aging and survival: the genetics of life span extension by dietary restriction. *Annual Review of Biochemistry, 77*, 727–754.

Manzanero, S., Erion, J. R., Santro, T., Steyn, F. J., Chen, C., Arumugam, T. V., et al. (2014). Intermittent fasting attenuates increases in neurogenesis after ischemia and reperfusion and improves recovery. *Journal of Cerebral Blood Flow and Metabolism, 34*, 897–905.

Mattison, J. A., Roth, G. S., Beasley, T. M., Tilmont, E. M., Handy, A. M., Herbert, R. L., et al. (2012). Impact of caloric restriction on health and survival in rhesus monkeys from the NIA study. *Nature, 489*, 318–321.

Mattson, M. P., Allison, D. B., Fontana, L., Harvie, M., Longo, V. D., Malaisse, W. J., et al. (2014). Meal frequency and timing in health and disease. *Proceedings of the National Academy of Sciences of the United States of America, 111*, 16647–16653.

Mattson, M. P. (2012). Energy intake and exercise as determinants of brain health and vulnerability to injury and disease. *Cell Metabolism, 16*, 706–722.

Meyer, B. J., Mann, N. J., Lewis, J. L., Milligan, G. C., Sinclair, A. J., & Howe, P. R. (2003). Dietary intakes and food sources of omega-6 and omega-3 polyunsaturated fatty acids. *Lipids, 38*, 391–398.

Mihaylova, M. M., Sabatini, D. M., & Yilmaz, O. H. (2014). Dietary and metabolic control of stem cell function in physiology and cancer. *Cell Stem Cell, 14*, 292–305.

Mouton, P. R., Chachich, M. E., Quigley, C., Spangler, E., & Ingram, D. K. (2009). Caloric restriction attenuates amyloid deposition in middle-aged dtg APP/PS1 mice. *Neuroscience Letters, 464*, 184–187.

Murphy, T., Dias, G. P., & Thuret, S. (2014). Effects of diet on brain plasticity in animal and human studies: mind the gap. *Neural Plasticity, 2014*, 563160.

Nagahama, Y., Nabatame, H., Okina, T., Yamauchi, H., Narita, M., Fujimoto, N., et al. (2003). Cerebral correlates of the progression rate of the cognitive decline in probable Alzheimer's disease. *European Neurology, 50*, 1–9.

Niculescu, M. D., Lupu, D. S., & Craciunescu, C. N. (2011). Maternal alpha-linolenic acid availability during gestation and lactation alters the postnatal hippocampal development in the mouse offspring. *International Journal of Developmental Neuroscience, 29*, 795–802.

Nilsson, A., Radeborg, K., Salo, I., & Bjorck, I. (2012). Effects of supplementation with n-3 polyunsaturated fatty acids on cognitive performance and cardiometabolic risk markers in healthy 51 to 72 years old subjects: a randomized controlled cross-over study. *Nutrition Journal, 11*, 99.

Palmer, T. D., Willhoite, A. R., & Gage, F. H. (2000). Vascular niche for adult hippocampal neurogenesis. *The Journal of Comparative Neurology, 425*, 479–494.

Park, H. R., Kong, K. H., Yu, B. P., Mattson, M. P., & Lee, J. (2012). Resveratrol inhibits the proliferation of neural progenitor cells and hippocampal neurogenesis. *The Journal of Biological Chemistry, 287*, 42588–42600.

Pasinetti, G. M., & Eberstein, J. A. (2008). Metabolic syndrome and the role of dietary lifestyles in Alzheimer's disease. *Journal of Neurochemistry, 106*, 1503–1514.

Peet, M., Murphy, B., Shay, J., & Horrobin, D. (1998). Depletion of omega-3 fatty acid levels in red blood cell membranes of depressive patients. *Biological Psychiatry, 43*, 315–319.

Perez, M. A., Terreros, G., Dagnino-Subiabre, A. (2013). Long-term omega-3 fatty acid supplementation induces anti-stress effects and improves learning in rats. *Behavioral and Brain Functions: BBF, 9*, 25.

Pfrieger, F. W. (2003). Cholesterol homeostasis and function in neurons of the central nervous system. *Cellular and Molecular Life Sciences, 60*, 1158–1171.

Qin, W., Zhao, W., Ho, L., Wang, J., Walsh, K., Gandy, S., et al. (2008). Regulation of forkhead transcription factor FoxO3a contributes to calorie restriction-induced prevention of Alzheimer's disease-type amyloid neuropathology and spatial memory deterioration. *Annals of the New York Academy of Sciences, 1147*, 335–347.

Rafalski, V. A., Mancini, E., & Brunet, A. (2012). Energy metabolism and energy-sensing pathways in mammalian embryonic and adult stem cell fate. *Journal of Cell Science, 125*, 5597–5608.

Redman, L. M., & Ravussin, E. (2011). Caloric restriction in humans: impact on physiological, psychological, and behavioral outcomes. *Antioxidants & Redox Signaling, 14*, 275–287.

Rendeiro, C., Vauzour, D., Kean, R. J., Butler, L. T., Rattray, M., Spencer, J. P., et al. (2012). Blueberry supplementation induces spatial memory improvements and region-specific regulation of hippocampal BDNF mRNA expression in young rats. *Psychopharmacology, 223*, 319–330.

Rendeiro, C., Vauzour, D., Rattray, M., Waffo-Teguo, P., Merillon, J. M., Butler, L. T., et al. (2013). Dietary levels of pure flavonoids improve spatial memory performance and increase hippocampal brain-derived neurotrophic factor. *PLoS One, 8*, e63535.

Rothman, S. M., Griffioen, K. J., Wan, R., & Mattson, M. P. (2012). Brain-derived neurotrophic factor as a regulator of systemic and brain energy metabolism and cardiovascular health. *Annals of the New York Academy of Sciences, 1264*, 49–63.

Ruitenberg, A., den Heijer, T., Bakker, S. L., van Swieten, J. C., Koudstaal, P. J., Hofman, A., et al. (2005). Cerebral hypoperfusion and clinical onset of dementia: the Rotterdam Study. *Annals of Neurology, 57*, 789–794.

Saharan, S., Jhaveri, D. J., & Bartlett, P. F. (2013). SIRT1 regulates the neurogenic potential of neural precursors in the adult subventricular zone and hippocampus. *Journal of Neuroscience Research, 91*, 642–659.

Scheltens, P., Kamphuis, P. J., Verhey, F. R., Olde Rikkert, M. G., Wurtman, R. J., Wilkinson, D., et al. (2010). Efficacy of a medical food in mild Alzheimer's disease: a randomized, controlled trial. *Alzheimer's & Dementia, 6*, 1–10 e11.

Schuchardt, J. P., Huss, M., Stauss-Grabo, M., & Hahn, A. (2010). Significance of long-chain polyunsaturated fatty acids (PUFAs) for the development and behaviour of children. *European Journal of Pediatrics, 169*, 149–164.

Seo, A. Y., Hofer, T., Sung, B., Judge, S., Chung, H. Y., & Leeuwenburgh, C. (2006). Hepatic oxidative stress during aging: effects of 8% long-term calorie restriction and lifelong exercise. *Antioxidants & Redox Signaling, 8*, 529–538.

Signer, R. A., & Morrison, S. J. (2013). Mechanisms that regulate stem cell aging and life span. *Cell Stem Cell, 12*, 152–165.

Silvers, K. M., & Scott, K. M. (2002). Fish consumption and self-reported physical and mental health status. *Public Health Nutrition, 5*, 427–431.

Singh, R., Lakhanpal, D., Kumar, S., Sharma, S., Kataria, H., Kaur, M., et al. (2012). Late-onset intermittent fasting dietary restriction as a potential intervention to retard age-associated brain function impairments in male rats. *Age (Dordrecht, Netherlands), 34*, 917–933.

Sinn, N., Milte, C., & Howe, P. R. (2010). Oiling the brain: a review of randomized controlled trials of omega-3 fatty acids in psychopathology across the lifespan. *Nutrients, 2*, 128–170.

Spalding, K. L., Bergmann, O., Alkass, K., Bernard, S., Salehpour, M., Huttner, H. B., et al. (2013). Dynamics of hippocampal neurogenesis in adult humans. *Cell, 153*, 1219–1227.

Spencer, J. P. (2010). The impact of fruit flavonoids on memory and cognition. *The British Journal of Nutrition, 104*(Suppl. 3), S40–S47.

Stranahan, A. M., & Mattson, M. P. (2012). Recruiting adaptive cellular stress responses for successful brain ageing. Nature reviews. *Neuroscience, 13*, 209–216.

Stringer, T. P., Guerrieri, D., Vivar, C., & van Praag, H. (2015). Plant-derived flavanol (-)epicatechin mitigates anxiety in association with elevated hippocampal monoamine and BDNF levels, but does not influence pattern separation in mice. *Translational Psychiatry, 5*, e493.

Su, H. M. (2010). Mechanisms of n-3 fatty acid-mediated development and maintenance of learning memory performance. *The Journal of Nutritional Biochemistry, 21*, 364–373.

Sydenham, E., Dangour, A. D., & Lim, W. S. (2012). Omega 3 fatty acid for the prevention of cognitive decline and dementia. *The Cochrane Database of Systematic Reviews, 6*, CD005379.

Tanskanen, A., Hibbeln, J. R., Hintikka, J., Haatainen, K., Honkalampi, K., & Viinamaki, H. (2001). Fish consumption, depression, and suicidality in a general population. *Archives of General Psychiatry, 58*, 512–513.

Titova, O. E., Sjogren, P., Brooks, S. J., Kullberg, J., Ax, E., Kilander, L., et al. (2013). Dietary intake of eicosapentaenoic and docosahexaenoic acids is linked to gray matter volume and cognitive function in elderly. *Age (Dordrecht, Netherlands), 35*, 1495–1505.

Vellas, B., Coley, N., Ousset, P. J., Berrut, G., Dartigues, J. F., Dubois, B., et al. (2012). Long-term use of standardised *Ginkgo biloba* extract for the prevention of Alzheimer's disease (GuidAge): a randomised placebo-controlled trial. *The Lancet. Neurology, 11*, 851–859.

Venna, V. R., Deplanque, D., Allet, C., Belarbi, K., Hamdane, M., & Bordet, R. (2009). PUFA induce antidepressant-like effects in parallel to structural and molecular changes in the hippocampus. *Psychoneuroendocrinology, 34*, 199–211.

Villeda, S. A., & Wyss-Coray, T. (2013). The circulatory systemic environment as a modulator of neurogenesis and brain aging. *Autoimmunity Reviews, 12*, 674–677.

Wang, J., Ho, L., Qin, W., Rocher, A. B., Seror, I., Humala, N., et al. (2005). Caloric restriction attenuates beta-amyloid neuropathology in a mouse model of Alzheimer's disease. *FASEB Journal, 19*, 659–661.

Wang, Z., Gu, J., Wang, X., Xie, K., Luan, Q., Wan, N., et al. (2013). Antidepressant-like activity of resveratrol treatment in the forced swim test and tail suspension test in mice: the HPA axis, BDNF expression and phosphorylation of ERK. *Pharmacology, Biochemistry, and Behavior, 112*, 104–110.

Witte, A. V., Fobker, M., Gellner, R., Knecht, S., & Floel, A. (2009). Caloric restriction improves memory in elderly humans. *Proceedings of the National Academy of Sciences of the United States of America, 106*, 1255–1260.

Wu, P., Shen, Q., Dong, S., Xu, Z., Tsien, J. Z., & Hu, Y. (2008). Calorie restriction ameliorates neurodegenerative phenotypes in forebrain-specific presenilin-1 and presenilin-2 double knockout mice. *Neurobiology of Aging, 29*, 1502–1511.

Zhao, Y. N., Li, W. F., Li, F., Zhang, Z., Dai, Y. D., Xu, A. L., et al. (2013). Resveratrol improves learning and memory in normally aged mice through microRNA-CREB pathway. *Biochemical and Biophysical Research Communications, 435*, 597–602.

CHAPTER

7

Aging

*R. König**, *P. Rotheneichner**, *J. Marschallinger*, *L. Aigner*, *S. Couillard-Despres*

Paracelsus Medical University, Salzburg, Austria

INTRODUCTION

Life expectancy has been continuously increasing over decades and no plateau is in sight at the moment. As a consequence, the proportion of the elderly in our societies steadily increases and latest projections infer that the number of individuals older than 60 years of age will increase from 810 million in 2012 to 2 billion worldwide by 2050 (United Nations Department of Economic and Social Affairs, 2012). With aging, physical and mental abilities decline markedly, translating into a loss in quality of life and a tremendous burden for the national health systems. During aging, organisms face a progressing loss of physiological functions affecting the whole body, including the brain. Advances in geroscience have identified a set of tightly interconnected genetic, molecular, and cellular changes involved in the process of aging, such as telomere attrition, mitochondrial dysfunction, cellular senescence, and loss of proteostasis (Kennedy et al., 2014; Lopez-Otin, Blasco, Partridge, Serrano, & Kroemer, 2013). Additionally, stem cell pool exhaustion and impaired regeneration are consistently been considered as hallmarks of aging (Kennedy et al., 2014; Lopez-Otin et al., 2013; Rando, 2006; Sharpless & DePinho, 2007).

Age-associated changes in the brain involve multiple molecular and cellular alterations undermining homeostasis and thus impairing brain functions in the elderly. Compromised homeostasis in the aged brain is additionally considered as a cause for an increased susceptibility and even the origin of age-related central nervous system diseases. The apparent increase of neuroinflammation (Lucin & Wyss-Coray, 2009; Sierra, Gottfried-Blackmore,

*These authors contributed equally.

Adult Neurogenesis in the Hippocampus
http://dx.doi.org/10.1016/B978-0-12-801977-1.00007-6

McEwen, & Bulloch, 2007), the partial disruption of the blood–brain barrier (BBB) (Bake, Friedman, & Sohrabji, 2009), and altered neuronal activity (Kanak, Rose, Zaveri, & Patrylo, 2013; Martinelli et al., 2013), together with a decrease in rates of neurogenesis (Kuhn, Dickinson-Anson, & Gage, 1996), are characteristic features of the aged brain.

We will review in the following sections the current knowledge on neurogenesis in the aging hippocampus, with a focus on the alterations within the aged hippocampal neurogenic niche and the intrinsic and extrinsic signals contributing to reduced neurogenesis in the aged brain. We examine possibilities for stimulating endogenous neurogenesis in the elderly and finally speculate on the functional consequences of reduced neurogenesis in the aged brain.

AGING AND NEUROGENESIS

An age-associated decrease in rates of adult hippocampal neurogenesis, detected by a reduced integration of radiolabeled thymidine in proliferating cells, has already been reported half a century ago by Altman and colleagues (Altman & Das, 1965). Decades later, others confirmed this finding using the newly available thymidine analogue 5-bromo-2′-deoxyuridine (BrdU) (Brown et al., 2003; Couillard-Despres et al., 2008; Kuhn et al., 1996; Marschallinger et al., 2014; Seki & Arai, 1995). In a pioneer work, for example, Kuhn and colleagues reported a more than fivefold decrease of the proliferation rate and number of PSA–NCAM-expressing neuronal progenitor cells (NPCs) in the dentate gyrus of 27-month-old as compared to 6-month-old rats. Thereafter, numerous studies investigated the age-related dynamics of neurogenesis in different mammalian species (Ben Abdallah, Slomianka, Vyssotski, & Lipp, 2010; Brown et al., 2003; Couillard-Despres et al., 2009; Gould, Reeves, et al., 1999; Kempermann, Kuhn, & Gage, 1998; Kronenberg et al., 2006; Leuner, Kozorovitskiy, Gross, & Gould, 2007; Miranda et al., 2012; Romine, Gao, Xu, So, & Chen, 2015; Siwak-Tapp, Head, Muggenburg, Milgram, & Cotman, 2007; Villeda et al., 2011). The data consistently demonstrated that adult hippocampal neurogenesis experiences a radical decline during life. Remarkably, this decline proceeds at the same pace in short-lived (eg, rodents) and long-lived (eg, primates, foxes) species (Amrein, Isler, & Lipp, 2011).

The existence of adult neurogenesis in the human brain is nowadays a well-accepted fact within the scientific community. Different methodological approaches have convincingly demonstrated ongoing adult neurogenesis in the human hippocampus and subventricular regions (SVZ) (Eriksson et al., 1998; Knoth et al., 2010; Lucassen, Stumpel, Wang, & Aronica, 2010; Moe et al., 2005; Spalding et al., 2013). Immunohistological postmortem analyses in the human dentate gyrus using markers such as phospho-histone 3 (PH3) (Lucassen et al., 2010), Ki-67, and PCNA (Knoth et al., 2010) confirmed the presence of proliferating cells, as well as of neuronal progenitors based on the detection of the expression of MCM2 (Lucassen et al., 2010) and doublecortin (DCX) (Knoth et al., 2010). Moreover, administration of BrdU, which labels dividing cells, demonstrated the existence of proliferating cells maturing and integrating as new neurons in the human hippocampus (Eriksson et al., 1998). In addition, multipotent progenitor cells residing in the human temporal lobes can be isolated which have the ability to proliferate and differentiate into neurons in vitro as originally described for rodents (Moe et al., 2005; Reynolds & Weiss, 1992).

Magnetic resonance spectroscopy (MRS) has also been suggested to detect human adult neurogenesis in vivo (Manganas et al., 2007). The abundance of a specific metabolite with a resonance peak at 1.28 ppm was correlated with the levels of neurogenesis in humans and reported to show the age-related decrease of neurogenesis (Manganas et al., 2007). The origin of the 1.28 ppm peak remains elusive and controversial. It has been shown to be unspecific to NPCs, but rather related to lipid-rich vesicles, such as detected in apoptotic cells (Ramm et al., 2009). An additional strategy for documenting the existence of adult hippocampal neurogenesis in humans recently employed by Spalding et al. (2013) was based on the ^{14}C isotope content within genomic DNA. According to this study, about 700 new neurons are exchanged daily in the adult human hippocampus, which corresponds to an annual turnover rate of 1.75%.

In contrast, reports on the impact of aging on human neurogenesis have been so far less consistent. A human postmortem histological study investigating the expression of DCX, a marker for NPCs, in the dentate gyrus of 54 individuals, between birth and 94 years of age, reported a decrease in the expression of DCX with aging that was similar to observations made in rodents (Knoth et al., 2010). According to this study, the number of DCX-expressing cells in humans of different ages followed an exponential decline over time. According to the estimates, roughly 20 DCX-expressing cells/mm^2 can be detected in the human dentate gyrus around 1 year of age, whereas at the age of 100 years, only 1 DCX-expressing cell/mm^2 would be detected. Furthermore, no $PCNA^+$ proliferating cells were found beyond the age of 40 years (Knoth et al., 2010). In contrast, the analysis of neurogenesis from Spalding et al. (2013) based on the ^{14}C content of genomic DNA revealed that only a very slow decline of hippocampal neuronal turnover rate can be recognized in adulthood.

Since neuroblasts and proliferating cells in the human hippocampus have been only detected in the dentate gyrus (Eriksson et al., 1998; Knoth et al., 2010), it can be assumed that the neuronal turnover rate in humans is also restricted to the dentate gyrus. Computational model-based analysis from Spalding and colleagues estimated that the fraction of neurons being exchanged in the hippocampus is about 35%, which corresponds to the neuronal fraction belonging to the dentate gyrus (Spalding et al., 2013). Spalding and coworkers thus assume that the neuronal population of dentate granular cells is subject to exchange. It must be taken into account that "turnover of neurons" indicates that some new neurons are generated, while others disappear at the population level. This does not mean, however, that a specific neuron is generated to directly replace a specific neuron lost.

Hippocampal Neurogenic Niche and Aging

Hippocampal adult neurogenesis comprises multiple steps starting from the activation of multipotent neural stem cells (NSCs) located in the subgranular lining of the dentate gyrus. These activated NSCs will give rise to fast-proliferating multipotent NPCs, most of which will be directed toward the neuronal lineage. While a large fraction of these generated progenitors undergoes programmed cell death, a selected population will survive and migrate within the granular cell layer to functionally integrate as new neurons in the hippocampal circuitry (reviewed in Vivar & van Praag, 2013). The net impact of aging on neurogenesis is the sum of the modulation of each step of the neurogenic process. Considering the complexity of neurogenesis per se, we are only starting to understand the impact of aging on the single components of neurogenesis.

The fate of NSCs during aging is a highly debated issue in the field of adult neurogenesis. Under physiological conditions, adult NSCs are relatively quiescent, ie, dividing only seldomly (Bonaguidi et al., 2011; Encinas & Sierra, 2012; Lugert et al., 2010; Morshead et al., 1994; Seri, Garcia-Verdugo, McEwen, & Alvarez-Buylla, 2001). On one side, several studies reported that the number of NSCs remains fairly constant from adulthood to old age in rodents, although the population of NSCs "retires" with time and becomes increasingly quiescent (Bouab, Paliouras, Aumont, Forest-Berard, & Fernandes, 2011; Hattiangady & Shetty, 2008; Lugert et al., 2010; Marschallinger et al., 2014). Hattiangady and Shetty, for example, showed that the number of Sox2-expressing cells in the subgranular zone, the residing site of NSCs, remained constant in aged rats (12 and 24 months old) as compared to young adult rats (4 months old). However, the proliferative activity of the Sox2-expressing cells was seen to diminish over time. Sox2 is a putative marker for NSCs within the subgranular zone (Suh et al., 2007).

In contrast, a substantial loss of NSCs within the dentate gyrus, together with a decreased proliferation of the remaining stem cells, was reported following fate mapping of nestin-expressing cells in transgenic young (3 months) and adult (16 months) mice (Walter, Keiner, Witte, & Redecker, 2011). The disagreement within the scientific community between stem cell depletion and stem cell quiescence was exemplified by two recent works, each supporting one of these two scenarios (reviewed in Kempermann, 2011). According to the disposable stem cell model (Encinas et al., 2011), quiescent NSCs stochastically enter the cell cycle and undergo a limited number of asymmetric cell divisions after which they terminally differentiate into mature astrocytes, thus leading to depletion of stem cells over time. In contrast, a second group reported an increase in quiescent NSCs, but a preserved capacity of self-renewal (Bonaguidi et al., 2011).

The conflicting data on NSC quiescence or depletion with aging result in part from the lack of reliable markers to unambiguously identify and distinguish adult NSCs from NPCs and surrounding astrocytes. In the study presented by Hattiangady and Shetty (2008), for example, NSCs have been identified by expression of the transcription factor Sox2. Noteworthy, these cell populations were found to be rather heterogeneous and Sox2 was found to be coexpressed with various cell type-specific markers, namely glial fibrillary acidic protein (GFAP) (astrocytes and NSCs), vimentin (radial glia), NG-2 (polydendrocytes), S-100β (mature astrocytes), RIP (mature oligodendrocytes), and DCX (immature neurons). Sox2 has also been detected in proliferating progenitors (Steiner et al., 2006) and in some mature astrocytes within the hippocampus (Komitova & Eriksson, 2004). Furthermore, NSC population cannot be solely detected based on the expression of nestin. While nestin is still the most common marker used to identify NSCs (eg, Lagace et al., 2007), it remains to be elucidated whether all NSCs contributing to neurogenesis can be identified solely on the basis of nestin expression (Decarolis et al., 2013).

In mice and rats, the majority of newborn neurons are eliminated by apoptosis within the first 8 weeks after final cell division, and only a small fraction (25–30%) of newly generated cells survive and integrate into the neuronal network (Brown et al., 2003; Couillard-Despres et al., 2009; Encinas et al., 2011; Kempermann, Jessberger, Steiner, & Kronenberg, 2004). Data on the consequences of aging on newly generated cell survival within the dentate gyrus are divergent. Some studies have documented a significant decline in the 4-week survival of newly generated $BrdU^+$ cells in old age (Kempermann et al., 1998; van Praag, Shubert, Zhao, & Gage, 2005), whereas others demonstrated that cell survival does not decrease as a function of age (Bondolfi, Ermini, Long, Ingram, & Jucker, 2004; McDonald & Wojtowicz, 2005; Rao, Hattiangady, Abdel-Rahman, Stanley, & Shetty, 2005). These contradictory results may, at least in part, be explained by methodological differences of BrdU administration, such as of various BrdU dosages, injection frequencies, and survival times used, which could lead to different interpretations on the survival of newborn cells (reviewed in Drapeau & Nora Abrous, 2008).

In addition to proliferation and survival aspects, Rao et al. (2005) observed that maturation of newly generated neurons is delayed in the aged hippocampus. Following analysis of BrdU$^+$/DCX$^+$ and BrdU$^+$/NeuN$^+$-labeled cells, Rao et al. (2005) demonstrated that the survival rate and migration of newly generated neurons were similar in young and aged rats. However, the onset of expression of the mature neuronal marker NeuN was delayed in the aged rats in newly generated cells, and likewise the early dendritic growth was retarded. Similarly, using retroviral labeling, Morgenstern, Lombardi, and Schinder (2008) have shown that neurons generated in the aged mouse dentate gyrus eventually reach a dendritic spine density similar to those born in young adulthood, however, with some delay.

The fate of newborn neurons within the dentate gyrus has been extensively investigated by lineage tracing using immunohistological detection of BrdU (or equivalent), by retroviral vector labeling, or by using transgenic technologies. On average, these studies concluded that in the adult dentate gyrus, the majority (approximately 70–85%) of newly generated cells differentiated into neurons 4 weeks after formation (Bizon, Lee, & Gallagher, 2004; Bondolfi et al., 2004; Brown et al., 2003; Couillard-Despres et al., 2005). In the aged brain, the proportion of newly generated cells undergoing glial differentiation has repeatedly been reported to increase as compared to the earlier time points (Bizon et al., 2004; Bondolfi et al., 2004; van Praag et al., 2005). Nevertheless, once integrated, newly generated granular neurons in young and old mice are morphologically comparable, exhibiting similar dendritic spine densities and electrophysiological properties (Couillard-Despres et al., 2006; Morgenstern et al., 2008; Rao et al., 2005; van Praag, 2008).

Nonneuronal Components of the Neurogenic Niche

The adult neurogenic niche comprises, in addition to neural stem and progenitor cells, a network of various cell types including endothelial cells, astrocytes, pericytes, and microglia, as shown in Fig. 7.1. These neighboring cells have the ability to regulate and support the generation of new neurons via cell–cell contacts and the secretion of soluble factors (Cotrina & Nedergaard, 2002; Mosher & Wyss-Coray, 2014; Oakley & Tharakan, 2014; Palmer, Willhoite, & Gage, 2000; Rodriguez et al., 2014; Seidenfaden, Desoeuvre, Bosio, Virard, & Cremer, 2006; Shen et al., 2004; Shen et al., 2008; Shihabuddin, Horner, Ray, & Gage, 2000; Song, Stevens, & Gage, 2002). Cell proliferation in the subgranular layer occurs mostly in close proximity to capillaries and astrocytes (Doetsch, 2003; Palmer et al., 2000). The close association of the niche with blood vessels secures sufficient supplies of nutrients and oxygen (Shen et al., 2004). In the aged brain, the support provided to stem and progenitor cells by the niche microenvironment decreases, resulting in lower rates of proliferation (Miranda et al., 2012). Nevertheless, NSCs

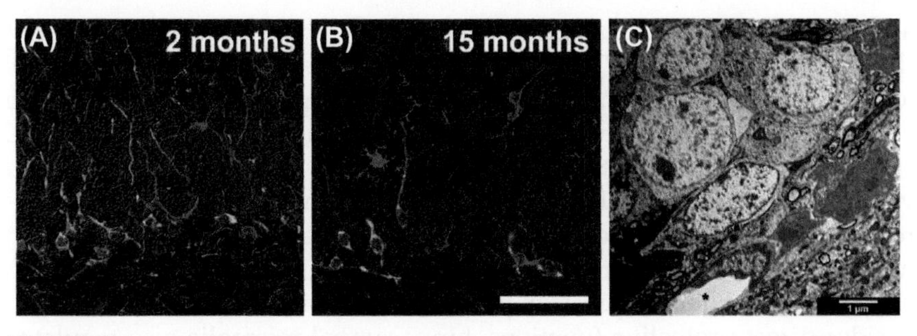

FIGURE 7.1 Immunohistological stainings and electron microscopy of the subgranular and granular zone of the mouse dentate gyrus. Representative confocal images from the dentate gyrus of (A) 2-month-old mouse and (B) 15-month-old mouse. Immunodetection of neuronal progenitors (anti-DCX, green), microglia (anti-Iba1, red), and cell nuclei (DAPI, blue); scale bar 50 μm. In older mice, the number of neuronal progenitors was significantly decreased in the subgranular zone, whereas the soma size of microglia was enlarged compared to the 2-month-old group. (C) Electron microscopy of the neurogenic niche within the dentate gyrus of a 2-month-old mouse. Blood vessels (asterisks) are associated with pericytes (red). Endothelial cells (yellow), NSCs (blue), neuroblasts (purple), microglia (green), mature granular neurons (orange), and astrocytes (not shown) reside within the neurogenic niche in close vicinity.

residing in the aged hippocampus maintain their ability to proliferate and differentiate into fully functional neurons (Ahlenius, Visan, Kokaia, Lindvall, & Kokaia, 2009).

The astrocytes constitute a key component of the neurogenic niche because of their pivotal role in homeostasis. Hippocampal astrocytes have been shown to regulate neurogenesis and promote the neuronal differentiation of NPCs via cell–cell contact and secretion of soluble factors, eg, FGF-2, IGF-1, VEGF, Wnt, and TGF-β (Ma, Ming, & Song, 2005; Song et al., 2002). In the course of aging, the secretion of proneurogenic factors by astrocytes declines significantly (Shetty, Hattiangady, & Shetty, 2005). Bernal and Peterson (2011) reported that astrocytes reduce their secretion of VEGF and FGF-2 during aging, which in turn compromises NSC maintenance and contributes to decreased neurogenesis. Furthermore, transgenic models in which astrocytes chronically secrete increased levels of TGF-β, as observed in the aged brain, showed a significant reduction of the rate of hippocampal neurogenesis (Buckwalter et al., 2006).

While the number of GFAP-expressing cells is maintained in the aged brain, dentate gyrus-specific morphological changes can be seen in astrocytes, such as elevated GFAP expression, hypertrophy of astrocytic processes as indicated by increased surface and volume, and increased GFAP and S100β immunoreactive profiles (Cotrina & Nedergaard, 2002; Rodriguez et al., 2014; Shetty et al., 2005). Astrocytes play a crucial role in glutamate homeostasis via glutamate reuptake and glutamine synthetase activity (Stobart & Anderson, 2013). This function is, however, impaired in hypertrophic reactive astrocytes as shown in situ (Ortinski et al., 2010) as well as suggested in vivo in aged hippocampal murine astrocytes (Rodriguez et al., 2014). The defective astrocytes may lead to deficits in local synaptic transmission in the aged hippocampus.

Microglia, which are the resident immune cells of the brain, constantly survey their environment for noxious substances or injury. On activation via environmental signals such as cytokines, chemokines, and pattern recognition signals, microglia begin to proliferate, migrate, change their morphology by thickening and retracting of their processes, and undergo changes in their gene expression profile (Graeber, 2010; Sierra et al., 2014). Besides direct protective functions, hippocampal microglia also regulate adult neurogenesis (Ekdahl, Claasen, Bonde, Kokaia, & Lindvall, 2003; Monje, Toda, & Palmer, 2003; Sierra et al., 2010). By phagocytizing apoptotic newborn cells in the dentate gyrus, microglia maintain the homeostasis of the neurogenic cascade (Sierra et al., 2010).

With aging, microglia become impaired and fail to execute their protective functions and might even be detrimental for the brain (Lucin & Wyss-Coray, 2009; Mosher & Wyss-Coray, 2014). The hallmarks of microglia aging have been recently defined by Mosher and Wyss-Coray (2014) as reduced proliferation, dystrophy, impaired movement, altered signaling, impaired proteostasis, and impaired phagocytosis. Altered signaling of aged microglia includes the increased production of proinflammatory cytokines such as interleukin (IL)-6, IL-1β, and tumor necrosis factor α (TNFα), and also an elevated expression of antiinflammatory cytokines such as IL-10 and transforming growth factor β1 (TGF-β1) (Dilger & Johnson, 2008; Sierra et al., 2007). Furthermore, impaired phagocytosis with aging (Mosher & Wyss-Coray, 2014) might result in impaired clearance of apoptotic newborn cells in the dentate gyrus, which in turn has been shown to influence adult hippocampal neurogenesis (Sierra et al., 2010).

Inadequate CX3CL1–CX3CR1 chemokine signaling was reported to be a critical contributor to the microglia-mediated impairment of neurogenesis in the aged brain (Bachstetter et al., 2011; Vukovic, Colditz, Blackmore, Ruitenberg, & Bartlett, 2012). The neuroprotective chemokine CX3CL1 (also known as fractalkine) is mainly expressed by neurons and targets the CX3CR1 receptor, which is exclusively expressed by microglia (Wolf, Yona, Kim, & Jung, 2013). This neuron–microglia signaling in the brain controls the extent of microglial activation by maintaining microglia in a resting phase (Wolf et al., 2013). CX3CR1 deficiency deregulates microglia function and results in excessive release of proinflammatory factors in various models of neurological disease (Cardona et al., 2006; Mizuno, Kawanokuchi, Numata, & Suzumura, 2003). In aged mice and rats, hippocampal CX3CL1 levels were reduced compared to young counterparts (Bachstetter et al., 2011; Vukovic et al., 2012), and this decline correlated with a decline in neural precursor cell activity. Chronic treatment with CX3CL1 increased the number of proliferating cells and the numbers of

DCX^+ neuronal progenitors in the dentate gyrus of aged (22 months old) but not young (3 months old) rats, whereas administration of a blocking antibody to CX3CR1 increased neurogenesis in young, but not in old rats (Bachstetter et al., 2011).

Endothelial cells are regarded as another central cell type of the neurogenic niche since they secrete strong proneurogenic factors (eg, VEGF, BDNF, IGF-1) (Leventhal, Rafii, Rafii, Shahar, & Goldman, 1999; Shen et al., 2004). In addition, together with the pericytes, they control the delivery of nutrients and oxygen to stem and progenitor cells closely apposed on blood capillary vessels (Palmer et al., 2000). A hallmark of the aging mammalian brain is the decrease in levels of growth hormone and proneurogenic factors secreted (eg, IGF-1), which is also paralleled by a decrease in the density of blood vessels (Sonntag, Lynch, Cooney, & Hutchins, 1997; reviewed, for instance, in Carter, Ramsey, & Sonntag, 2002; Trejo, Carro, Lopez-Lopez, & Torres-Aleman, 2004). Growth hormone and IGF-1 are known regulators of angiogenesis, but there is increasing evidence that an age-dependent decrease of these leads to a downregulation of adult hippocampal neurogenesis (Anderson, Aberg, Nilsson, & Eriksson, 2002; Sonntag et al., 1997). For instance, IGF-1 applied either in vitro or in vivo in rats significantly enhances proliferation of hippocampal NPCs and neurogenesis (Anderson et al., 2002).

Hormones secreted within the body, eg, corticosteroids from the adrenal glands, can be delivered over long distances to the neurogenic niche via blood vessels. The hypothalamic–pituitary–adrenal axis (HPA axis; also called the stress axis) is activated among others by stress, leading to a release into the blood of cortisol by the adrenal glands (Spiga, Walker, Terry, & Lightman, 2014). Cortisol, which can cross the BBB, will suppress further hormone secretion by acting in a negative feedback loop fashion on the hypothalamus and pituitary gland (Spiga et al., 2014). Importantly, cortisol is also acting on stem and progenitor cells in the dentate gyrus and inhibits their proliferation drastically (Cameron & Gould, 1994; Garcia, Steiner, Kronenberg, Bick-Sander, & Kempermann, 2004; McEwen, 1999). HPA deregulation leads to increased basal levels of cortisol in aged rats and humans, which could contribute to an age-associated decrease of neurogenesis (Lupien et al., 1994; Sapolsky, 1992).

A correlation between increased stress and diminished hippocampal cell proliferation was reported in different chronic stress models like social defeat in rodents (Lagace et al., 2010) or resident–intruder test in primates (Gould, Tanapat, McEwen, Flugge, & Fuchs, 1998). Supporting evidence of a direct contribution of corticosteroid levels on neurogenesis was provided by the observation that removal of adrenal glands in young and aged rodents led to a significant increase of cell proliferation and new granule neuron generation in the dentate gyrus (Cameron & McKay, 1999; Montaron et al., 1999).

The BBB is a structure selectively controlling the passage of blood-borne substances into the brain. This selectively permeable structure is composed of endothelial cells, pericytes, and astrocytes surrounding the blood vessels and prevents harmful or unwanted substances to be delivered into the brain (Bauer, Krizbai, Bauer, & Traweger, 2014; Lange et al., 2013; Trejo et al., 2004), as schematically shown in Fig. 7.2. Extensive evidence has been provided demonstrating the leakiness of the BBB in the aged brain (Bauer et al., 2014; Simpson et al., 2010). Reports have, for example, described an increased extravasation of plasma proteins or intravenously

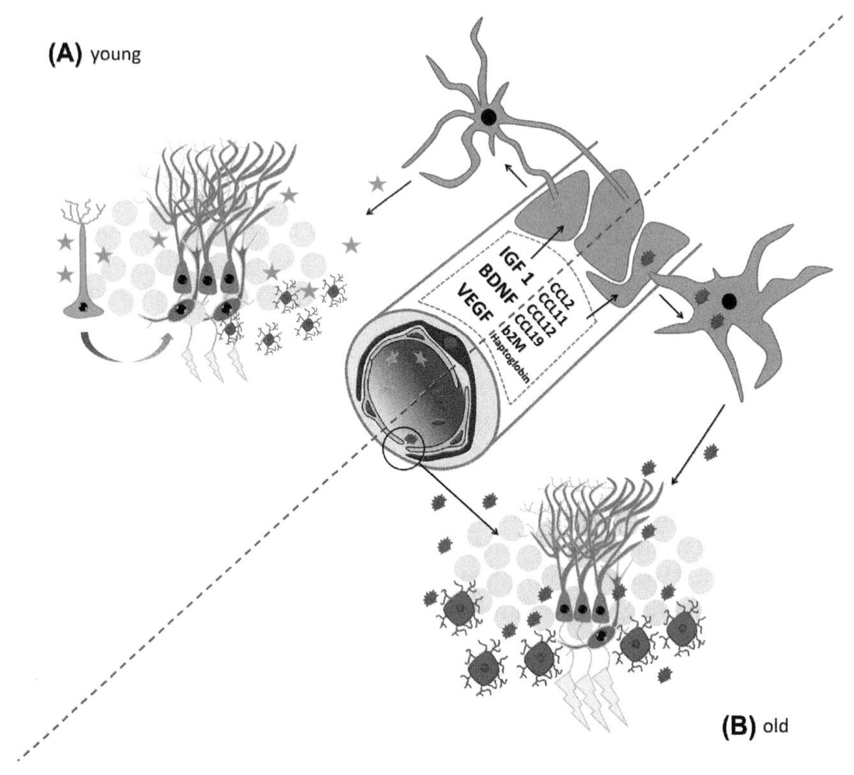

FIGURE 7.2 **Schematic representation of the young and the old neurovascular unit and hippocampal neurogenic niche.** (A) The intact blood–brain barrier (BBB) of the young rodent brains. Astrocytes (blue) and endothelial cells (beige) deliver proneurogenic factors (eg, IGF-1, BDNF, and VEGF; green stars). Continuous proliferation of stem cells (green) leads to the addition of NPCs (orange), which will slowly mature as adult granular cells (gray). Microglia (red) and pericytes (purple) are also integral component of the niche. (B) Defective BBB in the old rodent brain. Blood-derived aging factors (eg, CCL2, CCL11, CCL12, CCL19, b2M, and haptoglobin; red spots) can infiltrate in the neurogenic compartments. Microglial cells and astrocytes become activated. In addition, mature granular cells in the aged hippocampus are reported to change their electrophysiological properties, as indicated by enhanced lightning symbols.

injected Evan's blue dye, high levels of IgG in brains of senescent compared to young adult rodents, as well as in dynamic contrast-enhanced MRI studies in vivo revealing a disrupted BBB in aged humans (Bake et al., 2009; Bake & Sohrabji, 2004; Heye, Culling, Valdes Hernandez Mdel, Thrippleton, & Wardlaw, 2014). Detailed analysis on the ultrastructural level of the BBB in humans revealed no age-dependent variation of endothelial vesicles or gap junctions, but a decrease in pericytes, suggesting a lack of compensation of transient leaks (Stewart et al., 1987).

Neuroinflammation and Neurogenesis in the Aged Brain

The aging brain, including the hippocampus, is characterized by a significant glial activation, an increase in complement factors, and in inflammatory mediators (Lucin & Wyss-Coray, 2009; Streit, Miller, Lopes, & Njie, 2008). Human and mouse brains show age-related increases in the expression of genes related to cellular stress and inflammation, while the expression of genes related to synaptic function, vesicle transport, growth factors, and trophic support decreases (Lee, Weindruch, & Prolla, 2000; Lu et al., 2004). The causes for increased inflammation during aging are still not clarified, but DNA damage caused by increased reactive oxygen species (ROS) might play an important role (Lu et al., 2004).

Inflammation (M1, M2) is generally mediated by proinflammatory cytokines such as IL-1β, IL- 6, and TNFα, in addition to lipidic mediators such as prostaglandins and leukotrienes (Sierra et al., 2014). In the brain, microglia are the main cell populations responsible for the neuroinflammatory response, but other resident cell types, including astrocytes, endothelial cells, mast cells, perivascular and meningeal macrophages, and even neurons, can produce proinflammatory mediators (Jeong, Ji, Min, & Joe, 2013). It should be noted that aged microglia express higher levels of proinflammatory cytokines such as IL-1β, IL-6, and TNFα and show a greater response to LPS inflammatory challenge than their younger counterparts (Sierra et al., 2007).

An inflammation-dependent reduction of adult hippocampal neurogenesis has been demonstrated using systemic and intrahippocampal administration of lipopolysaccharide (LPS) (Ekdahl et al., 2003; Monje et al., 2003). The administration of indomethacin, a nonsteroidal antiinflammatory drug (NSAID), which inhibits the synthesis of proinflammatory prostaglandins, was able to block the LPS-associated decrease of neurogenesis. Along the same line, an age-related increase of proinflammatory cytokines could block neurogenesis in the aging brain. Application of IL-1β and transgenic IL-1β overexpression decreases hippocampal cell proliferation and decreased the number of DCX-labeled neuronal progenitors detected in young mice (Koo & Duman, 2008; Wu et al., 2012). Furthermore, inhibition of IL-1 by blocking IL-1β cleavage by the caspase-1 inhibitor Ac-YVAD-CMK partially restored the number of newborn neurons in aged mice (Gemma et al., 2007). Yet, the actual mechanism of action of IL-1β on neurogenesis in aged mice remains to be fully deciphered.

Similarly, elevated levels of leukotrienes, ie, eicosanoid lipid mediators of inflammation, have been implicated in the aging process of the hippocampus (Chinnici, Yao, & Pratico, 2007; Qu, Uz, & Manev, 2000). Microglia express the leukotriene receptors CysLTR1 and CysLTR2 as well as the putative leukotriene receptor GPR17, which mediate proinflammatory effects of leukotrienes (Lecca et al., 2008; Zhang et al., 2014; Zhang et al., 2013). Antagonizing CysLTR1 and GPR17 with montelukast reduced microglia activation by diminishing proinflammatory cytokine expression in vitro (Zhang et al., 2014). Furthermore, we demonstrated that adult neural progenitor cells cultivated as neurospheres express the leukotriene receptors, in particular GPR17, and reported a dose-dependent increase of cell proliferation for progenitors exposed to the leukotriene receptor antagonist montelukast (Huber et al., 2011). Remarkably, chronic administration of montelukast in aged rats induced a significant increase in hippocampal neurogenesis, in particular NPC proliferation, which results in an increased number of DCX^+ cells. Treatment with the leukotriene receptor antagonist further changed the morphology of aged hippocampal microglia toward an antiinflammatory phenotype and, most importantly, restored spatial learning and memory in aged rats (Marschallinger et al., 2015).

Stimulation of Neurogenesis in Aging

Electroconvulsive therapy (ECT), with its experimental animal counterpart named electroconvulsive shock (ECS), is used in the clinics to treat severe forms of depression in humans (Group, 2003). It has been demonstrated that ECS robustly stimulates hippocampal neurogenesis in mammals by acting on proliferation, survival, and synaptogenesis (Chen, Madsen, Wegener, & Nyengaard, 2009; Nakamura et al., 2013; Perera et al., 2007; Rotheneichner et al., 2014; Scott, Wojtowicz, & Burnham, 2000; Weber et al., 2013; Zhao, Warner-Schmidt, Duman, & Gage, 2012), without altering cell fates or inducing tissue damage (Madsen et al., 2000; Perera et al., 2007). It remains to be elucidated to what extent ECS neurogenic activity is involved in the antidepressive mode of action or represents an independent process. Selective serotonin reuptake inhibitors (SSRIs), which are often used as first-line antidepressant therapy, were also shown to enhance neurogenesis by increasing proliferation of NPCs and survival rates of immature neurons (Malberg, Eisch, Nestler, & Duman, 2000; Marcussen, Flagstad, Kristjansen, Johansen, & Englund, 2008). Whereas the action of SSRIs appears to be limited to the neuronal progenitors, ECT/ECS appears to stimulate in addition the proliferation of both stem and progenitor cells (Encinas, Vaahtokari, & Enikolopov, 2006; Nakamura et al., 2013; Rotheneichner et al., 2014). In addition, SSRI stimulation on neurogenesis was reported to be effective only in young mice (Couillard-Despres et al., 2009). In contrast, ECS was also very potent in the aged mouse brain and promoted a relative increase of 600% of cell proliferation in the aged hippocampus (Rotheneichner et al., 2014).

Recent findings suggest that several factors delivered by the blood system to the brain have the capacity to stimulate or inhibit hippocampal neurogenesis in the aged brain (Villeda et al., 2011). The parabiosis model, in which two animals are joined surgically to share their circulatory systems, represents an ideal model for studying the influence of blood-borne factors from one animal on the body homeostasis of its connected counterpart (Eggel & Wyss-Coray, 2014). The parabiosis model has been used to address the contribution of the aged systemic milieu on neurogenesis and cognitive function (Villeda et al., 2011). In heterochronic parabionts, for example, the circulatory systems of young (3–4 months) and aged (18–20 months) mice were conjugated for 4 weeks. Remarkably, young heterochronic mice exposed to an aged systemic milieu (as well as young mice simply receiving injections of plasma from aged mice) exhibited reduced hippocampal neurogenesis and were impaired in their learning and memory performances. Conversely, old heterochronic parabionts exposed to a young systemic environment showed increased rates of hippocampal neurogenesis, as indicated by an increase in proliferating cells, Sox2-expressing progenitors, and DCX-expressing NPCs. Furthermore, in the heterochronic parabiosis model (as well as by infusions of young plasma in aged mice) exposure to younger blood reverted age-related structural alterations (eg, impaired synaptic plasticity) as well as cognitive impairments (Villeda et al., 2014), indicating that young blood provided late in life is capable of rejuvenating the aged brain.

Villeda and colleagues identified six chemokines, namely CCL2, CCL11 (eotaxin), CCL12, CCL19, haptoglobin, and b2m, whose plasma levels correlated with the reduced neurogenesis in heterochronic parabionts and aged mice. The concentration of these six cytokines was also found to be increased in plasma and cerebral spinal fluid of healthy aging humans. Further evidence that these chemokines are involved in the aging processes was provided by this study in which peripheral administration of eotaxin in young mice resulted in a decreased adult neurogenesis and impaired learning and memory. In addition, a recent heterochronic parabiosis study reported rejuvenating effects of young blood on adult neurogenesis and identified GDF11 as a candidate mediating this effect (Katsimpardi et al., 2014). Together, these results suggest two strategies for reversing age-related alterations on neurogenesis and cognition: administer "proyouthful" factors enriched in the young blood and/or counterpoise proaging factors from aged blood.

NEUROGENESIS AND COGNITION IN AGING

In the adult mammalian brain, the hippocampus is a key structure for the formation of certain types of memory, such as episodic memory and spatial memory (Squire, 1992). Being involved in emotional and cognitive processes, the hippocampal neuronal circuit is central for our everyday behaviors and defines, in some ways, who we are. Whether the reported reduction of hippocampal neurogenesis has an impact on our cognition and mood remains to be elucidated. Moreover, how direct or indirect is the influence of adult-born neurons on cognitive performances and mood in elderly humans? There is a substantial body of evidence supporting that, at least in young rodents, adult-generated neurons in the dentate gyrus play a crucial role in memory processing, such as spatial learning and trace memories (Ambrogini et al., 2000; Gould et al., 1999; Gould & Tanapat, 1999; Lemaire, Koehl, Le Moal, & Abrous, 2000; Shors et al., 2001; Shors, Townsend, Zhao, Kozorovitskiy, & Gould, 2002). In young rats, most of the newborn neuroblasts differentiate into granular cells within the dentate gyrus, which implied that they will on maturation send axonal projections to the CA3 subfield and integrate functionally into the existing circuitry of the hippocampal formation as schematized in Fig. 7.3A (Cameron & McKay, 1999; Hastings & Gould, 1999; Kaplan & Hinds, 1977; Markakis & Gage, 1999; van Praag et al., 2002). These young adult-born neurons exhibit, during a fairly narrow time window found between 4 and 6 weeks of age, enhanced synaptic plasticity, as evidenced by increased LTP amplitude and decreased LTP induction thresholds (Ge, Yang, Hsu, Ming, & Song, 2007). However, following complete maturation, the electrophysiological properties of newly generated neurons within the granular layer become indistinguishable from the properties of their neighboring mature neurons (Ge et al., 2007; Lemaire et al., 2012).

Adult-born neurons in the dentate gyrus are proposed to keep the hippocampal circuits plastic in young rodents and therefore are linked to cognitive plasticity. Proliferation and survival of newly generated hippocampal neurons can be promoted by exposure to an enriched environment (Ambrogini et al., 2000; Gould, Beylin et al., 1999; Gould & Tanapat, 1999). This increased number of adult-born hippocampal neurons was associated with improved memory encoding and pattern separation in mice and rats (Cameron & McKay, 1999; Kempermann & Gage, 2002; Kempermann, Gast, & Gage, 2002; Nilsson, Perfilieva, Johansson, Orwar, & Eriksson, 1999). Furthermore, manipulations to deplete adult neurogenesis, such as low-dose X-irradiation, decreased the amount of adult-born neurons and strongly impaired pattern separation (Clelland et al., 2009; Nakashiba et al., 2012). This growing evidence for an association between adult-born neurons in young animals and specific cognitive performance suggests that recently generated neurons in adulthood may play a significant role

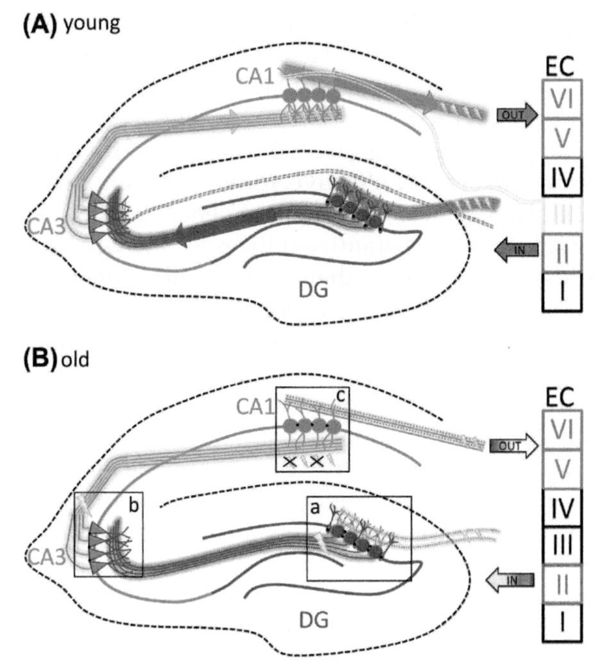

FIGURE 7.3 **(A) Schematic representation of the "young" trisynaptic pathway (entorhinal cortex–dentate gyrus–CA3–CA1–entorhinal cortex).** Adult-born neurons (gray cells) integrate almost exclusively into the granular cell layer, becoming indistinguishable from surrounding mature granular cells (gray circles) according to electrophysiology. Granular cells receive synaptic inputs through the perforant path (gray lines) from layer II of the entorhinal cortex. Granular cells send their projections via the mossy fibers (red arrow) to the pyramidal cells in CA3 where they form the next synaptic relay. CA3 neurons also receive few direct projections from entorhinal cortex layer II through the perforant path (dashed gray lines). Pyramidal CA3 neurons project via Schaffer collaterals (green arrow) to CA1 pyramidal neurons. The main output of the trisynaptic pathway is constituted by these CA1 neurons, which send projections back into the entorhinal cortex deep layers. In addition, the CA1 subfield receives direct input from the entorhinal cortex layer III by the temporoammonic pathway (yellow lines). The dentate granule cells also project to the mossy cells in the hilus and hilar interneurons, which send excitatory and inhibitory projections, respectively, back to the granule cells. Yellow lightning symbols indicate strong signal strength in input and output of the young trisynaptic pathway. **(B) Schematic representation of the "aged" trisynaptic pathway describing age-related changes.** (Inset a) Due to the age-associated reduction of afferent fibers from the entorhinal cortex layer II (dashed gray lines), lower fiber potential amplitudes reach the dentate gyrus granular cells. This weakened incoming stimulus is partially compensated by an increased postsynaptic sensitivity, increased electric coupling (black dots), and a lower discharge threshold of mature granular cells. (Inset b) In the CA3 subfield, the firing rates of CA3 pyramidal neurons are slightly higher, and these neurons show increased numbers of gap junction with age. (Inset c) While no age-associated changes in the presynaptic fiber potentials from the incoming Schaffer collaterals were consistently reported, the CA1 neurons show a reduction in the field excitatory postsynaptic potential. Hence, the net output of the trisynaptic pathway to the deep layers of the entorhinal cortex is reduced with age. This age-associated loss in signal strength is indicated by crossed lightning symbols, and may account for memory deficits in the elderly.

in processes of plasticity (Bizon & Gallagher, 2005; Couillard-Despres, 2013; Mongiat & Schinder, 2014). Hence, the processing of new memories could be facilitated by increased plasticity in hippocampal networks resulting from the integration of new neurons and therefore modification of already existing circuits. This plasticity is defined by the ability of the network to form new, or more effective, routes of communication between cells (Bizon & Gallagher, 2005; Meyer, Bonhoeffer, & Scheuss, 2014).

Morphological studies substantiated the role of newly born neurons in plasticity processes by showing that the functional integration of adult-born neurons into the dentate gyrus network entails a competition with existing synapses and thereby alters preestablished connections (Toni et al., 2008; Toni et al., 2007). Hence, it is hypothesized that the newly generated neurons, which are abundant in the dentate gyrus of young adult mammals, provide a fundamentally different mechanism of plasticity as compared to the canonical synaptic plasticity of mature granule cells (Gould et al., 1999). Intriguingly, while some studies report that this form of neurogenesis-based plasticity is required for learning (reviewed in Drapeau & Nora Abrous, 2008), others demonstrated that adult neurogenesis also promotes forgetting (Akers et al., 2014; Mongiat & Schinder, 2014). Models have been proposed to reconcile the apparent discrepancy reported on the role of adult hippocampal neurogenesis in which the destabilization of preexisting synaptic connections by arriving new neurons may mask previously encoded memories (Akers et al., 2014; Crick & Miranker, 2006; Mongiat & Schinder, 2014).

In detail, Akers et al. (2014) further substantiated the evidence for a competitive circuit modification promoted by hippocampal neurogenesis. A pronounced reduction in neurogenesis is observed in mice from day 17 (infant) to day 60 of age (young adult). During this decline, the neurogenesis levels were found to be inversely related to memory persistence; ie, adult mice showed better memory stability along with reduced neurogenesis, compared to infant mice. This inverse relationship was also reported in other species. Hence, in contextual fear conditioning tests, memories were more persistent in infant species with low postnatal hippocampal neurogenesis, like in infant guinea pigs and degus. In contrast, voluntary running increased neurogenesis and promoted forgetting of a context fear memory in adult mice, thereby counteracting the age-related stronger persistence described above (Akers et al., 2014).

Taken together, evidence accumulates that adult hippocampal neurogenesis facilitates learning as well as forgetting by increasing the neuronal circuit plasticity. The consequence of this hypothesis is that lower neurogenesis detected in the aged brain leads to scanty plasticity and the observed loss of specific cognitive performances. Thus, elderly people, with reduced capacities in their hippocampal neurogenic niche, do not profit from a lack of forgetting in a positive manner, but rather suffer from a lack of plasticity. In addition, Akers et al. (2014) selectively targeted neurogenesis in the infant mice to support this hypothesis. In their report, the running-induced increase in neurogenesis was prevented in transgenic mice bearing a suicide gene for NPCs. This system is based on the use of the promoter/enhancer element of nestin to drive the expression of a modified herpes simplex virus thymidine kinase (HSV-tk) suicide gene (TK+ mice) (Niibori et al., 2012). In these TK+ mice, the administration of ganciclovir ablates selectively proliferating NPCs expressing the tk transgene. Surprisingly, in infant mice with access to a running wheel, the administration of ganciclovir led to increased memory persistence (Akers et al., 2014). Therefore, the age-related stabilization or destabilization observed could be directly linked to neurogenesis. If this relationship can be translated to aged human hippocampi, losing neurogenesis would imply losing plasticity and finally impair memory function negatively.

Although attractive, the neurogenesis-based cognitive plasticity model carries some substantial flaws within. A central weakness of the model resides in the fact that the most important reduction of neurogenesis rates occurs early in life (Eriksson et al., 1998; Fahrner et al., 2007; Gould, Reeves et al., 1999), whereas age-related cognitive deficits in humans, affecting, for example, the mental speed (Salthouse, 1996), executive function (Connelly, Hasher, & Zacks, 1991; Schretlen et al., 2000), and episodic memory (Buckner, 2004; Nyberg, Lovden, Riklund, Lindenberger, & Backman, 2012; Salthouse, 2003), appear in late adulthood. As a consequence, if neurogenesis and cognition were directly linked with cognitive performances, most mammals would show a dramatic decline in their cognitive functions in early life, rather than with old age (Amrein & Lipp, 2009; Gould, Reeves et al., 1999; Kuhn et al., 1996). In humans, for example, one would expect a dramatic decline of specific cognitive abilities long before puberty, instead of the episodic memory decline reported to begin around the age of 50–60 years on a population basis (Nyberg et al., 2012).

The discrepancy existing regarding the role of neurogenesis in human cognition might result from the fact that the vast majority of functional studies have aimed to increase or decrease neurogenesis rates in cohorts of young adult rodents. However, can the observations made in young mice or rats directly be projected to the elderly brain of long-living mammals? One aspect arguing against a simple extrapolation from mice to humans is the existence of mammalian species devoid, or with only low amounts, of adult-born neurons in their hippocampi, which nevertheless demonstrate remarkably good memory and navigational capacities (Amrein, Dechmann, Winter, & Lipp, 2007; Amrein & Lipp, 2009; Amrein, Slomianka, Poletaeva, Bologova, & Lipp, 2004; Bartkowska, Djavadian, Taylor, & Turlejski, 2008; Johannes & York, 2005). For example, adult Sorex shrews (Lipotyphla) do not generate new granular cells in their dentate gyri, albeit injections of 5HT1A receptor agonists induce hippocampal neurogenesis, suggesting the presence of dormant progenitor cells (Bartkowska et al., 2008). In addition, no hippocampal cell proliferation could be found in 9 out of 12 bat species (Amrein et al., 2007). These observations indicate that adult hippocampal neurogenesis is not essential for learning and navigation in all mammals, yet it might still act as a facilitator in some.

From a comparative point of view, it is very challenging to attempt to transpose findings on the functional relevance of hippocampal neurogenesis on cognitive performances from small, short-living rodents to long-living mammals, like humans. Hence, high rates of neurogenesis, directly linked to hippocampus-dependent learning, might be a predominant feature of short-living mammals (Amrein et al., 2007). Especially in young adult mice and rats, high rates of neurogenesis may promote efficient and fast learning, unlearning (forgetting) and relearning. This might be of particular importance for highly predated rodents that must permanently identify dangers within changing contexts. These continuous environmental demands may create a particularly high need for adult hippocampal neurogenesis (Gage, Kempermann, & Song, 2007). In comparison, long-living mammals like most bats and certainly humans might profit more from a fine tuning of previously established and stable networks rather than from constant neuronal turnover (Changeux & Danchin, 1976).

After comparison of different species, including seven different laboratory strains of mice (BALB, C57BL/6, CD1, outbred) and rats (F344, Sprague–Dawley, Wistar), three nonhuman primates (marmoset and two macaque species), and one red fox, Amrein et al. (2011) reported an overall exponential decline in proliferation that is chronologically equal between these species. They concluded that the absolute age is the critical factor regulating adult hippocampal neurogenesis, independent of early ontogenetic processes and the individual life span. Amrein et al. (2011) strongly pointed out that other factors, like ontogeny and ecological demands, primarily influence the regulation of neuronal differentiation, but not the rate of proliferation.

While an age-dependent connection of decreased neurogenesis to the decrease of some cognitive performances is under question, age-related reduction in the hippocampal volume is frequently put in parallel to age-associated cognitive impairments in humans. Notably, the decrease of volume of the human hippocampus as measured by magnetic resonance imaging at the age of 20 and 80 years corresponds to an annual reduction of approximately 0.8–2.0% (Du et al., 2006; Ezekiel et al., 2004; Fjell, McEvoy, Holland, Dale, & Walhovd, 2014; Jack et al., 1998; Raz et al., 2005; Scahill et al., 2003). Similarly, the entorhinal cortex, which is the major input and output structure of the hippocampal formation, and the primary nodal point of cortico-hippocampal circuits, loses about 0.3–2.4% of its volume per year in humans (Du et al., 2006; Du et al., 2003; Ezekiel et al., 2004; Fjell et al., 2014; Fjell et al., 2009; Raz et al., 2005). In contrast, the overall reduction of gross brain volume is now well documented, with an annual decrease of 0.2–0.5% (Enzinger et al., 2005; Ezekiel et al., 2004; Fotenos, Snyder, Girton, Morris, & Buckner, 2005; Hedman, van Haren, Schnack, Kahn, & Hulshoff Pol, 2012; Scahill et al., 2003).

However, these snapshot volume measurements at the two ends of the adulthood spectrum mask another reality, ie, the net loss of volume in the various regions does not follow the same kinetic. Hence, while the total cortical volume declines linearly from 20 years of age onward, the hippocampal volume remains remarkably stable up to 60 years and drops drastically afterward in humans (Fjell et al., 2014; Fjell et al., 2013). Similarly, based on cross-sectional estimates and confirmed by longitudinal data, Fjell et al. (2014) demonstrated elegantly that the rate of the cortical thinning within the entorhinal cortex is minimal until the age of 50, but greatly accelerates afterward.

While the temporal pattern of hippocampal atrophy points to its relevance to age-related cognitive impairments, its neurobiological basis is still under discussion. For example, episodic memory declines from about the age of 50–60 years onward on a population basis in humans (Nyberg et al., 2012), fitting well the onset of age-related loss of hippocampal volume (Fjell et al., 2013). Nevertheless, the numbers of neurons within the dentate gyrus and the Ammon's horn and within the entorhinal cortex were reported to remain stable during normal aging, although neuronal numbers within the hilus and the subiculum decrease in humans and nonhuman primates (Gazzaley, Thakker, Hof, & Morrison, 1997; Hara et al., 2011; Merrill, Roberts, & Tuszynski, 2000; Nakajima, Soc, & Ono, 1995; Rapp, 1995; West, 1993; West, Coleman, Flood, & Troncoso, 1994). Hence, no correlation can be made between an age-associated decrease in the neurogenesis rate and a decreased hippocampal volume.

Taken that the number of hippocampal neurons remains roughly stable during aging, other structural elements must be involved in the hippocampal volume reduction and potentially in cognitive decline. The perforant path, eg, the white matter connecting the entorhinal cortex with the dentate gyrus and other subfields of the hippocampal formation (Fig. 7.3A), constitutes an example. In humans, age-related memory loss has been in part correlated with volume loss of the perforant path (Cabalka, Hyman, Goodlett, Ritchie, & Van Hoesen, 1992; Flood, Buell, Horwitz, & Coleman, 1987; Hyman, Van Hoesen, Damasio, & Barnes, 1984; Morrison & Hof, 1997; Yassa, Muftuler, & Stark, 2010). In aged rats, a decrease in the amplitude of the presynaptic fiber potential together with a reduction of the number of perforant path synapses on the granule cells hippocampi suggests an overall reduction in the number of perforant path axon collaterals connecting to dentate gyrus (Barnes & McNaughton, 1980; Bondareff & Geinisman, 1976; Geinisman, deToledo-Morrell, Morrell, Persina, &

Rossi, 1992; Lister & Barnes, 2009). As schematically represented in Fig. 7.3B an age-related loss of perforant path signal strength, due to the loss of synapses on granule cells, has an impact on the signal transduction and processing throughout the trisynaptic path.

While cell numbers cannot explain the reduced hippocampal volume of elderly humans, reduced connectivity might partially. The relation between gray and white matter loss during human aging has received particular attention in recent years (Peters & Rosene, 2003). Several reports demonstrated the presence of white matter volume loss in the absence of gray matter reduction (Allen, Bruss, Brown, & Damasio, 2005; Bartzokis et al., 2003; Guttmann et al., 1998; Jernigan et al., 2001; Resnick, Pham, Kraut, Zonderman, & Davatzikos, 2003; reviewed in Gunning-Dixon, Brickman, Cheng, & Alexopoulos, 2009). Therefore, age-related loss of volume within the hippocampal formation may be in part due to a loss of connectivity. Furthermore, age-related decreases in structural connectives might explain some of the age-dependent cognitive impairments in humans.

Remarkably, our aged hippocampi may have compensatory mechanisms to counteract for age-related loss of synaptic connections. Thus, an age-related increase in excitability of mature granular cells of the dentate gyrus has been suggested to compensate the age-related decrease in perforant path signal strength (Burke & Barnes, 2006; Lister & Barnes, 2009) and may balance at the same time lower rates of neurogenesis in long-living organisms. For example, mature dentate gyrus granular cells residing in the aged rat have lower discharge thresholds compared to young animals (Barnes & McNaughton, 1980). This adaptation was shown by intracellular recordings and makes granule cells of the aged brain more sensitive for smaller excitatory events on their presynapses (Barnes & McNaughton, 1980). Furthermore, mature granular cells within the dentate gyrus in the aged brain exhibit increased unitary excitatory postsynaptic potential sizes, due to an increase in α-amino-3-hydroxy-5-methyl-4-isoxaole propionic acid receptor currents (Barnes & McNaughton, 1980; Barnes, Rao, Foster, & McNaughton, 1992; Burke & Barnes, 2006). Together with the increased electric coupling (Forster, 1991) the morphological adaptation of granular cells in the aged hippocampus may partially compensate the weaker input from the entorhinal cortex. Age-related changes in adult granular cells might preserve, to a certain extent, the processing efficacy through the trisynaptic pathway.

Adult-born immature neurons might have a significant impact on the sensitivity and response intensity of the dentate gyrus to incoming signals. Notably, the newly added neurons are tonically activated by GABA (γ-aminobutyric acid) during the first week after being generated (Esposito et al., 2005; Ge et al., 2006; Zhao, Teng, Summers, Ming, & Gage, 2006). The timing of synaptic integration coincides with the transition of GABAergic input from being excitatory to being inhibitory and with the onset of glutamatergic synaptic inputs. Until approximately 6 weeks of age, adult-born dentate gyrus granular cells exhibit characteristics that could contribute to increased excitability, such as high membrane resistance and high resting potentials (Esposito et al., 2005; Ge et al., 2007; Schmidt-Hieber, Jonas, & Bischofberger, 2004). Beyond 8 weeks of age, the basic physiological properties and synaptic plasticity become indistinguishable from those of mature granular cells (Ge et al., 2007).

Due to their strategic position within the hippocampus, adult-born granular cells could have a significant impact on the sensitivity and response intensity, even in their immature stages. In addition, with the increased coupling with mature granular cells, as well as their increased presynaptic sensitivity, immature granular cells may amplify the inputs from the perforant path. Therefore, the addition of immature granular cells in the dentate gyrus per se could be seen as a functional modifier for mature hypoactive networks, as previously demonstrated in vitro (Stephens, Toda, Palmer, Demarse, & Ormerod, 2012). In aging, the quantitative decrease of adult-born neurons might be partly compensated by their slower maturation of the newly generated neurons. This would, as a consequence, also extend the duration of the highly excitability stage of immature granular cells (Nyffeler, Yee, Feldon, & Knuesel, 2010; Rao et al., 2005).

In summary, in the aged brain of long-living mammals, the age-associated decrease in neurogenesis and immature granular cells appears to have a lower impact as compared to their short-living relatives and to be compensated over long time periods. Nevertheless, neurogenesis offers plastic cells that can efficiently fine-tune, shape, and maintain neuronal circuits for cognitive processes previously established.

References

Ahlenius, H., Visan, V., Kokaia, M., Lindvall, O., & Kokaia, Z. (2009). Neural stem and progenitor cells retain their potential for proliferation and differentiation into functional neurons despite lower number aged brain. *Journal of Neuroscience, 29*, 4408–4419.

Akers, K. G., Martinez-Canabal, A., Restivo, L., Yiu, A. P., De Cristofaro, A., Hsiang, H. L., et al. (2014). Hippocampal neurogenesis regulates forgetting during adulthood and infancy. *Science, 344*, 598–602.

Allen, J. S., Bruss, J., Brown, C. K., & Damasio, H. (2005). Normal neuroanatomical variation due to age: the major lobes and a parcellation of the temporal region. *Neurobiology Aging, 26*, 1245–1260; discussion 1279–1282.

Altman, J., & Das, G. D. (1965). Autoradiographic and histological evidence of postnatal hippocampal neurogenesis in rats. *Journal of Comparative Neurology, 124*, 319–335.

Ambrogini, P., Cuppini, R., Cuppini, C., Ciaroni, S., Cecchini, T., Ferri, P., et al. (2000). Spatial learning affects immature granule cell survival in adult rat dentate gyrus. *Neuroscience Letters, 286*, 21–24.

Amrein, I., Dechmann, D. K., Winter, Y., & Lipp, H. P. (2007). Absent or low rate of adult neurogenesis in the hippocampus of bats (Chiroptera). *PLoS One, 2*, e455.

Amrein, I., Isler, K., & Lipp, H. P. (2011). Comparing adult hippocampal neurogenesis in mammalian species and orders: influence of chronological age and life history stage. *European Journal of Neuroscience, 34*, 978–987.

Amrein, I., & Lipp, H. P. (2009). Adult hippocampal neurogenesis of mammals: evolution and life history. *Biology Letters, 5*, 141–144.

Amrein, I., Slomianka, L., Poletaeva, I. I., Bologova, N. V., & Lipp, H. P. (2004). Marked species and age-dependent differences in cell proliferation and neurogenesis in the hippocampus of wild-living rodents. *Hippocampus, 14*, 1000–1010.

Anderson, M. F., Aberg, M. A., Nilsson, M., & Eriksson, P. S. (2002). Insulin-like growth factor-I and neurogenesis in the adult mammalian brain. *Brain Research. Development Brain Research, 134*, 115–122.

Bachstetter, A. D., Morganti, J. M., Jernberg, J., Schlunk, A., Mitchell, S. H., Brewster, K. W., et al. (2011). Fractalkine and CX 3 CR1 regulate hippocampal neurogenesis in adult and aged rats. *Neurobiology Aging, 32*, 2030–2044.

Bake, S., Friedman, J. A., & Sohrabji, F. (2009). Reproductive age-related changes in the blood–brain barrier: expression of IgG and tight junction proteins. *Microvascular Research, 78*, 413–424.

Bake, S., & Sohrabji, F. (2004). 17beta-estradiol differentially regulates blood–brain barrier permeability in young and aging female rats. *Endocrinology, 145*, 5471–5475.

Barnes, C. A., & McNaughton, B. L. (1980). Physiological compensation for loss of afferent synapses in rat hippocampal granule cells during senescence. *Journal of Physiology, 309*, 473–485.

Barnes, C. A., Rao, G., Foster, T. C., & McNaughton, B. L. (1992). Region-specific age effects on AMPA sensitivity: electrophysiological evidence for loss of synaptic contacts in hippocampal field CA1. *Hippocampus, 2*, 457–468.

Bartkowska, K., Djavadian, R. L., Taylor, J. R., & Turlejski, K. (2008). Generation recruitment and death of brain cells throughout the life cycle of Sorex shrews (Lipotyphla). *European Journal of Neuroscience, 27*, 1710–1721.

Bartzokis, G., Cummings, J. L., Sultzer, D., Henderson, V. W., Nuechterlein, K. H., & Mintz, J. (2003). White matter structural integrity in healthy aging adults and patients with Alzheimer disease: a magnetic resonance imaging study. *Archives of Neurology, 60*, 393–398.

Bauer, H. C., Krizbai, I. A., Bauer, H., & Traweger, A. (2014). "You Shall Not Pass"-tight junctions of the blood–brain barrier. *Frontiers in Neuroscience, 8*, 392.

Ben Abdallah, N. M., Slomianka, L., Vyssotski, A. L., & Lipp, H. P. (2010). Early age-related changes in adult hippocampal neurogenesis in C57 mice. *Neurobiology of Aging, 31*, 151–161.

Bernal, G. M., & Peterson, D. A. (2011). Phenotypic and gene expression modification with normal brain aging in GFAP-positive astrocytes and neural stem cells. *Aging Cell, 10*, 466–482.

Bizon, J. L., & Gallagher, M. (2005). More is less: neurogenesis and age-related cognitive decline in Long-Evans rats. *Science of Aging Knowledge Environment, 2005*, re2.

Bizon, J. L., Lee, H. J., & Gallagher, M. (2004). Neurogenesis in a rat model of age-related cognitive decline. *Aging Cell, 3*, 227–234.

Bonaguidi, M. A., Wheeler, M. A., Shapiro, J. S., Stadel, R. P., Sun, G. J., Ming, G. L., et al. (2011). In vivo clonal analysis reveals self-renewing and multipotent adult neural stem cell characteristics. *Cell, 145*, 1142–1155.

Bondareff, W., & Geinisman, Y. (1976). Loss of synapses in the dentate gyrus of the senescent rat. *American Journal of Anatomy, 145*, 129–136.

Bondolfi, L., Ermini, F., Long, J. M., Ingram, D. K., & Jucker, M. (2004). Impact of age and caloric restriction on neurogenesis in the dentate gyrus of C57BL/6 mice. *Neurobiology Aging, 25*, 333–340.

Bouab, M., Paliouras, G. N., Aumont, A., Forest-Berard, K., & Fernandes, K. J. (2011). Aging of the subventricular zone neural stem cell niche: evidence for quiescence-associated changes between early and mid-adulthood. *Neuroscience, 173*, 135–149.

Brown, J. P., Couillard-Despres, S., Cooper-Kuhn, C. M., Winkler, J., Aigner, L., & Kuhn, H. G. (2003). Transient expression of doublecortin during adult neurogenesis. *Journal of Comparative Neurology, 467*, 1–10.

Buckner, R. L. (2004). Memory and executive function in aging and AD: multiple factors that cause decline and reserve factors that compensate. *Neuron, 44*, 195–208.

Buckwalter, M. S., Yamane, M., Coleman, B. S., Ormerod, B. K., Chin, J. T., Palmer, T., et al. (2006). Chronically increased transforming growth factor-beta1 strongly inhibits hippocampal neurogenesis in aged mice. *American Journal of Pathology, 169*, 154–164.

Burke, S. N., & Barnes, C. A. (2006). Neural plasticity in the ageing brain. *Nature Reviews Neuroscience, 7*, 30–40.

Cabalka, L. M., Hyman, B. T., Goodlett, C. R., Ritchie, T. C., & Van Hoesen, G. W. (1992). Alteration in the pattern of nerve terminal protein immunoreactivity in the perforant pathway in Alzheimer's disease and in rats after entorhinal lesions. *Neurobiology Aging, 13*, 283–291.

Cameron, H. A., & Gould, E. (1994). Adult neurogenesis is regulated by adrenal steroids in the dentate gyrus. *Neuroscience, 61*, 203–209.

Cameron, H. A., & McKay, R. D. (1999). Restoring production of hippocampal neurons in old age. *Nature Neuroscience, 2*, 894–897.

Cardona, A. E., Pioro, E. P., Sasse, M. E., Kostenko, V., Cardona, S. M., Dijkstra, I. M., et al. (2006). Control of microglial neurotoxicity by the fractalkine receptor. *Nature Neuroscience, 9*, 917–924.

Carter, C. S., Ramsey, M. M., & Sonntag, W. E. (2002). A critical analysis of the role of growth hormone and IGF-1 in aging and lifespan. *Trends in Genetics, 18*, 295–301.

Changeux, J. P., & Danchin, A. (1976). Selective stabilisation of developing synapses as a mechanism for the specification of neuronal networks. *Nature, 264*, 705–712.

Chen, F., Madsen, T. M., Wegener, G., & Nyengaard, J. R. (2009). Repeated electroconvulsive seizures increase the total number of synapses in adult male rat hippocampus. *European Neuropsychopharmacology, 19*, 329–338.

Chinnici, C. M., Yao, Y., & Pratico, D. (2007). The 5-lipoxygenase enzymatic pathway in the mouse brain: young versus old. *Neurobiology Aging, 28*, 1457–1462.

Clelland, C. D., Choi, M., Romberg, C., Clemenson, G. D., Jr., Fragniere, A., Tyers, P., et al. (2009). A functional role for adult hippocampal neurogenesis in spatial pattern separation. *Science, 325*, 210–213.

Connelly, S. L., Hasher, L., & Zacks, R. T. (1991). Age and reading: the impact of distraction. *Psychology and Aging, 6*, 533–541.

Cotrina, M. L., & Nedergaard, M. (2002). Astrocytes in the aging brain. *Journal of Neuroscience Research, 67*, 1–10.

Couillard-Despres, S. (2013). Hippocampal neurogenesis and ageing. *Current Topics in Behavioral Neurosciences, 15*, 343–355.

Couillard-Despres, S., Finkl, R., Winner, B., Ploetz, S., Wiedermann, D., Aigner, R., et al. (2008). In vivo optical imaging of neurogenesis: watching new neurons in the intact brain. *Molecular Imaging, 7*, 28–34.

Couillard-Despres, S., Winner, B., Karl, C., Lindemann, G., Schmid, P., Aigner, R., et al. (2006). Targeted transgene expression in neuronal precursors: watching young neurons in the old brain. *European Journal of Neuroscience, 24*, 1535–1545.

Couillard-Despres, S., Winner, B., Schaubeck, S., Aigner, R., Vroemen, M., Weidner, N., et al. (2005). Doublecortin expression levels in adult brain reflect neurogenesis. *European Journal of Neuroscience, 21*, 1–14.

Couillard-Despres, S., Wuertinger, C., Kandasamy, M., Caioni, M., Stadler, K., Aigner, R., et al. (2009). Ageing abolishes the effects of fluoxetine on neurogenesis. *Molecular Psychiatry, 14*, 856–864.

Crick, C., & Miranker, W. (2006). Apoptosis, neurogenesis, and information content in Hebbian networks. *Biological Cybernetics, 94*, 9–19.

Decarolis, N. A., Mechanic, M., Petrik, D., Carlton, A., Ables, J. L., Malhotra, S., et al. (2013). In vivo contribution of nestin- and GLAST-lineage cells to adult hippocampal neurogenesis. *Hippocampus, 23*, 708–719.

Dilger, R. N., & Johnson, R. W. (2008). Aging, microglial cell priming, and the discordant central inflammatory response to signals from the peripheral immune system. *Journal of Leukocyte Biology, 84*, 932–939.

Doetsch, F. (2003). A niche for adult neural stem cells. *Current Opinion in Genetics and Development, 13*, 543–550.

Drapeau, E., & Nora Abrous, D. (2008). Stem cell review series: role of neurogenesis in age-related memory disorders. *Aging Cell, 7*, 569–589.

Du, A. T., Schuff, N., Chao, L. L., Kornak, J., Jagust, W. J., Kramer, J. H., et al. (2006). Age effects on atrophy rates of entorhinal cortex and hippocampus. *Neurobiology Aging, 27*, 733–740.

Du, A. T., Schuff, N., Zhu, X. P., Jagust, W. J., Miller, B. L., Reed, B. R., et al. (2003). Atrophy rates of entorhinal cortex in AD and normal aging. *Neurology, 60*, 481–486.

Eggel, A., & Wyss-Coray, T. (2014). A revival of parabiosis in biomedical research. *Swiss Medical Weekly, 144*, w13914.

Ekdahl, C. T., Claasen, J. H., Bonde, S., Kokaia, Z., & Lindvall, O. (2003). Inflammation is detrimental for neurogenesis in adult brain. *Proceedings of the National Academy of Sciences of the United States of America, 100*, 13632–13637.

Encinas, J. M., Michurina, T. V., Peunova, N., Park, J. H., Tordo, J., Peterson, D. A., et al. (2011). Division-coupled astrocytic differentiation and age-related depletion of neural stem cells in the adult hippocampus. *Cell Stem Cell, 8*, 566–579.

Encinas, J. M., & Sierra, A. (2012). Neural stem cell deforestation as the main force driving the age-related decline in adult hippocampal neurogenesis. *Behavioural Brain Research, 227*, 433–439.

Encinas, J. M., Vaahtokari, A., & Enikolopov, G. (2006). Fluoxetine targets early progenitor cells in the adult brain. *Proceedings of the National Academy of Sciences of the United States of America, 103*, 8233–8238.

Enzinger, C., Fazekas, F., Matthews, P. M., Ropele, S., Schmidt, H., Smith, S., et al. (2005). Risk factors for progression of brain atrophy in aging: six-year follow-up of normal subjects. *Neurology, 64*, 1704–1711.

Eriksson, P. S., Perfilieva, E., Bjork-Eriksson, T., Alborn, A. M., Nordborg, C., Peterson, D. A., et al. (1998). Neurogenesis in the adult human hippocampus. *Nature Medicine, 4*, 1313–1317.

Esposito, M. S., Piatti, V. C., Laplagne, D. A., Morgenstern, N. A., Ferrari, C. C., Pitossi, F. J., et al. (2005). Neuronal differentiation in the adult hippocampus recapitulates embryonic development. *Journal of Neuroscience, 25*, 10074–10086.

Ezekiel, F., Chao, L., Kornak, J., Du, A. T., Cardenas, V., Truran, D., et al. (2004). Comparisons between global and focal brain atrophy rates in normal aging and Alzheimer disease: boundary shift integral versus tracing of the entorhinal cortex and hippocampus. *Alzheimer Disease and Associated Disorders, 18*, 196–201.

Fahrner, A., Kann, G., Flubacher, A., Heinrich, C., Freiman, T. M., Zentner, J., et al. (2007). Granule cell dispersion is not accompanied by enhanced neurogenesis in temporal lobe epilepsy patients. *Experimental Neurology, 203*, 320–332.

Fjell, A. M., McEvoy, L., Holland, D., Dale, A. M., & Walhovd, K. B. (2014). What is normal in normal aging? Effects of aging, amyloid and Alzheimer's disease on the cerebral cortex and the hippocampus. *Progress in Neurobiology, 117*, 20–40.

Fjell, A. M., Walhovd, K. B., Fennema-Notestine, C., McEvoy, L. K., Hagler, D. J., Holland, D., et al. (2009). One-year brain atrophy evident in healthy aging. *Journal of Neuroscience, 29*, 15223–15231.

Fjell, A. M., Westlye, L. T., Grydeland, H., Amlien, I., Espeseth, T., Reinvang, I., et al. (2013). Critical ages in the life course of the adult brain: nonlinear subcortical aging. *Neurobiology Aging, 34*, 2239–2247.

Flood, D. G., Buell, S. J., Horwitz, G. J., & Coleman, P. D. (1987). Dendritic extent in human dentate gyrus granule cells in normal aging and senile dementia. *Brain Research, 402*, 205–216.

Foster, T. C., Barnes, C. A., Rao, G., McNaughton, B. L. (1991). Increase in perforant path quantal size in aged F-344 rats. *Neurobiology of Aging, 12*(5), 441–448.

Fotenos, A. F., Snyder, A. Z., Girton, L. E., Morris, J. C., & Buckner, R. L. (2005). Normative estimates of cross-sectional and longitudinal brain volume decline in aging and AD. *Neurology, 64*, 1032–1039.

Gage, F. H., Kempermann, G., & Song, H. (2007). Adult neurogenesis. In *Cold Spring Harbor monograph series*.

Garcia, A., Steiner, B., Kronenberg, G., Bick-Sander, A., & Kempermann, G. (2004). Age-dependent expression of glucocorticoid- and mineralocorticoid receptors on neural precursor cell populations in the adult murine hippocampus. *Aging Cell, 3*, 363–371.

Gazzaley, A. H., Thakker, M. M., Hof, P. R., & Morrison, J. H. (1997). Preserved number of entorhinal cortex layer II neurons in aged macaque monkeys. *Neurobiology Aging, 18*, 549–553.

Ge, S., Goh, E. L., Sailor, K. A., Kitabatake, Y., Ming, G. L., & Song, H. (2006). GABA regulates synaptic integration of newly generated neurons in the adult brain. *Nature, 439*, 589–593.

Geinisman, Y., deToledo-Morrell, L., Morrell, F., Persina, I. S., & Rossi, M. (1992). Age-related loss of axospinous synapses formed by two afferent systems in the rat dentate gyrus as revealed by the unbiased stereological dissector technique. *Hippocampus, 2*, 437–444.

Gemma, C., Bachstetter, A. D., Cole, M. J., Fister, M., Hudson, C., & Bickford, P. C. (2007). Blockade of caspase-1 increases neurogenesis in the aged hippocampus. *European Journal of Neuroscience, 26*, 2795–2803.

Ge, S., Yang, C. H., Hsu, K. S., Ming, G. L., & Song, H. (2007). A critical period for enhanced synaptic plasticity in newly generated neurons of the adult brain. *Neuron, 54*, 559–566.

Gould, E., Beylin, A., Tanapat, P., Reeves, A., & Shors, T. J. (1999). Learning enhances adult neurogenesis in the hippocampal formation. *Nature Neuroscience, 2*, 260–265.

Gould, E., Reeves, A. J., Fallah, M., Tanapat, P., Gross, C. G., & Fuchs, E. (1999). Hippocampal neurogenesis in adult Old World primates. *Proceedings of the National Academy of Sciences of the United States of America, 96*, 5263–5267.

Gould, E., & Tanapat, P. (1999). Stress and hippocampal neurogenesis. *Biological Psychiatry, 46*, 1472–1479.

Gould, E., Tanapat, P., Hastings, N. B., & Shors, T. J. (1999). Neurogenesis in adulthood: a possible role in learning. *Trends in Cognitive Sciences, 3*, 186–192.

Gould, E., Tanapat, P., McEwen, B. S., Flugge, G., & Fuchs, E. (1998). Proliferation of granule cell precursors in the dentate gyrus of adult monkeys is diminished by stress. *Proceedings of the National Academy of Sciences of the United States of America, 95*, 3168–3171.

Graeber, M. B. (2010). Changing face of microglia. *Science, 330*, 783–788.

Group, T. U. E. (2003). Efficacy and safety of electroconvulsive therapy in depressive disorders: a systematic review and meta-analysis. *Lancet, 361*, 799–808.

Gunning-Dixon, F. M., Brickman, A. M., Cheng, J. C., & Alexopoulos, G. S. (2009). Aging of cerebral white matter: a review of MRI findings. *International Journal of Geriatric Psychiatry, 24*, 109–117.

Guttmann, C. R., Jolesz, F. A., Kikinis, R., Killiany, R. J., Moss, M. B., Sandor, T., et al. (1998). White matter changes with normal aging. *Neurology, 50*, 972–978.

Hara, Y., Park, C. S., Janssen, W. G., Punsoni, M., Rapp, P. R., & Morrison, J. H. (2011). Synaptic characteristics of dentate gyrus axonal boutons and their relationships with aging, menopause, and memory in female rhesus monkeys. *Journal of Neuroscience, 31*, 7737–7744.

Hastings, N. B., & Gould, E. (1999). Rapid extension of axons into the CA3 region by adult-generated granule cells. *Journal of Comparative Neurology, 413*, 146–154.

Hattiangady, B., & Shetty, A. K. (2008). Aging does not alter the number or phenotype of putative stem/progenitor cells in the neurogenic region of the hippocampus. *Neurobiology Aging, 29*, 129–147.

Hedman, A. M., van Haren, N. E., Schnack, H. G., Kahn, R. S., & Hulshoff Pol, H. E. (2012). Human brain changes across the life span: a review of 56 longitudinal magnetic resonance imaging studies. *Human Brain Mapping, 33*, 1987–2002.
Heye, A. K., Culling, R. D., Valdes Hernandez Mdel, C., Thrippleton, M. J., & Wardlaw, J. M. (2014). Assessment of blood–brain barrier disruption using dynamic contrast-enhanced MRI. A systematic review. *Neuroimage Clinical, 6*, 262–274.
Huber, C., Marschallinger, J., Tempfer, H., Furtner, T., Couillard-Despres, S., Bauer, H. C., et al. (2011). Inhibition of leukotriene receptors boosts neural progenitor proliferation. *Cellular Physiology and Biochemistry, 28*, 793–804.
Hyman, B. T., Van Hoesen, G. W., Damasio, A. R., & Barnes, C. L. (1984). Alzheimer's disease: cell-specific pathology isolates the hippocampal formation. *Science, 225*, 1168–1170.
Jack, C. R., Jr., Petersen, R. C., Xu, Y., O'Brien, P. C., Smith, G. E., Ivnik, R. J., et al. (1998). Rate of medial temporal lobe atrophy in typical aging and Alzheimer's disease. *Neurology, 51*, 993–999.
Jeong, H. K., Ji, K., Min, K., & Joe, E. H. (2013). Brain inflammation and microglia: facts and misconceptions. *Experimental Neurobiology, 22*, 59–67.
Jernigan, T. L., Archibald, S. L., Fennema-Notestine, C., Gamst, A. C., Stout, J. C., Bonner, J., et al. (2001). Effects of age on tissues and regions of the cerebrum and cerebellum. *Neurobiology Aging, 22*, 581–594.
Johannes, T., & York, W. (2005). Hierarchical strategy for relocating food targets in flower bats: spatial memory versus cue-directed search. *Animal Behaviour, 69*, 315–327.
Kanak, D. J., Rose, G. M., Zaveri, H. P., & Patrylo, P. R. (2013). Altered network timing in the CA3-CA1 circuit of hippocampal slices from aged mice. *PLoS One, 8*, e61364.
Kaplan, M. S., & Hinds, J. W. (1977). Neurogenesis in the adult rat: electron microscopic analysis of light radioautographs. *Science, 197*, 1092–1094.
Katsimpardi, L., Litterman, N. K., Schein, P. A., Miller, C. M., Loffredo, F. S., Wojtkiewicz, G. R., et al. (2014). Vascular and neurogenic rejuvenation of the aging mouse brain by young systemic factors. *Science, 344*, 630–634.
Kempermann, G. (2011). The pessimist's and optimist's views of adult neurogenesis. *Cell, 145*, 1009–1011.
Kempermann, G., & Gage, F. H. (2002). Genetic determinants of adult hippocampal neurogenesis correlate with acquisition, but not probe trial performance, in the water maze task. *European Journal of Neuroscience, 16*, 129–136.
Kempermann, G., Gast, D., & Gage, F. H. (2002). Neuroplasticity in old age: sustained fivefold induction of hippocampal neurogenesis by long-term environmental enrichment. *Annals of Neurology, 52*, 135–143.
Kempermann, G., Jessberger, S., Steiner, B., & Kronenberg, G. (2004). Milestones of neuronal development in the adult hippocampus. *Trends in Neurosciences, 27*, 447–452.
Kempermann, G., Kuhn, H. G., & Gage, F. H. (1998). Experience-induced neurogenesis in the senescent dentate gyrus. *Journal of Neuroscience, 18*, 3206–3212.
Kennedy, B. K., Berger, S. L., Brunet, A., Campisi, J., Cuervo, A. M., Epel, E. S., et al. (2014). Geroscience: linking aging to chronic disease. *Cell, 159*, 709–713.
Knoth, R., Singec, I., Ditter, M., Pantazis, G., Capetian, P., Meyer, R. P., et al. (2010). Murine features of neurogenesis in the human hippocampus across the lifespan from 0 to 100 years. *PLoS One, 5*, e8809.
Komitova, M., & Eriksson, P. S. (2004). Sox-2 is expressed by neural progenitors and astroglia in the adult rat brain. *Neuroscience Letters, 369*, 24–27.
Koo, J. W., & Duman, R. S. (2008). IL-1beta is an essential mediator of the antineurogenic and anhedonic effects of stress. *Proceedings of the National Academy of Sciences of the United States of America, 105*, 751–756.
Kronenberg, G., Bick-Sander, A., Bunk, E., Wolf, C., Ehninger, D., & Kempermann, G. (2006). Physical exercise prevents age-related decline in precursor cell activity in the mouse dentate gyrus. *Neurobiology Aging, 27*, 1505–1513.

Kuhn, H. G., Dickinson-Anson, H., & Gage, F. H. (1996). Neurogenesis in the dentate gyrus of the adult rat: age-related decrease of neuronal progenitor proliferation. *Journal of Neuroscience, 16,* 2027–2033.

Lagace, D. C., Donovan, M. H., DeCarolis, N. A., Farnbauch, L. A., Malhotra, S., Berton, O., et al. (2010). Adult hippocampal neurogenesis is functionally important for stress-induced social avoidance. *Proceedings of the National Academy of Sciences of the United States of America, 107,* 4436–4441.

Lagace, D. C., Whitman, M. C., Noonan, M. A., Ables, J. L., DeCarolis, N. A., Arguello, A. A., et al. (2007). Dynamic contribution of nestin-expressing stem cells to adult neurogenesis. *Journal of Neuroscience, 27,* 12623–12629.

Lange, S., Trost, A., Tempfer, H., Bauer, H. C., Bauer, H., Rohde, E., et al. (2013). Brain pericyte plasticity as a potential drug target in CNS repair. *Drug Discov Today, 18,* 456–463.

Lecca, D., Trincavelli, M. L., Gelosa, P., Sironi, L., Ciana, P., Fumagalli, M., et al. (2008). The recently identified P2Y-like receptor GPR17 is a sensor of brain damage and a new target for brain repair. *PLoS One, 3,* e3579.

Lee, C. K., Weindruch, R., & Prolla, T. A. (2000). Gene-expression profile of the ageing brain in mice. *Nature Genetics, 25,* 294–297.

Lemaire, V., Koehl, M., Le Moal, M., & Abrous, D. N. (2000). Prenatal stress produces learning deficits associated with an inhibition of neurogenesis in the hippocampus. *Proceedings of the National Academy of Sciences of the United States of America, 97,* 11032–11037.

Lemaire, V., Tronel, S., Montaron, M. F., Fabre, A., Dugast, E., & Abrous, D. N. (2012). Long-lasting plasticity of hippocampal adult-born neurons. *Journal of Neuroscience, 32,* 3101–3108.

Leuner, B., Kozorovitskiy, Y., Gross, C. G., & Gould, E. (2007). Diminished adult neurogenesis in the marmoset brain precedes old age. *Proceedings of the National Academy of Sciences of the United States of America, 104,* 17169–17173.

Leventhal, C., Rafii, S., Rafii, D., Shahar, A., & Goldman, S. A. (1999). Endothelial trophic support of neuronal production and recruitment from the adult mammalian subependyma. *Molecular and Cellular Neuroscience, 13,* 450–464.

Lister, J. P., & Barnes, C. A. (2009). Neurobiological changes in the hippocampus during normative aging. *Archives of Neurology, 66,* 829–833.

Lopez-Otin, C., Blasco, M. A., Partridge, L., Serrano, M., & Kroemer, G. (2013). The hallmarks of aging. *Cell, 153,* 1194–1217.

Lucassen, P. J., Stumpel, M. W., Wang, Q., & Aronica, E. (2010). Decreased numbers of progenitor cells but no response to antidepressant drugs in the hippocampus of elderly depressed patients. *Neuropharmacology, 58,* 940–949.

Lucin, K. M., & Wyss-Coray, T. (2009). Immune activation in brain aging and neurodegeneration: too much or too little? *Neuron, 64,* 110–122.

Lugert, S., Basak, O., Knuckles, P., Haussler, U., Fabel, K., Gotz, M., et al. (2010). Quiescent and active hippocampal neural stem cells with distinct morphologies respond selectively to physiological and pathological stimuli and aging. *Cell Stem Cell, 6,* 445–456.

Lu, T., Pan, Y., Kao, S. Y., Li, C., Kohane, I., Chan, J., et al. (2004). Gene regulation and DNA damage in the ageing human brain. *Nature, 429,* 883–891.

Lupien, S., Lecours, A. R., Lussier, I., Schwartz, G., Nair, N. P., & Meaney, M. J. (1994). Basal cortisol levels and cognitive deficits in human aging. *Journal of Neuroscience, 14,* 2893–2903.

Madsen, T. M., Treschow, A., Bengzon, J., Bolwig, T. G., Lindvall, O., & Tingstrom, A. (2000). Increased neurogenesis in a model of electroconvulsive therapy. *Biological Psychiatry, 47,* 1043–1049.

Malberg, J. E., Eisch, A. J., Nestler, E. J., & Duman, R. S. (2000). Chronic antidepressant treatment increases neurogenesis in adult rat hippocampus. *Journal of Neuroscience, 20,* 9104–9110.

Ma, D. K., Ming, G. L., & Song, H. (2005). Glial influences on neural stem cell development: cellular niches for adult neurogenesis. *Current Opinion in Neurobiology, 15,* 514–520.

Manganas, L. N., Zhang, X., Li, Y., Hazel, R. D., Smith, S. D., Wagshul, M. E., et al. (2007). Magnetic resonance spectroscopy identifies neural progenitor cells in the live human brain. *Science, 318*, 980–985.

Marcussen, A. B., Flagstad, P., Kristjansen, P. E., Johansen, F. F., & Englund, U. (2008). Increase in neurogenesis and behavioural benefit after chronic fluoxetine treatment in Wistar rats. *Acta Neurologica Scandinavica, 117*, 94–100.

Markakis, E. A., & Gage, F. H. (1999). Adult-generated neurons in the dentate gyrus send axonal projections to field CA3 and are surrounded by synaptic vesicles. *Journal of Comparative Neurology, 406*, 449–460.

Marschallinger, J., Schäffner, I., Klein, B., Gelfert, R., Rivera, F. J., Illes, S., et al. (2015). Structural and functional rejuvenation of the aged brain by an approved anti-asthmatic drug. *Nature Communications, 27*(6), 8466.

Marschallinger, J., Krampert, M., Couillard-Despres, S., Heuchel, R., Bogdahn, U., & Aigner, L. (2014). Age-dependent and differential effects of Smad7DeltaEx1 on neural progenitor cell proliferation and on neurogenesis. *Experimental Gerontology, 57*, 149–154.

Martinelli, P., Sperduti, M., Devauchelle, A. D., Kalenzaga, S., Gallarda, T., Lion, S., et al. (2013). Age-related changes in the functional network underlying specific and general autobiographical memory retrieval: a pivotal role for the anterior cingulate cortex. *PLoS One, 8*, e82385.

McDonald, H. Y., & Wojtowicz, J. M. (2005). Dynamics of neurogenesis in the dentate gyrus of adult rats. *Neuroscience Letters, 385*, 70–75.

McEwen, B. S. (1999). Stress and hippocampal plasticity. *Annu Rev Neurosci, 22*, 105–122.

Merrill, D. A., Roberts, J. A., & Tuszynski, M. H. (2000). Conservation of neuron number and size in entorhinal cortex layers II, III, and V/VI of aged primates. *Journal of Comparative Neurology, 422*, 396–401.

Meyer, D., Bonhoeffer, T., & Scheuss, V. (2014). Balance and stability of synaptic structures during synaptic plasticity. *Neuron, 82*, 430–443.

Miranda, C. J., Braun, L., Jiang, Y., Hester, M. E., Zhang, L., Riolo, M., et al. (2012). Aging brain microenvironment decreases hippocampal neurogenesis through Wnt-mediated survivin signaling. *Aging Cell, 11*, 542–552.

Mizuno, T., Kawanokuchi, J., Numata, K., & Suzumura, A. (2003). Production and neuroprotective functions of fractalkine in the central nervous system. *Brain Research, 979*, 65–70.

Moe, M. C., Varghese, M., Danilov, A. I., Westerlund, U., Ramm-Pettersen, J., Brundin, L., et al. (2005). Multipotent progenitor cells from the adult human brain: neurophysiological differentiation to mature neurons. *Brain, 128*, 2189–2199.

Mongiat, L. A., & Schinder, A. F. (2014). Neuroscience. A price to pay for adult neurogenesis. *Science, 344*, 594–595.

Monje, M. L., Toda, H., & Palmer, T. D. (2003). Inflammatory blockade restores adult hippocampal neurogenesis. *Science, 302*, 1760–1765.

Montaron, M. F., Petry, K. G., Rodriguez, J. J., Marinelli, M., Aurousseau, C., Rougon, G., et al. (1999). Adrenalectomy increases neurogenesis but not PSA-NCAM expression in aged dentate gyrus. *European Journal of Neuroscience, 11*, 1479–1485.

Morgenstern, N. A., Lombardi, G., & Schinder, A. F. (2008). Newborn granule cells in the ageing dentate gyrus. *Journal of Physiology, 586*, 3751–3757.

Morrison, J. H., & Hof, P. R. (1997). Life and death of neurons in the aging brain. *Science, 278*, 412–419.

Morshead, C. M., Reynolds, B. A., Craig, C. G., McBurney, M. W., Staines, W. A., Morassutti, D., et al. (1994). Neural stem cells in the adult mammalian forebrain: a relatively quiescent subpopulation of subependymal cells. *Neuron, 13*, 1071–1082.

Mosher, K. I., & Wyss-Coray, T. (2014). Microglial dysfunction in brain aging and Alzheimer's disease. *Biochemical Pharmacology, 88*, 594–604.

Nakajima, Soc, J. S., & Ono, T. (1995). Emotion, memory and behavior: studies on human and nonhuman primates. In *Taniguchi symposia on brain sciences*.

Nakamura, K., Ito, M., Liu, Y., Seki, T., Suzuki, T., & Arai, H. (2013). Effects of single and repeated electroconvulsive stimulation on hippocampal cell proliferation and spontaneous behaviors in the rat. *Brain Research, 1491*, 88–97.

Nakashiba, T., Cushman, J. D., Pelkey, K. A., Renaudineau, S., Buhl, D. L., McHugh, T. J., et al. (2012). Young dentate granule cells mediate pattern separation, whereas old granule cells facilitate pattern completion. *Cell, 149*, 188–201.

Niibori, Y., Yu, T. S., Epp, J. R., Akers, K. G., Josselyn, S. A., & Frankland, P. W. (2012). Suppression of adult neurogenesis impairs population coding of similar contexts in hippocampal CA3 region. *Nature Communications, 3*, 1253.

Nilsson, M., Perfilieva, E., Johansson, U., Orwar, O., & Eriksson, P. S. (1999). Enriched environment increases neurogenesis in the adult rat dentate gyrus and improves spatial memory. *Journal of Neurobiology, 39*, 569–578.

Nyberg, L., Lovden, M., Riklund, K., Lindenberger, U., & Backman, L. (2012). Memory aging and brain maintenance. *Trends in Cognitive Sciences, 16*, 292–305.

Nyffeler, M., Yee, B. K., Feldon, J., & Knuesel, I. (2010). Abnormal differentiation of newborn granule cells in age-related working memory impairments. *Neurobiology Aging, 31*, 1956–1974.

Oakley, R., & Tharakan, B. (2014). Vascular hyperpermeability and aging. *Aging and Disease, 5*, 114–125.

Ortinski, P. I., Dong, J., Mungenast, A., Yue, C., Takano, H., Watson, D. J., et al. (2010). Selective induction of astrocytic gliosis generates deficits in neuronal inhibition. *Nature Neuroscience, 13*, 584–591.

Palmer, T. D., Willhoite, A. R., & Gage, F. H. (2000). Vascular niche for adult hippocampal neurogenesis. *Journal of Comparative Neurology, 425*, 479–494.

Perera, T. D., Coplan, J. D., Lisanby, S. H., Lipira, C. M., Arif, M., Carpio, C., et al. (2007). Antidepressant-induced neurogenesis in the hippocampus of adult nonhuman primates. *Journal of Neuroscience, 27*, 4894–4901.

Peters, A., & Rosene, D. L. (2003). In aging, is it gray or white? *Journal of Comparative Neurology, 462*, 139–143.

van Praag, H. (2008). Neurogenesis and exercise: past and future directions. *Neuromolecular Medicine, 10*, 128–140.

van Praag, H., Schinder, A. F., Christie, B. R., Toni, N., Palmer, T. D., & Gage, F. H. (2002). Functional neurogenesis in the adult hippocampus. *Nature, 415*, 1030–1034.

van Praag, H., Shubert, T., Zhao, C., & Gage, F. H. (2005). Exercise enhances learning and hippocampal neurogenesis in aged mice. *Journal of Neuroscience, 25*, 8680–8685.

Qu, T., Uz, T., & Manev, H. (2000). Inflammatory 5-LOX mRNA and protein are increased in brain of aging rats. *Neurobiology Aging, 21*, 647–652.

Ramm, P., Couillard-Despres, S., Plotz, S., Rivera, F. J., Krampert, M., Lehner, B., et al. (2009). A nuclear magnetic resonance biomarker for neural progenitor cells: is it all neurogenesis? *Stem Cells, 27*, 420–423.

Rando, T. A. (2006). Stem cells, ageing and the quest for immortality. *Nature, 441*, 1080–1086.

Rao, M. S., Hattiangady, B., Abdel-Rahman, A., Stanley, D. P., & Shetty, A. K. (2005). Newly born cells in the ageing dentate gyrus display normal migration, survival and neuronal fate choice but endure retarded early maturation. *European Journal of Neuroscience, 21*, 464–476.

Rapp, P. R. (1995). Cognitive neuroscience perspectives on aging in non-human primates. In T. Nakajima, & T. Ono (Eds.), *Emotion, memory and behavior: Studies on human and nonhuman primates* (p. 197).

Raz, N., Lindenberger, U., Rodrigue, K. M., Kennedy, K. M., Head, D., Williamson, A., et al. (2005). Regional brain changes in aging healthy adults: general trends, individual differences and modifiers. *Cerebral Cortex, 15*, 1676–1689.

Resnick, S. M., Pham, D. L., Kraut, M. A., Zonderman, A. B., & Davatzikos, C. (2003). Longitudinal magnetic resonance imaging studies of older adults: a shrinking brain. *Journal of Neuroscience, 23*, 3295–3301.

Reynolds, B. A., & Weiss, S. (1992). Generation of neurons and astrocytes from isolated cells of the adult mammalian central nervous system. *Science, 255*, 1707–1710.

Rodriguez, J. J., Yeh, C. Y., Terzieva, S., Olabarria, M., Kulijewicz-Nawrot, M., & Verkhratsky, A. (2014). Complex and region-specific changes in astroglial markers in the aging brain. *Neurobiology Aging, 35*, 15–23.

Romine, J., Gao, X., Xu, X. M., So, K. F., & Chen, J. (2015). The proliferation of amplifying neural progenitor cells is impaired in the aging brain and restored by the mTOR pathway activation. *Neurobiology Aging, 36*.

Rotheneichner, P., Lange, S., O'Sullivan, A., Marschallinger, J., Zaunmair, P., Geretsegger, C., et al. (2014). Hippocampal neurogenesis and antidepressive therapy: shocking relations. *Neural Plasticity, 2014*, 723915.

Salthouse, T. A. (1996). The processing-speed theory of adult age differences in cognition. *Psychological Review, 103*, 403–428.

Salthouse, T. A. (2003). Memory aging from 18 to 80. *Alzheimer Disease and Associated Disorders, 17*, 162–167.

Sapolsky, R. M. (1992). Do glucocorticoid concentrations rise with age in the rat? *Neurobiology Aging, 13*, 171–174.

Scahill, R. I., Frost, C., Jenkins, R., Whitwell, J. L., Rossor, M. N., & Fox, N. C. (2003). A longitudinal study of brain volume changes in normal aging using serial registered magnetic resonance imaging. *Archives of Neurology, 60*, 989–994.

Schmidt-Hieber, C., Jonas, P., & Bischofberger, J. (2004). Enhanced synaptic plasticity in newly generated granule cells of the adult hippocampus. *Nature, 429*, 184–187.

Schretlen, D., Pearlson, G. D., Anthony, J. C., Aylward, E. H., Augustine, A. M., Davis, A., et al. (2000). Elucidating the contributions of processing speed, executive ability, and frontal lobe volume to normal age-related differences in fluid intelligence. *Journals of the Internaitonal Neuropsychological Society, 6*, 52–61.

Scott, B. W., Wojtowicz, J. M., & Burnham, W. M. (2000). Neurogenesis in the dentate gyrus of the rat following electroconvulsive shock seizures. *Experimental Neurology, 165*, 231–236.

Seidenfaden, R., Desoeuvre, A., Bosio, A., Virard, I., & Cremer, H. (2006). Glial conversion of SVZ-derived committed neuronal precursors after ectopic grafting into the adult brain. *Molecular and Cellular Neuroscience, 32*, 187–198.

Seki, T., & Arai, Y. (1995). Age-related production of new granule cells in the adult dentate gyrus. *Neuroreport, 6*, 2479–2482.

Seri, B., Garcia-Verdugo, J. M., McEwen, B. S., & Alvarez-Buylla, A. (2001). Astrocytes give rise to new neurons in the adult mammalian hippocampus. *Journal of Neuroscience, 21*, 7153–7160.

Sharpless, N. E., & DePinho, R. A. (2007). How stem cells age and why this makes us grow old. *Nature Reviews Molecular Cell Biology, 8*, 703–713.

Shen, Q., Goderie, S. K., Jin, L., Karanth, N., Sun, Y., Abramova, N., et al. (2004). Endothelial cells stimulate self-renewal and expand neurogenesis of neural stem cells. *Science, 304*, 1338–1340.

Shen, Q., Wang, Y., Kokovay, E., Lin, G., Chuang, S. M., Goderie, S. K., et al. (2008). Adult SVZ stem cells lie in a vascular niche: a quantitative analysis of niche cell-cell interactions. *Cell Stem Cell, 3*, 289–300.

Shetty, A. K., Hattiangady, B., & Shetty, G. A. (2005). Stem/progenitor cell proliferation factors FGF-2, IGF-1, and VEGF exhibit early decline during the course of aging in the hippocampus: role of astrocytes. *Glia, 51*, 173–186.

Shihabuddin, L. S., Horner, P. J., Ray, J., & Gage, F. H. (2000). Adult spinal cord stem cells generate neurons after transplantation in the adult dentate gyrus. *Journal of Neuroscience, 20*, 8727–8735.

Shors, T. J., Miesegaes, G., Beylin, A., Zhao, M., Rydel, T., & Gould, E. (2001). Neurogenesis in the adult is involved in the formation of trace memories. *Nature, 410*, 372–376.

Shors, T. J., Townsend, D. A., Zhao, M., Kozorovitskiy, Y., & Gould, E. (2002). Neurogenesis may relate to some but not all types of hippocampal-dependent learning. *Hippocampus, 12*, 578–584.

Sierra, A., Beccari, S., Diaz-Aparicio, I., Encinas, J. M., Comeau, S., & Tremblay, M. E. (2014). Surveillance, phagocytosis, and inflammation: how never-resting microglia influence adult hippocampal neurogenesis. *Neural Plasticity, 2014*, 610343.

Sierra, A., Encinas, J. M., Deudero, J. J., Chancey, J. H., Enikolopov, G., Overstreet-Wadiche, L. S., et al. (2010). Microglia shape adult hippocampal neurogenesis through apoptosis-coupled phagocytosis. *Cell Stem Cell, 7*, 483–495.

Sierra, A., Gottfried-Blackmore, A. C., McEwen, B. S., & Bulloch, K. (2007). Microglia derived from aging mice exhibit an altered inflammatory profile. *Glia, 55*, 412–424.

Simpson, J. E., Wharton, S. B., Cooper, J., Gelsthorpe, C., Baxter, L., Forster, G., et al. (2010). Alterations of the blood–brain barrier in cerebral white matter lesions in the ageing brain. *Neuroscience Letters, 486*, 246–251.

Siwak-Tapp, C. T., Head, E., Muggenburg, B. A., Milgram, N. W., & Cotman, C. W. (2007). Neurogenesis decreases with age in the canine hippocampus and correlates with cognitive function. *Neurobiology of Learning and Memory, 88*, 249–259.

Song, H., Stevens, C. F., & Gage, F. H. (2002). Astroglia induce neurogenesis from adult neural stem cells. *Nature, 417*, 39–44.

Sonntag, W. E., Lynch, C. D., Cooney, P. T., & Hutchins, P. M. (1997). Decreases in cerebral microvasculature with age are associated with the decline in growth hormone and insulin-like growth factor 1. *Endocrinology, 138*, 3515–3520.

Spalding, K. L., Bergmann, O., Alkass, K., Bernard, S., Salehpour, M., Huttner, H. B., et al. (2013). Dynamics of hippocampal neurogenesis in adult humans. *Cell, 153*, 1219–1227.

Spiga, F., Walker, J. J., Terry, J. R., & Lightman, S. L. (2014). HPA axis-rhythms. *Comprehensive Physiology, 4*, 1273–1298.

Squire, L. R. (1992). Memory and the hippocampus: a synthesis from findings with rats, monkeys, and humans. *Psychological Review, 99*, 195–231.

Steiner, B., Klempin, F., Wang, L., Kott, M., Kettenmann, H., & Kempermann, G. (2006). Type-2 cells as link between glial and neuronal lineage in adult hippocampal neurogenesis. *Glia, 54*, 805–814.

Stephens, C. L., Toda, H., Palmer, T. D., Demarse, T. B., & Ormerod, B. K. (2012). Adult neural progenitor cells reactivate superbursting in mature neural networks. *Experimental Neurology, 234*, 20–30.

Stewart, P. A., Magliocco, M., Hayakawa, K., Farrell, C. L., Del Maestro, R. F., Girvin, J., et al. (1987). A quantitative analysis of blood–brain barrier ultrastructure in the aging human. *Microvascular Research, 33*, 270–282.

Stobart, J. L., & Anderson, C. M. (2013). Multifunctional role of astrocytes as gatekeepers of neuronal energy supply. *Frontiers in Cell Neuroscience, 7*, 38.

Streit, W. J., Miller, K. R., Lopes, K. O., & Njie, E. (2008). Microglial degeneration in the aging brain–bad news for neurons? *Frontiers in Bioscience, 13*, 3423–3438.

Suh, H., Consiglio, A., Ray, J., Sawai, T., D'Amour, K. A., & Gage, F. H. (2007). In vivo fate analysis reveals the multipotent and self-renewal capacities of Sox2+ neural stem cells in the adult hippocampus. *Cell Stem Cell, 1*, 515–528.

Toni, N., Laplagne, D. A., Zhao, C., Lombardi, G., Ribak, C. E., Gage, F. H., et al. (2008). Neurons born in the adult dentate gyrus form functional synapses with target cells. *Nature Neuroscience, 11*, 901–907.

Toni, N., Teng, E. M., Bushong, E. A., Aimone, J. B., Zhao, C., Consiglio, A., et al. (2007). Synapse formation on neurons born in the adult hippocampus. *Nature Neuroscience, 10*, 727–734.

Trejo, J. L., Carro, E., Lopez-Lopez, C., & Torres-Aleman, I. (2004). Role of serum insulin-like growth factor I in mammalian brain aging. *Growth Hormone and IGF Research, 14*(Suppl. A), S39–S43.

United Nations Department of Economic and Social Affairs. (2012). *Population facts*. http://www.un.org/en/development/desa/population/publications/pdf/popfacts/popfacts_2012-4.pdf.
Villeda, S. A., Luo, J., Mosher, K. I., Zou, B., Britschgi, M., Bieri, G., et al. (2011). The ageing systemic milieu negatively regulates neurogenesis and cognitive function. *Nature, 477*, 90–94.
Villeda, S. A., Plambeck, K. E., Middeldorp, J., Castellano, J. M., Mosher, K. I., Luo, J., et al. (2014). Young blood reverses age-related impairments in cognitive function and synaptic plasticity in mice. *Nature Medicine, 20*, 659–663.
Vivar, C., & van Praag, H. (2013). Functional circuits of new neurons in the dentate gyrus. *Frontiers in Neural Circuits, 7*, 15.
Vukovic, J., Colditz, M. J., Blackmore, D. G., Ruitenberg, M. J., & Bartlett, P. F. (2012). Microglia modulate hippocampal neural precursor activity in response to exercise and aging. *Journal of Neuroscience, 32*, 6435–6443.
Walter, J., Keiner, S., Witte, O. W., & Redecker, C. (2011). Age-related effects on hippocampal precursor cell subpopulations and neurogenesis. *Neurobiology Aging, 32*, 1906–1914.
Weber, T., Baier, V., Lentz, K., Herrmann, E., Krumm, B., Sartorius, A., et al. (2013). Genetic fate mapping of type-1 stem cell-dependent increase in newborn hippocampal neurons after electroconvulsive seizures. *Hippocampus, 23*, 1321–1330.
West, M. J. (1993). Regionally specific loss of neurons in the aging human hippocampus. *Neurobiology Aging, 14*, 287–293.
West, M. J., Coleman, P. D., Flood, D. G., & Troncoso, J. C. (1994). Differences in the pattern of hippocampal neuronal loss in normal ageing and Alzheimer's disease. *Lancet, 344*, 769–772.
Wolf, Y., Yona, S., Kim, K. W., & Jung, S. (2013). Microglia, seen from the CX3CR1 angle. *Frontiers in Cell Neuroscience, 7*, 26.
Wu, M. D., Hein, A. M., Moravan, M. J., Shaftel, S. S., Olschowka, J. A., & O'Banion, M. K. (2012). Adult murine hippocampal neurogenesis is inhibited by sustained IL-1beta and not rescued by voluntary running. *Brain, Behavior, and Immunity, 26*, 292–300.
Yassa, M. A., Muftuler, L. T., & Stark, C. E. (2010). Ultrahigh-resolution microstructural diffusion tensor imaging reveals perforant path degradation in aged humans in vivo. *Proceedings of the National Academy of Sciences of the United States of America, 107*, 12687–12691.
Zhang, X. Y., Chen, L., Yang, Y., Xu, D. M., Zhang, S. R., Li, C. T., et al. (2014). Regulation of rotenone-induced microglial activation by 5-lipoxygenase and cysteinyl leukotriene receptor 1. *Brain Research, 1572*, 59–71.
Zhang, X. Y., Wang, X. R., Xu, D. M., Yu, S. Y., Shi, Q. J., Zhang, L. H., et al. (2013). HAMI 3379, a CysLT2 receptor antagonist, attenuates ischemia-like neuronal injury by inhibiting microglial activation. *Journal of Pharmacology and Experimental Therapeutics, 346*, 328–341.
Zhao, C., Teng, E. M., Summers, R. G., Jr., Ming, G. L., & Gage, F. H. (2006). Distinct morphological stages of dentate granule neuron maturation in the adult mouse hippocampus. *Journal of Neuroscience, 26*, 3–11.
Zhao, C., Warner-Schmidt, J., Duman, R. S., & Gage, F. H. (2012). Electroconvulsive seizure promotes spine maturation in newborn dentate granule cells in adult rat. *Developmental Neurobiology, 72*, 937–942.

CHAPTER

8

Adult Neurogenesis, Chronic Stress and Depression

P.J. Lucassen
University of Amsterdam, Amsterdam, The Netherlands

C.A. Oomen
Radboud University Medical Centre, Nijmegen, The Netherlands

M. Schouten
University of Amsterdam, Amsterdam, The Netherlands

J.M. Encinas
Achucarro Basque Center for Neuroscience, Bizkaia, Spain

C.P. Fitzsimons
University of Amsterdam, Amsterdam, The Netherlands

INTRODUCTION

Stress occurs whenever an endogenous or exogenous challenge is perceived as unpleasant, aversive, or threatening for the homeostasis or survival of an individual. Following exposure to stress, various sensory and cognitive signals that trigger specific processes in the body and brain converge, helping the individual to suppress several ongoing processes and refocus attention to cope with the stressor. Even though stress is often considered a "modern disease," the stress response itself occurs in many organisms and has been conserved throughout evolution. It enables individuals to adapt to challenges in their environment and regain homeostasis.

Stress is a very broad term and no single entity. Stress can be psychological in nature and, e.g., be triggered by interpersonal or financial

http://dx.doi.org/10.1016/B978-0-12-801977-1.00008-8

problems or result from a loss of control or unpredictability considering the outcome of a given situation (Ursin & Eriksen, 2004). Also, more biological challenges such as blood loss, dehydration or inflammation, elicit stress responses. Stress can further be acute (e.g., when facing a predator) or chronic (living in poverty or in a broken family). Important elements that determine the nature and impact of a stressor are its (un)predictability, (un)controllability, intensity, and the context in which stress occurs. The perception and interpretation of a stressor, as well as the magnitude and duration of an individual's response to a given stressor, depend largely on his/her genetic background, sex, personality, and early life history (Joels, Karst, Krugers, & Lucassen, 2007; Joels, Sarabdjitsingh, Karst, 2012; Kim et al., 2013; Koolhaas et al., 2011; Lucassen, Fitzsimons, et al., 2013; Lucassen, Naninck, et al., 2013).

The physiological responses to stress can be divided into a fast and a more delayed response. In the classic neuroendocrine stress circuit, several limbic and hypothalamic brain regions integrate a variety of inputs, and together determine the magnitude and specificity of the behavioral, neural, and hormonal responses of the individual to a particular stressor (Joels & Baram, 2009; Joels et al., 2012).

The first phase of the stress response involves a rapid activation of the autonomic nervous system that causes epinephrine and norepinephrine release from the adrenal medulla. These hormones quickly elevate basal metabolic rate, blood pressure, and respiration in seconds to minutes and increase blood flow to the organs essential for the "fight-or-flight" response, such as heart, lungs, and muscles.

The second, slower response to stress involves activation of the hypothalamic–pituitary–adrenal (HPA) axis that controls the release of glucocorticoid (GC) hormones (corticosterone in rodents and cortisol in humans) from the adrenal cortex. As transcriptional regulators, GCs generally act in a slow, genomic manner, but faster GC actions have also been described (Joels et al., 2012). Furthermore, other signaling pathways, such as the gonadal axis and the metabolic and immune system, act in concert with the HPA axis and together they help to redirect energy resources such that attention can be focused on the most urgent and important elements, whereas "maintenance" functions like food digestion or reproduction are temporarily suppressed.

Activation of the HPA axis is triggered by corticotropin-releasing hormone (CRH) in the paraventricular nucleus (PVN) that in turn induces adrenocorticotropic hormone (ACTH) release from the pituitary, which causes the release of GCs from the adrenal. Regulation occurs through negative feedback of GCs that bind to the high-affinity mineralocorticoid receptor (MR) and lower affinity glucocorticoid receptors (GR) (de Kloet, Joëls, & Holsboer, 2005). The GR helps to maintain GC levels within physiological limits (Erdmann, Berger, & Schütz, 2008; de Kloet et al., 2005; Kretz, Reichardt, Schütz, & Bock, 1999) and aberrant GR expression has, e.g., been implicated in hypercortisolism, stress resistance, anxiety, and depression (de Kloet et al., 2005; Ridder et al., 2005; Wei et al., 2007). Furthermore, GC plasma levels are under strict circadian and ultradian control (Liston et al., 2013; Qian, Droste, Lightman, Reul, & Linthorst, 2012) and together with GR and MR function determine an individual's sensitivity and responsivity to stress (Harris, Holmes, de Kloet, Chapman, & Seckl, 2013; Medina et al., 2013; Pruessner et al., 2010; Sousa, Cerqueira, & Almeida, 2008).

On their release in the periphery, GCs affect numerous important functions such as energy, inflammation, and lipid metabolism, among others. Thus, an imbalance in stress hormone regulation can have deleterious consequences, particularly for the brain, where GCs modulate memory, fear, and attention (de Kloet et al., 2005). Whereas acute stress is generally adaptive, chronic stress may alter the MR/GR balance (Harris et al., 2013; de Kloet et al., 2005; Qi et al., 2013) or HPA feedback and result in (prolonged) overexposure of the brain and body to stress hormones, and thus to changes in many of the functions, processes, and behaviors affected by GCs as noted above.

The abundant presence of GRs, particularly in the hippocampus, makes this brain structure very sensitive to stress (de Kloet et al., 2005; Lucassen et al., 2014; Swaab, Bao, & Lucassen, 2005; Wang et al., 2013). GRs have considerable genetic diversity in humans and changes in MR and GR (variants) have been implicated in stress disorders such as major depression disorder (MDD), in stress responsivity and in the associated reductions in hippocampal volume (Alt et al., 2010; Czéh & Lucassen, 2007; Klok, Alt, et al., 2011; Klok, Giltay, et al., 2011; Medina et al., 2013; Qi et al., 2013; Ridder et al., 2005; Sapolsky, 2000; Sinclair, Tsai, Woon, & Weickert, 2011; Spijker et al., 2011; Vinkers et al., 2014; Wang, Joëls, Swaab, & Lucassen, 2012). In functional terms, chronic stress has been associated with reductions in hippocampal excitability, long-term potentiation, and hippocampal memory, but positive effects that depend on the timing, type, and controllability of a stressor have also been described (Joels et al., 2007, 2012). Morphological consequences of chronic stress commonly include hippocampal volume reductions, dendritic atrophy, and reductions in neurogenesis (for a review, see Lucassen et al., 2014, and references therein).

STRESS-RELATED CHANGES IN MAJOR DEPRESSION

Stress is one of the most common risk factors for the development of mood disorders such as MDD, which is thought to result from interactions between genetic predispositions and environmental interactions (Karg, Burmeister, Shedden, & Sen, 2011; Risch et al., 2009). Especially stressful life events experienced during early childhood or adolescence can program plasticity and increase the risk for MDD (Bilbo & Schwarz, 2009;

Bland et al., 2010; Boku et al., 2014; Brunson et al., 2005; Co et al., 2003; Heim, Newport, Mletzko, Miller, & Nemeroff, 2008; Koehl, van der Veen, Gonzales, Piazza, & Abrous, 2012; Korosi et al., 2012; Leuner, 2010; Lucassen, Naninck, et al., 2013; Mirescu, Peters, & Gould, 2004; Risch et al., 2009). Indeed, in a large proportion of depressed patients, the HPA axis activation and GC feedback resistance are common, as reflected by the high percentage of dexamethasone nonsuppressors in this population, hypertrophy of the adrenals and pituitary, increased plasma levels of cortisol, and increases in CRH and AVP expression in the PVN (Lucassen et al., 2014; Raadsheer et al., 1995; Swaab et al., 2005). Notably, depressed subjects show a remarkable heterogeneity in neuroendocrine function and the proportion of depressed individuals demonstrating overt HPA axis abnormalities may range from 35% to 65%.

Maladaptive responses to stress and the associated GC hypersecretion can induce hyperemotional states, mood dysfunction, and cognitive impairments in depressed patients. This is often paralleled by volume changes in various brain regions, including the hippocampus. In the literature, variations in hippocampal volume have been reported in depression, which can relate to differences in disease duration, anatomical delineation, lateralization, early life conditions, and genotype. In general, amygdalar, prefrontal, and hippocampal changes in MDD are well-replicated findings in psychiatry (Cobb et al., 2013; Czéh, Fuchs, Wiborg, & Simon, 2015; Drevets, Price, & Furey, 2008; Kempton et al., 2011; Lorenzetti, Allen, Fornito, & Yücel, 2009). Whether hippocampal volume loss reflects a cause or a consequence of MDD remains unclear but lower hippocampal volumes in patients can be predicted by a more extensive depressive episode duration and recurrence, the size of their integrated cortisol responses, and a history of early life stress. Also, a smaller hippocampal volume could predispose for the development of psychopathology (Czéh & Lucassen, 2007; Lucassen et al., 2014; Sapolsky, 2000).

Preclinical and postmortem studies indicate that chronic stress and depression affect different hippocampal subfields, and different structural substrates, to a different extent. In addition to sex-specific changes in GR expression (Medina et al., 2013; Wang et al., 2012), differences across its transversal and longitudinal axis and connectivity changes were found. Detailed high-field MRI measurements (Huang et al., 2013) revealed that the mean volumes of the DG and CA1–3 subregions were smaller in non-medicated or recently unmedicated depressed patients than in healthy controls. Along the longitudinal axis, a smaller volume was mainly found posteriorly, i.e., in the hippocampal body and tail, rather than anteriorly. Of interest, both the subfield and the posterior hippocampal volume reductions seen in unmedicated depression were absent after antidepressant treatment. The posterior hippocampus may differ from the anterior part when studying volume changes or treatment outcome (Kheirbek & Hen, 2011; MacQueen, Yucel, Taylor, Macdonald, & Joffe, 2008).

In a postmortem study, total hippocampal volume in depression was decreased with increasing duration of depressive illness. There was no significant difference between depressed and control cases in the total number or density of neurons or glia in the CA1, CA2/3, hilus, or DG subregion (Cobb et al., 2013). However, both granule cell and glial cell numbers increased with age in depressed patients on medication, which may reflect proliferative effects of antidepressants (see below) and suggest that GC-induced volume reductions parallel to increased cellular densities are best explained by assuming cell shrinkage and changes in neuropil rather than cell loss (Lucassen et al., 2014; Stockmeier et al., 2004; Swaab et al., 2005). The finding that the loss of brain volume in Cushing's syndrome is reversible after correction of hypercortisolism is consistent with this concept (Bordeau et al., 2002; Starkman et al., 1999). Another option is that volume changes may relate to (lasting) changes in structural plasticity and/or neurogenesis (Czéh & Lucassen, 2007; Jacobs, van Praag, & Gage, 2000; Kempermann, Krebs, & Fabel, 2008; Sapolsky, 2000), as will be discussed below.

STRUCTURAL PLASTICITY AND ADULT NEUROGENESIS

Traditionally, MDD and stress-related disturbances were explained by neurochemical (mainly monoaminergic) imbalances, thought to take place mainly at the synaptic level. More recent studies have now indicated that impairments in structural plasticity also contribute to, e.g., the volumetric changes and pathophysiology of these disorders. Various candidate cellular substrates, such as dendritic retraction, spine alterations, neuronal loss, or glial changes, that are stress sensitive, have been proposed. However, it remains elusive whether changes in these substrates can be considered truly "pathological" or whether they reflect more dynamic adaptations to stress, that can, at least to some extent, be transient and/or reversible, as discussed before (Czéh & Lucassen, 2007; Heine, Maslam, Zareno, Joels, & Lucassen, 2004; Lucassen et al., 2014; Sapolsky, 2000).

One highly plastic and dynamic cellular substrate is adult neurogenesis (AN). AN refers to neural stem cells that continue to produce new neurons in the adult brain (see Fig. 8.1 for a schematic representation). New neuron formation in the adult hippocampus received considerable attention during recent years as it has been implicated, e.g., in (aspects of) mood, epilepsy, cognition, and pattern separation (Abrous & Wojtowicz, 2015; Aimone, Deng, Gage, 2011; Bielefeld et al., 2014; Cho et al., 2015; Clelland et al., 2009; Déry, Goldstein, & Becker, 2015; Jessberger et al., 2009; Jessberger & Gage, 2014; Oomen, Bekinschtein, Kent, Saksida, & Bussey, 2014; Oomen et al., 2013; Richetin et al., 2015; Sahay et al., 2011; Sierra et al., 2015; Tronel et al., 2015). Also, the adult-generated cells are regulated

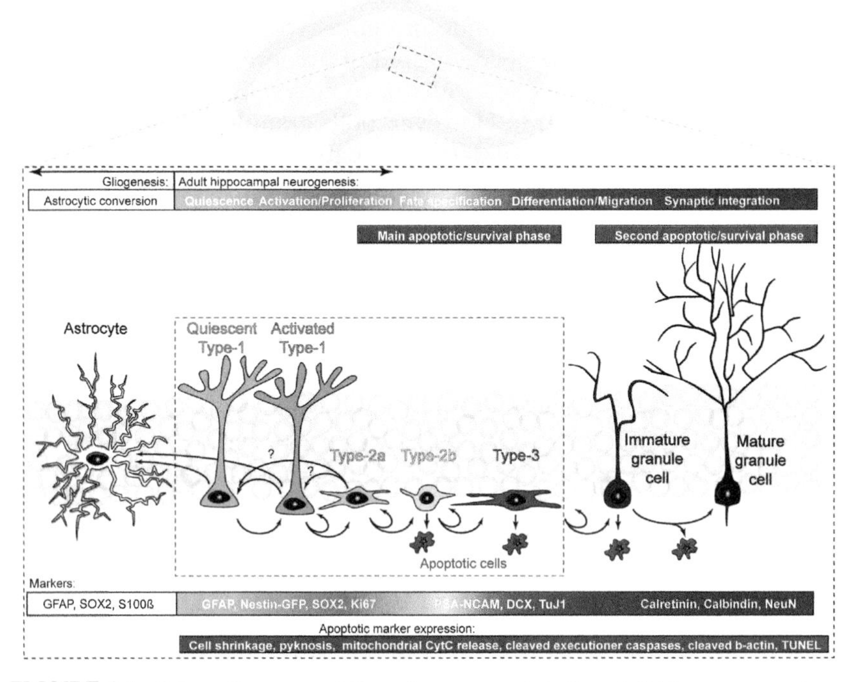

FIGURE 8.1 Schematic representation of neurogenesis in the adult hippocampus, showing the different stages of the neurogenic cascade, the different cell types, and the expression of some of the cell type-specific markers typically used to characterize them. *Arrows* indicate possible transitions between cell types originating from a type 1 neuronal stem cell (NSC) with self-renewal potential. The scheme is based on Encinas et al. (2006) and Sierra et al. (2010).

by (early) stress and various molecular factors. They express GRs (see Fig. 8.1 and Anacker et al., 2013; Gould, McEwen, Tanapat, Galea, & Fuchs, 1997; Heine, Maslam, Joëls, & Lucassen, 2004; Heine, Maslam, Zareno, et al., 2004; Lehmann, Brachman, Martinowich, Schloesser, & Herkenham, 2013; Lemaire, Koehl, Le Moal, & Abrous, 2000; Lucassen, Meerlo, et al., 2010; Lucassen, Fitzsimons, et al., 2013; Lucassen, Naninck, et al., 2013; Montaron et al., 2003; Pham, Nacher, Hof, & McEwen, 2003; Schoenfeld & Gould, 2013; Schouten, Buijink, Lucassen, & Fitzsimons, 2012; Schouten et al., 2015; Song et al., 2012; Wong & Herbert, 2004, 2005) and have been implicated in cognition, anxiety, pattern separation, behavioral flexibility, and the pathophysiology of depression (Clelland et al., 2009; Déry et al., 2015; Duman, 2004; Eisch & Petrik, 2012; Fitzsimons et al., 2013; Jacobs et al., 2000; Kempermann, 2008; Kempermann et al., 2008; Kempermann & Kronenberg, 2003; Lucassen, Meerlo, et al., 2010; Lucassen et al., 2014; Lucassen, Stumpel, Wang, Aronica, 2010; Opendak & Gould, 2015; Petrik, Lagace, & Eisch, 2012; Revest et al., 2009).

The neurogenic hypothesis of depression was postulated based on three observations. First, stress is a risk factor for the development of depression and associated with reductions in hippocampal volume. Stress also reduces neurogenesis in the hippocampus. Second, most antidepressants generally require a period of several weeks to exert their beneficial action, similar to the time frame that newborn neuronal precursors need to integrate and contribute as functional neurons to the adult DG circuitry. Third, antidepressant treatment and the resulting increases in serotonin levels, e.g., promote cell proliferation and generation of new neurons in the hippocampus (Boldrini et al., 2012; Malberg & Duman, 2003; Sahay & Hen, 2007; Wu et al., 2014) that has further been implicated in stress response regulation (Anacker & Pariante, 2012; Lucassen, Fitzsimons, et al., 2013; Opendak & Gould, 2011; Snyder, Soumier, Brewer, Pickel, & Cameron, 2011; Surget et al., 2011).

It is important to note that exceptions exist too, where neurogenesis or newborn cell survival is not stimulated by antidepressants, an effect that appears to depend on animal species and mouse strain, sex, age, and early life history, and on the pharmacology of the antidepressant, or may become apparent only under disease-specific conditions, or only in more anxious mouse strains (Couillard-Despres et al., 2009; Cowen, Takase, Fornal, Jacobs, 2008; David et al., 2009; Hodes, Hill-Smith, Suckow, Cooper, Lucki, 2009; Hodes, Yang, Van Kooy, Santollo, & Shors, 2009; Holick, Lee, Hen, & Dulawa, 2008; Huang, Bannerman, Flint, 2008; Klomp, Václavů, Meerhoff, Reneman, & Lucassen, 2014; Marlatt, Lucassen, & van Praag, 2010; Mendez-David et al., 2014; Navailles, Hof, & Schmauss, 2008; Santarelli et al., 2003; Snyder et al., 2009). Also, anatomical differences within the hippocampus are involved (Kheirbek & Hen, 2011) while indirect effects of serotonin, e.g., on other forms of plasticity, or in interaction with other drugs, may also influence antidepressant effects on neurogenesis and brain plasticity (Maya Vetencourt et al., 2008; Wu & Castrén, 2009). Moreover, a reduction in neurogenesis per se, i.e., other than induced by stress, does not cause a "depressed" phenotype (Henn & Vollmayr, 2004; Lucassen, Fitzsimons, et al., 2013; Snyder et al., 2011; Vollmayr, Simonis, Weber, Gass, & Henn, 2003). Later studies have suggested that neurogenesis is implicated in antidepressant drug actions in rodents and primates and that blocking the proneurogenic effects of antidepressants prevented the behavioral effects of these drugs measured in some animal models (see also Figs. 8.1 and 8.2, and Santarelli et al., 2003; Sahay & Hen, 2007; Surget et al., 2008; Perera et al., 2011).

So far, however, a coherent functional theory is still lacking as to how a limited number of new hippocampal neurons in only a subregion of the DG can contribute to brain features and functions that are as general as mood or depressive symptoms. This is besides the cognitive deficits, which are related to, but not specific for, mood disorders. Although a reduced rate

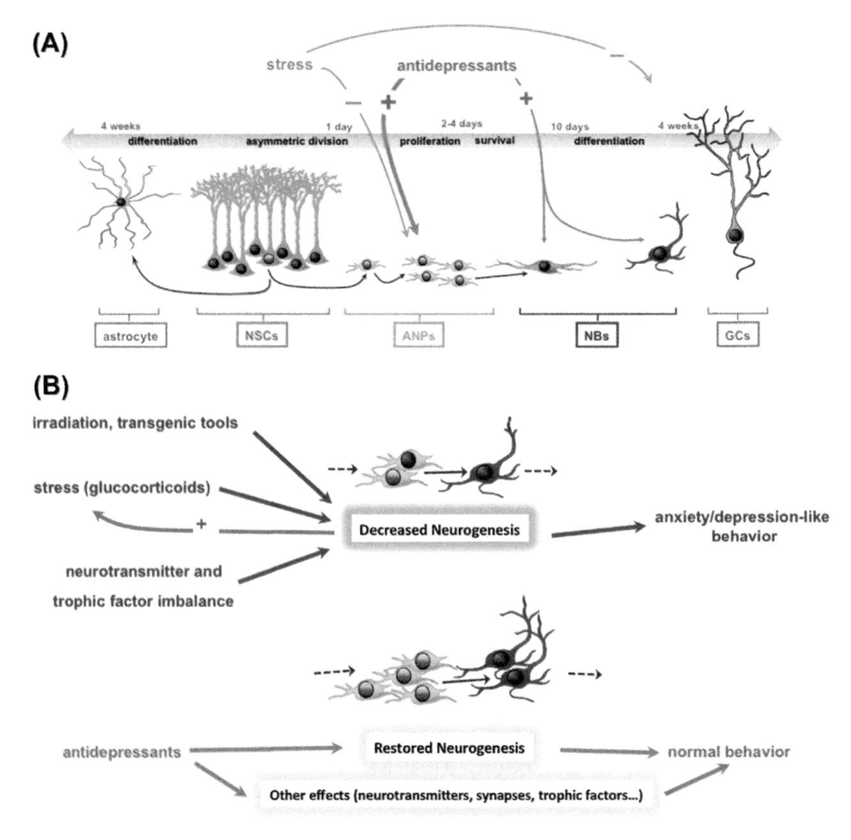

FIGURE 8.2 Schematic representation of the different stages of neurogenesis and their relationship/sensitivity to stress and depression. (A) Model of adult hippocampal neurogenesis. Neural stem cells (NSCs) are activated and enter the cell cycle to produce amplifying neuronal precursors (ANPs). These can proliferate and die, or differentiate into neuroblasts (NBs) and eventually into granule cells (GCs). Stress can reduce newborn cell proliferation and also their survival and differentiation (see also Heine, Maslam, Joëls, et al., 2004; Heine, Maslam, Zareno, et al., 2004 and Wong and Herbert, 2004). Many antidepressants, including treatment with antiglucocorticoids, exert the opposite effect. (B) An acute decrease in neurogenesis per se is not sufficient to provoke depression or depression-like behavior, but when neurogenesis is decreased by an imbalance of neurotransmitters or trophic factors, or induced with transgenic tools or, e.g., irradiation, it can increase the response to stress, which by itself can further diminish neurogenesis (blue +). As a result, depression-like behavior can emerge in experimental animals. Most antidepressants increase neurogenesis, a response that has been shown to be required for their (functional) antidepressant action, most likely in combination with other effects at the neurotransmitter, trophic factor, and/or synaptic levels. Chronic exposure to stress, or to stress early in life, can lastingly reduce AN, resulting in a diminished neurogenic reserve and a reduced cognitive potential and flexibility, associated with depression (Kheirbek, Klemenhagen, Sahay, & Hen, 2012).

of neurogenesis may reflect impaired hippocampal plasticity, reductions in adult neurogenesis alone are unlikely to produce depression. Lasting reductions in the turnover rate of DG granule cells, however, will, over time, clearly alter the average age and overall composition of the DG cell population and thereby influence properties of the hippocampal circuit.

Neural stem cells present in the hippocampus and their progeny go through different stages of proliferation (apoptotic), cell selection, fate specification, migration, and neuronal differentiation before they are eventually integrated as new functional neurons into the preexisting adult hippocampal network (Fig. 8.1) (Abrous et al., 2005; Jessberger & Gage, 2014; Kempermann, 2012; Opendak & Gould, 2015; Zhao, Deng, & Gage, 2008). Neurogenesis and related structural plasticity have also been reported in other brain structures, such as the amygdala, striatum, hypothalamus, and neocortex, but to a more limited extent, with differences between species and often occurring in response to specific challenges or injury. Whether the cellular/structural plasticity in these other brain regions is also regulated by similar environmental factors, such as stress, is less well studied (Amrein, Isler, & Lipp, 2011; Cavegn et al., 2013; Gould, 2007).

AN is dynamically regulated by various hormonal and environmental factors and drugs and declines with age. Neurogenesis in the hippocampal DG is potently stimulated by exercise and environmental enrichment, notably parallel to changes in hippocampal function (Holmes, Galea, Mistlberger, & Kempermann, 2004; Kannangara et al., 2011; Kempermann, 2012; Kempermann et al., 2010; Kobilo et al., 2011; Vivar, Potter, & van Praag, 2013; Voss et al., 2013). Exercise also exerts effects on the stress axis itself (Droste et al., 2003) and on growth factor levels (Marlatt, Potter, Lucassen, & van Praag, 2012; Vivar et al., 2013). Whereas rewarding experiences can also stimulate neurogenesis, aversive experiences like stress generally decrease proliferation and neurogenesis in the hippocampus (Balu & Lucki, 2009; Lucassen, Meerlo, et al., 2010; Lucassen et al., 2014).

STRESS REGULATION OF NEUROGENESIS

Stress is one of the best known environmental inhibitors of AN. Both psychosocial (Czéh et al., 2002; Gould et al., 1997) and physical stressors (Malberg & Duman, 2003; Pham et al., 2003; Vollmayr et al., 2003) can suppress one or more phases of the neurogenesis process (see Figs. 8.2 and 8.3 and Czéh et al., 2001, 2006, 2002; Lucassen, Meerlo, et al., 2010; Lucassen et al., 2014; Mirescu & Gould, 2006). In classical studies, predator stress produced significant reductions in hippocampal proliferation (Czéh et al., 2001; Gould et al., 1997; see Lucassen et al., 2014, and Czéh et al., 2015, for recent reviews) and both acute and chronic unpredictable stress can suppress proliferation while also other stressors, including physical restraint,

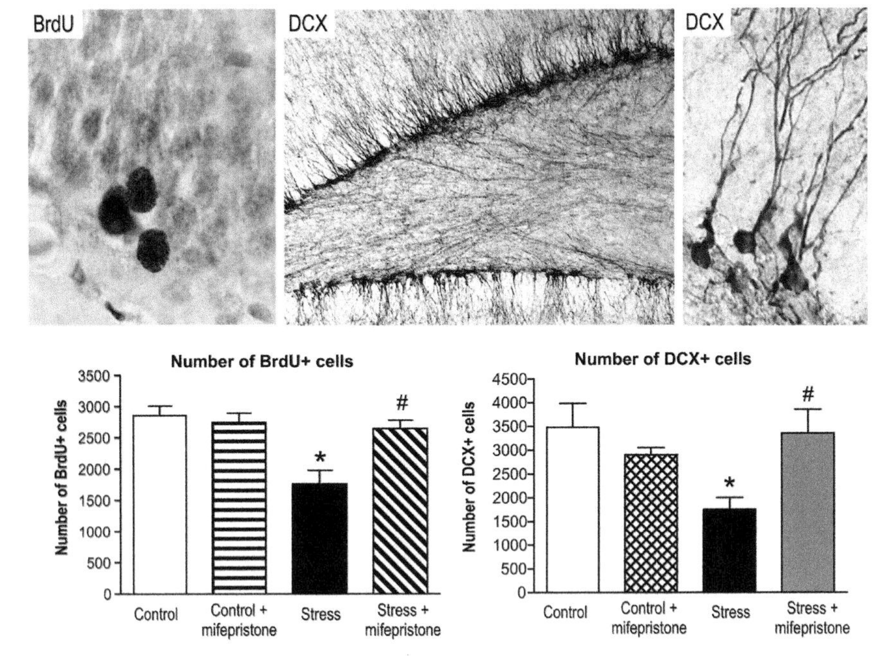

FIGURE 8.3 Effects of prolonged stress or glucocorticoid treatment on neurogenesis and the rescue effect of GR blockade. Top panels show examples of BrdU+ and Doublecortin (DCX)+ immunostained cells in the rat hippocampal dentate gyrus (DG). DCX-positive somata are located in the SGZ with their extensions (arrowheads) passing through the GCL. Lower panels display BrdU- (left graph) and DCX-positive cell numbers (right graph) in a 1 in 10 series of hippocampal sections of rats subjected to 21 days of chronic unpredictable stress (Oomen et al., 2007) similar results have been found after chronic corticosterone treatment (Mayer et al., 2006). The significant reduction in both BrdU- (21-day-old cells) and DCX-positive cell numbers after chronic stress or corticosterone treatment is normalized by 4 days of high dose treatment with the GR antagonist mifepristone, whereas the drug alone has no effect (see Mayer et al., 2006, and Oomen et al., 2007; Zalachoras et al., 2013, for details). * indicates $p<0.001$ relative to the control group in the left graph and $p<0.05$ in the right graph. Similar recovery is found when mifepristone is applied for only 1 day (Hu et al., 2012).

social defeat, inescapable foot shock, sleep deprivation, and inflammation can decrease the numbers of new neurons in the dentate gyrus (Czéh et al., 2002; Dagyte et al., 2009; Gould et al., 1997; Heine, Maslam, Joëls, et al., 2004; Heine, Maslam, Zareno, et al., 2004; Hulshof, Novati, Luiten, den Boer, & Meerlo, 2012; Lucassen, Meerlo, et al., 2010; Lucassen, Stumpel, et al., 2010; Pham et al., 2003; Schoenfeld & Gould, 2013; Wu et al., 2014). Interestingly, also an absence of an effect or even increases in neurogenesis have been reported after stress in some instances, but here, differences in temporal design may have been involved, or repeated stressors could have become predictable or were relatively mild and may actually have enriched an otherwise "boring" environment and could thus have been perceived as rewarding experiences (Parihar, Hattiangady, Kuruba, Shuai, & Shetty, 2011).

Also the duration of the stress period and the anatomical location may be of influence as acute stress could, e.g., induce proliferation in the dorsal hippocampus under specific conditions (Kirby et al., 2013; Kheirbek & Hen, 2011).

When no other transmitter systems are altered, the stressor is unpredictable and its nature severe, then reductions in neurogenesis are commonly seen. In fact, multiple stages of the neurogenic process are then affected, including proliferation of the neural stem cells and amplifying progenitor cells, as well as subsequent neuronal differentiation phase and dendritic expansion. Stress not only reduces proliferation and neurogenesis in many different species, it may also shift neural stem cells away from neuronal differentiation and instead "redirect" them toward the generation of oligodendrocytes (Chetty et al., 2014). Although not studied in great detail yet, such stress-induced fate shifts may have important functional consequences, e.g., for the myelination of axons and/or mossy fibers, and hence network connectivity.

Thus, while different types of stress can trigger different behavioral and functional responses, GCs are considered instrumental in mediating the effects of stress, e.g., on new neuron production. Chronic exogenous administration of GCs to animals affects cell proliferation, neuronal differentiation, and cell survival, as well as the production of oligodendrocytes and (micro-)glia responses and behavior (Bland et al., 2010; Butovsky et al., 2006; Chetty et al., 2014; Ekdahl, Kokaia, & Lindvall, 2009; Hu et al., 2012; Lehmann et al., 2013; Mayer et al., 2006). Moreover, the reductions in neurogenesis after stress, and many of the related molecular alterations as well (Datson et al., 2012), can be prevented by, e.g., blocking GCs from the adrenal or HPA mediators, using, e.g., CRH or GR antagonists (Alonso et al., 2004). Also, a short treatment for 1 or 3 days with a GR antagonist already normalized stress or GC-induced reductions in hippocampal neurogenesis (Hu et al., 2012; Mayer et al., 2006; Oomen, Mayer, de Kloet, Joels, & Lucassen, 2007; Zalachoras et al., 2013) (see also Fig. 8.3).

Although more information has become available on molecular control and timing of stem cell regulation (Anacker et al., 2013;Fitzsimons et al., 2014, 2013; Miller et al., 2013; Schouten et al., 2012, 2015), the precise mechanisms by which GCs decrease neurogenesis remains unknown. NMDA receptors, GRs, and MRs are present on the new cells, albeit in different ratios over time, that likely act in concert to mediate effects of stress on neurogenesis (Montaron et al., 2003; Wong & Herbert, 2004, 2005). Notably, GR knockdown, selectively in cells of the hippocampal neurogenic niche, accelerates their neuronal differentiation and migration, induces ectopic positioning, alters their dendritic complexity, and increases their dendritic spines and basal excitability but impairs contextual freezing during fear conditioning (Fitzsimons et al., 2013). Hence, GR expression in, and thus stress sensitivity of, the newborn hippocampal cells is important for their structural and functional integration in the hippocampal circuit.

The precursors are further located closely to blood vessels, which is of relevance as it is indeed this population that is particularly sensitive to stress (Heine, Zareno, Maslam, Joels, & Lucassen, 2005). Astrocytes are also important as this cell type supports the survival of developing neurons, possesses GR, and are affected by some types of stress (Banasr & Duman, 2008; Czéh, Simon, Schmelting, Hiemke, & Fuchs, 2006; Oomen et al., 2009) and changes in this cell population can contribute to depressive-like behavior (Banasr & Duman, 2008).

Stress further slows down neuronal differentiation, as evidenced by the upregulation of markers indicating cell cycle arrest (Heine, Maslam, Joëls, et al., 2004) and related changes in granule cell dendritic trees. Stress and GCs also reduce the survival of neurons produced prior to the stressful experience. While the underlying mechanism is largely unknown, this is thought to be mediated by inhibitory effects on neurotrophins such as brain-derived neurotrophic factor (BDNF) (Schmidt & Duman, 2007). The reduction in newborn cell survival likely also involves microglia, which are known to phagocytose newborn neuronal precursors (Ekdahl et al., 2009; Guadagno et al., 2013; Hinwood, Morandini, Day, & Walker, 2012; Morris, Clark, Zinn, & Vissel, 2013; Sierra et al., 2010). Indeed, stress influences microglia and their responsivity, which may modulate their efficiency in cleaning up debris. Although under normal conditions, "resting" or unchallenged microglia do not trigger apoptosis of those cells that they efficiently phagocytose in the hippocampal neurogenic niche (Sierra et al., 2010), microglia could help reduce new neuron survival under acute and chronic stress (Jakubs et al., 2008), by becoming activated and releasing cytokines and chemokines with neurotoxic effects (Johansson, Price, & Modo, 2008; Koo & Duman, 2008; Spulber, Oprica, Bartfai, Winblad, & Schultzberg, 2008). Also, they could undergo changes in intrinsic properties such as motility and morphology (Walker, Nilsson, & Jones, 2013). Whether they have the capacity to switch to a model in which they can actually induce apoptosis of neuronal precursors in the hippocampal neurogenic niche, as has been shown in vitro via TNFa (Guadagno et al., 2013), remains to be determined.

Although a role for (nor)adrenaline has not been studied in detail with respect to stress-induced changes in neurogenesis, an important difference between several studies is whether or not GC levels remain elevated after the exposure to the stressor has ended. In some psychosocial stress models, the GC "milieu" is altered in a lasting manner and GC levels remain elevated for prolonged periods of time, which has stronger inhibitory effects on AN than apparently severe, but predictable, physical stressors, such as restraint (Wong & Herbert, 2004). Several examples of a persistent and lasting inhibition of AN after an initial stressor exist, despite a later normalization of GC levels (e.g., Czéh et al., 2002; Mirescu & Gould, 2006; Schoenfeld & Gould, 2013). Also, GC levels can remain elevated after the onset of the first, often psychosocial, stressor that suppresses neurogenesis for prolonged periods. In other, milder models, stress hormone levels generally normalize, yet neurogenesis remains reduced (Schoenfeld & Gould, 2013; Van Bokhoven et al., 2011). This suggests that while GCs are involved in the initial suppression of proliferation, they are not always necessary for the maintenance of this effect.

When studying the effects of stress on adult neurogenesis under laboratory conditions, it is important to realize that many variables influence the outcome of such studies, as interindividual and gender differences in stress coping, handling, time of day at sacrifice, and previous exposure to stressful learning tasks can influence stress responses and neurogenesis (e.g., Ehninger & Kempermann, 2006; Holmes et al., 2004). In addition to stress hormones like GCs, also other mediators of the stress system that interact with neurogenesis are changed. Models employing repeated injections with exogenous GCs to imitate the hypercortisolism found, e.g., in depression, exert negative feedback at the level of the pituitary and inhibit the endogenous production of GCs by the adrenals. As a result, ACTH and CRH levels are very low in GC-treated rodents, a condition which is in contrast to the endogenous HPA axis activation seen in chronically stressed animals and patients where CRH, ACTH, and GCs are elevated. A large number of other factors may also contribute to the stress-induced inhibition of AN, such as the stress-induced increase in glutamate release via NMDA receptor activation (Gould et al., 1997; Nacher & McEwen, 2006; Schoenfeld & Gould, 2013).

Stress further affects various neurotransmitters implicated in the regulation of neurogenesis: GABA (Ge, Yang, Hsu, Ming, & Song, 2007), serotonin (Djavadian, 2004), noradrenaline (Joca et al., 2007), acetylcholine (Bruel-Jungerman, Lucassen, & Francis, 2011), and dopamine, e.g., (Domínguez-Escribà et al., 2006; Takamura et al., 2014). GABA deserves special mention as it has been recently reported to be a key regulator for the recruitment and activation of hippocampal neural stem cells. The balance of activation and quiescence would be controlled by tonic release of GABA by neighboring interneurons. Higher levels would promote quiescence while lower levels would promote activation of neural stem cells, via GABAA receptors expressed by them (Song et al., 2012). Also other neurotransmitter systems such as the cannabinoids, opioids, nitric oxide, various neuropeptides, and also gonadal steroids may contribute (see, e.g., Balu & Lucki, 2009; Galea, 2008). Importantly, stress reduces the expression of several growth and neurotrophic factors, such as BDNF, insulin-like growth factor-1, nerve growth factor, epidermal growth factor, and vascular endothelial growth factor, that can influence neurogenesis (see, e.g., Schmidt & Duman, 2007; Wilson, 2014).

Chronic stress can also affect proliferation of glial cells as, e.g., seen in the medial prefrontal cortex of rats after social defeat, after chronic unpredictable stress, or after chronic GC administration. Similarly, prolonged and elevated GC treatment inhibited NG2-positive cell proliferation, reflecting changes in oligodendrocyte precursors. Chronic stress also promotes structural remodeling of microglia and can enhance the release of proinflammatory cytokines from microglia, and can even mediate aspects of depressive-like behavior (Kreisel et al., 2014).

Astrocytes are also key components of the "neurogenic niche" that provides the necessary local microenvironment for neurogenesis. For instance, astrocytes participate physically in the establishment of synapses between newborn and preexistent neurons, and the inhibition of glutamate reuptake by astrocytes significantly impairs postsynaptic currents and facilitates paired-pulse facilitation in adult-born neurons (Krzisch et al., 2015). Furthermore, astrocytes promote neuronal differentiation by secreting ephrin-B2 acting on ephrin-B4 receptors present in neural stem cells and modulating β-catenin in a Wnt-independent manner (Ashton et al., 2012). Since astrocytes also contain GRs and can be regulated by stress, this implies that stress can also modulate neural progenitors through interactions with astrocytes (Vallieres et al., 2002; Wang et al., 2013).

Stress-induced suppression of adult neurogenesis has been associated with impaired performance on various cognitive tasks that require the hippocampus, such as spatial navigation learning and object memory (Braun & Jessberger, 2014; Oomen et al., 2014). It should be noted that these effects are typically observed within a shorter time frame than what would be expected for the involvement and integration of newborn neurons, which should be taken into account in the experimental design. In addition, additional younger, immature, and excitable neurons, as well as the older populations of DG cells, exist that are sensitive too, and could contribute to hippocampal performance. As a more specific test for the new neurons in the DG per se, pattern separation has received a lot of attention recently (see above).

An extensive literature further indicates that AN is sensitive to stress exposure during the early life period. As this is beyond the scope of this chapter, early life stress will only be discussed briefly. The set point of HPA axis activity, and possibly also of neurogenesis regulation, is, on the one hand, programmed by genotype, but can be further modified by early development, and, e.g., by epigenetic changes (Lucassen, Naninck, et al., 2013). In humans, early life stressors (ELS) are among the strongest predisposing factors for developing psychopathology and cognitive decline later in life (Baram et al., 2012; Heim et al., 2008; Loman, Gunnar, & Early Experience Stress and Neurobehavioral Development Center, 2010; Maselko, Kubzansky, Lipsitt, & Buka, 2011; Risch et al., 2009; Teicher, Anderson, & Polcari, 2012) and in experimental conditions where emotional and cognitive functions are altered after ELS (Aisa, Tordera, Lasheras, Del Río, & Ramírez, 2007; Baram et al., 2012; Brunson et al., 2005; Ivy et al., 2010; Oomen et al., 2010). Also neurogenesis is sensitive to stress during the perinatal period, the effects of which often depend on sex and on the developmental stage during which the organism experienced stress (Coe et al., 2003; Galea, 2008; Kim et al., 2013; Korosi et al., 2012; Lemaire et al., 2000; Loi, Koricka, Lucassen, & Joëls, 2014; Lucassen et al., 2009; Lucassen, Naninck, et al., 2013; Mirescu et al., 2004; Naninck et al., 2015), although exceptions exist as well (Tauber et al., 2008).

NEUROGENESIS AND MAJOR DEPRESSION

Given the clear association among stress, hippocampal volume reductions, and major depression, it came as no surprise that antidepressants could affect hippocampal neurogenesis in several animal models (for a recent overview, see Czéh et al., 2015). Given the technical limitations to visualize neurogenesis in vivo, only few studies have so far addressed this issue in human brain tissue. Reif, Fritzen, Finger, Strobel, and Lauer (2006) failed to find differences in neural stem cell proliferation in postmortem brain samples between patients suffering from MDD, bipolar disorder, or schizophrenia and control subjects. Antidepressants did not alter these numbers but changes were found in schizophrenia. More recent studies (Boldrini et al., 2012, 2009; Lucassen et al., 2014; Lucassen, Stumpel, et al., 2010) (see Fig. 8.4) compared progenitor and dividing cells with different immunocytochemical markers and found that in hippocampal tissues of untreated depressed subjects, the numbers of progenitor cells were significantly decreased. Both treatments with selective serotonin-reuptake inhibitors (SSRIs) and tricyclic antidepressants (TCAs) increased the number of nestin-positive progenitors and TCAs had a robust stimulatory effect on the number of Ki-67-reactive dividing cells. These changes were reported in middle aged (Boldrini et al., 2012, 2009) but not in older depressed patients (Lucassen, Meerlo, et al., 2010; Lucassen, Stumpel, et al., 2010), possibly because of limited power or due to age-related differences in plasticity in these patients (see also Felice et al., 2015).

In a postmortem study on MDD patients, the volume of the histologically defined DG was determined and found to be in fact 68% larger in SSRI-treated depressed subjects, while SSRI treatment substantially increased neural progenitor cells in the dentate gyrus. A recent study by Huang et al. (2013) found smaller DG volumes using MRI in unmedicated depressed patients, which was confirmed at postmortem analysis, which is consistent with the neurogenic hypothesis of depression. Interestingly, both subfield and posterior hippocampal volume reductions were only seen in unmedicated depression but were absent in patients treated with antidepressants. Although it is so far not simple to detect ongoing neurogenesis in vivo (Manganas et al., 2007), these data are consistent with preclinical studies demonstrating subregional specific and opposite effects of stress or depression and antidepressant treatment.

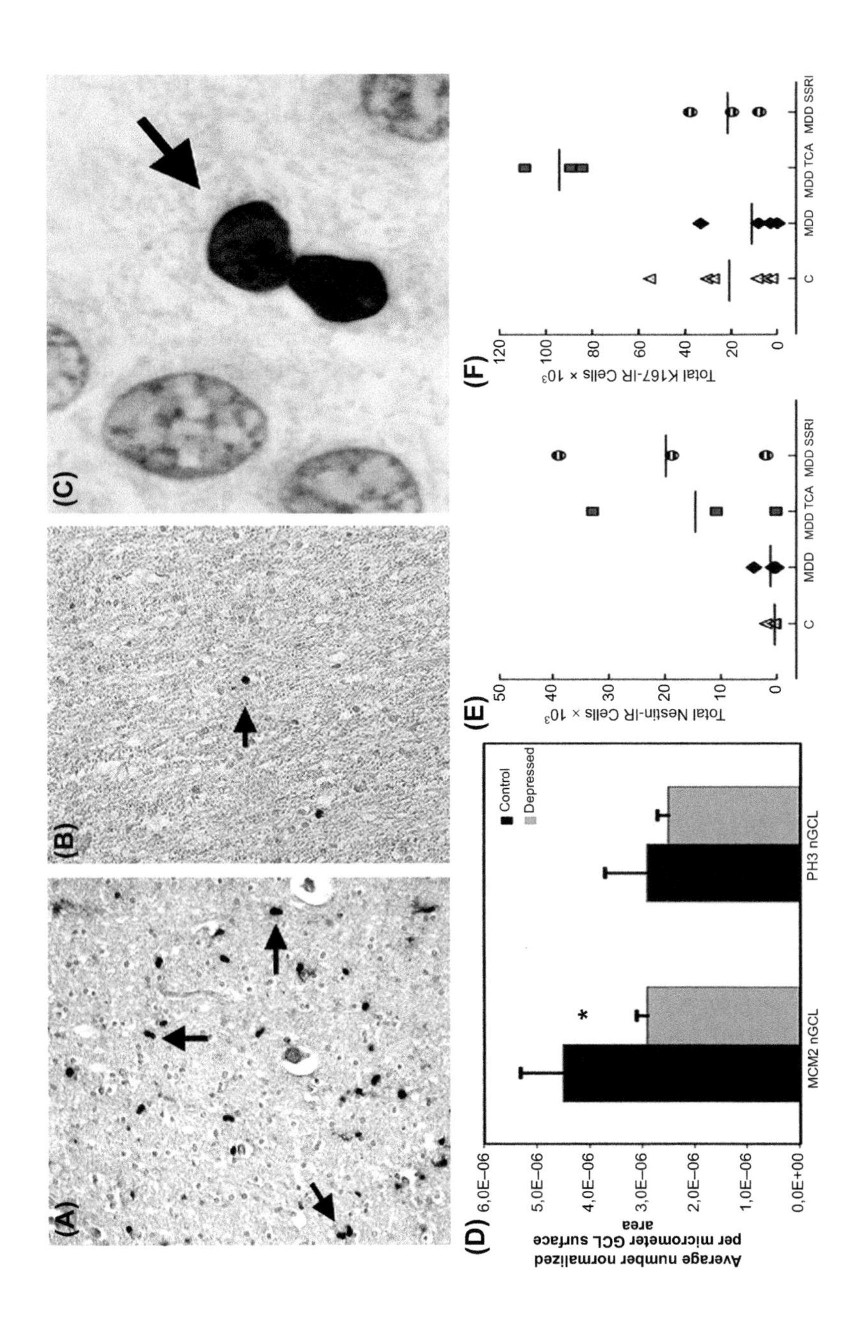
(A)
(B)
(C)
(D)
Average number normalized per micrometer GCL surface area
6,0E–06
5,0E–06
4,0E–06
3,0E–06
2,0E–06
1,0E–06
0,0E+00
*
MCM2 nGCL
PH3 nGCL
Control
Depressed
(E)
Total Nestin-IR Cells × 10³
50
40
30
20
10
0
C
MDD
MDD TCA
MDD SSRI
(F)
Total K167-IR Cells × 10³
120
100
80
60
40
20
0
C
MDD
MDD TCA
MDD SSRI

Although AN may thus not be essential for the development of depression, it may be required for clinically effective antidepressant treatment (Perera et al., 2011; Santarelli et al., 2003; Surget et al., 2008, 2011). Hence, stimulation of neurogenesis has been regarded as a promising strategy for identifying new antidepressant targets (see also Figs. 8.1 and 8.2). Accordingly, when tested in chronic stress paradigms, several candidate antidepressant compounds, such as corticotrophin-releasing factor (CRF1), vasopressin (V1b), or glucocorticoid receptor antagonists (Alonso et al., 2004; Oomen et al., 2007; Surget et al., 2008), tianeptine (Czéh et al., 2001) or selective neurokinin-1 (NK1) receptor antagonists (Czéh et al., 2005) could indeed normalize the inhibitory effects of stress on proliferation or neurogenesis.

Hippocampal volume loss is well documented in various psychopathologies and also in patients with Cushing's disease and in subjects treated with synthetic GCs (Bourdeau et al., 2002; for a review, see Lucassen et al., 2014, and references therein). In addition to neurochemical changes, structural connectivity and plasticity changes, including neurogenesis, may contribute to its etiology as well. However, it remains elusive how exactly a subpopulation of newborn neurons can contribute to general features such as stress regulation, mood, and depression (Anacker & Pariante, 2012; Lehmann et al., 2013; Lucassen, Fitzsimons, et al., 2013; Lucassen et al., 2014; Opendak & Gould, 2011; Snyder et al., 2011).

Although a reduced rate of neurogenesis may reflect impaired hippocampal plasticity, reductions in adult neurogenesis per se, i.e., induced by irradiation, but without the concomitant presence of stress, are unlikely to

FIGURE 8.4 Changes in cell proliferation in postmortem human brain tissue of depressed and antidepressant-treated patients. (A) Cells immunopositive for the cell cycle marker minichromosome maintenance protein 2 (MCM2) that is involved in the control of DNA replication. Many MCM2-immunopositive cells and doublets (arrows) are observed in cortical tissue of a 2-year-old subject that served as positive control for the immunocytochemical procedure. (B) As expected, MCM2-ir cell numbers are strongly reduced to very low numbers (arrow) in a 69-year-old control subject. (C) Detail of an MCM2-ir doublet of two cells that appear to have recently separated in the hippocampus of a depressed patient (arrow), cresyl violet counterstain. (D) Graphs depicting numbers of MCM2 and phosphorylated histone H3 (PH3)-immunopositive cells (the latter marker reflecting the late G2 and M phases of cell division). PH3-immunoreactive cells in the subgranular zone and granular cell layer of the dentate gyrus, normalized to the surface area of the GCL. A significant reduction was found for MCM2, but not PH3, in elderly (average age of 68 years) depressed patients compared to controls. (E) Neural progenitor and (F) dividing cells (nestin and Ki-67-positive, respectively) are increased in the dentate gyrus of younger (average ages of 40 and 54 years) patients with major depressive disorder (MDD) who were treated with antidepressants compared to untreated MDDs and control subjects. Progenitor numbers (E, nestin-positive) were higher in MDD patients treated with tricyclics (TCA) or with selective serotonin reuptake inhibitors (SSRI), compared to untreated MDD and control cases, whereas the numbers of dividing cells (F, Ki-67-ir) were higher only in the TCA but not the SSRI-treated group. *Reproduced from Lucassen P.J., Pruessner J., Sousa N., Almeida O.F., Van Dam A.M., Rajkowska G., et al. (2014). Neuropathology of stress.* Acta Neuropathologica, 127, *109–135.*

produce depression (Henn & Vollmayr, 2004). Lasting and stress-related reductions in DG neurogenesis will, however, alter turnover rate, average cellular age, and overall composition of the DG, and in the long term modify DG volume, and thereby influence the properties and vulnerability of the hippocampal circuit (Teicher et al., 2012). Indeed, hippocampal volume changes often coincide with stressful episodes in depressed patients, correlating with cognitive impairments.

The hippocampus is further thought to provide negative feedback control of the HPA axis, in which neurogenesis is at least partly implicated. Initial disturbances in hippocampal neurogenesis or output may thus disturb feedback and hence amplify HPA axis dysregulation that is common in approximately 50% of the depressed patients. Since massive cell loss could not be demonstrated in the hippocampus, the observed hippocampal volume changes could be due to (atrophy of) the somatodendritic or synaptic components, but also glia or changes in fluid balance may be involved (Czéh et al., 2006; Lucassen et al., 2014). Interestingly, AN might form a link between depression and stress. When neurogenesis is ablated, either by X-irradiation or by selective killing of mitotically active GFAP-expressing cells (neural stem cells) in mice, stressors induce a significantly stronger effect, measured both by blood levels of cortisone and by behavioral outcome in tests such as novelty-suppressed feeding, forced swimming, and the sucrose preference test (Snyder et al., 2011). Thus AN has been postulated to act as a buffer to both stress responses and depressive behavior.

CONCLUSIONS

Different types of stress and glucocorticoid and antidepressant treatment often interfere with one or more phases of the neurogenetic process. While inhibitory effects of acute stress can normalize after a recovery period, e.g., reductions in neurogenesis that are caused by chronic, severe, and/or unpredictable stress may last longer and have functional consequences. While neurogenesis has been implicated in cognition, mood, and anxiety regulation as well as in the therapeutic effects of antidepressant drugs, its exact role, i.e., cause or consequence, in relation to depression remains poorly understood. A reduced rate of neurogenesis may reflect impaired hippocampal plasticity, but reductions in AN alone are unlikely to produce depression and may require additional stress exposure, e.g., during critical developmental periods. Lasting reductions in turnover rate of DG granule cells, e.g., influenced by genotype or programmed by early life events, may alter the overall composition of the DG cell population. This, in turn, could modify stress responsivity and thereby influence functioning of the adult hippocampal circuit as well as the vulnerability for developing brain disorders.

Acknowledgments

P.J.L. is supported by grants from the HersenStichting Nederland, NWO PrioMedChild, and Amsterdam Brain and Cognition (ABC). C.P.F. and M.S. are supported by ISAO and by the Innovational Research Incentive Scheme VIDI from The Netherlands Organization for Scientific Research (NWO). J.M.E. is supported by MINECO (Spanish Ministry for Economy and Competitiveness) and IKERBASQUE (The Basque Science Foundation).

References

Abrous, D. N., Koehl, M., & Le Moal, M. (2005). Adult neurogenesis: from precursors to network and physiology. *Physiological Reviews, 85*, 523–569.

Abrous, D. N., & Wojtowicz, J. M. (June 1, 2015). Interaction between neurogenesis and hippocampal memory system: new vistas. *Cold Spring Harbor Perspectives in Biology, 7*(6). pii:a018952.

Aimone, J. B., Deng, W., & Gage, F. H. (May 26, 2011). Resolving new memories: a critical look at the dentate gyrus, adult neurogenesis, and pattern separation. *Neuron, 70*(4), 589–596.

Aisa, B., Tordera, R., Lasheras, B., Del Río, J., & Ramírez, M. J. (2007). Cognitive impairment associated to HPA axis hyperactivity after maternal separation in rats. *Psychoneuroendocrinology, 32*, 256–266.

Alonso, R., Griebel, G., Pavone, G., Stemmelin, J., Le Fur, G., & Soubrié, P. (2004). Blockade of CRF(1) or V(1b) receptors reverses stress-induced suppression of neurogenesis in a mouse model of depression. *Molecular Psychiatry, 9*(3), 278–286, 224.

Alt, S. R., Turner, J. D., Klok, M. D., Meijer, O. C., Lakke, E. A., Derijk, R. H., et al. (2010). Differential expression of glucocorticoid receptor transcripts in major depressive disorder is not epigenetically programmed. *Psychoneuroendocrinology, 35*(4), 544–556.

Amrein, I., Isler, K., & Lipp, H. P. (September 2011). Comparing adult hippocampal neurogenesis in mammalian species and orders: influence of chronological age and life history stage. *European Journal of Neuroscience, 34*(6), 978–987.

Anacker, C., Cattaneo, A., Luoni, A., Musaelyan, K., Zunszain, P. A., Milanesi, E., et al. (2013). Glucocorticoid-related molecular signaling pathways regulating hippocampal neurogenesis. *Neuropsychopharmacology, 38*, 872–883.

Anacker, C., & Pariante, C. M. (2012). Can adult neurogenesis buffer stress responses and depressive behaviour? *Molecular Psychiatry, 17*, 9–10.

Ashton, R. S., Conway, A., Pangarkar, C., Bergen, J., Lim, K. I., Shah, P., et al. (2012). Astrocytes regulate adult hippocampal neurogenesis through ephrin-B signaling. *Nature Neuroscience, 15*(10), 1399–1406.

Balu, D. T., & Lucki, I. (2009). Adult hippocampal neurogenesis: regulation, functional implications, and contribution to disease pathology. *Neuroscience and Biobehavioral Reviews, 33*, 232–252.

Banasr, M., & Duman, R. S. (2008). Glial loss in the prefrontal cortex is sufficient to induce depressive-like behaviors. *Biological Psychiatry, 64*, 863–870.

Baram, T. Z., Davis, E. P., Obenaus, A., Sandman, C. A., Small, S. L., Solodkin, A., et al. (2012). Fragmentation and unpredictability of early-life experience in mental disorders. *American Journal of Psychiatry, 169*, 907–915.

Bielefeld, P., van Vliet, E. A., Gorter, J. A., Lucassen, P. J., & Fitzsimons, C. P. (January 2014). Different subsets of newborn granule cells: a possible role in epileptogenesis? *European Journal of Neuroscience, 39*(1), 1–11.

Bilbo, S. D., & Schwarz, J. M. (2009). Early-life programming of later-life brain and behavior: a critical role for the immune system. *Frontiers in Behavioral Neuroscience, 3*, 1–14.

Bland, S. T., Beckley, J. T., Young, S., Tsang, V., Watkins, L. R., Maier, S. F., et al. (2010). Enduring consequences of early-life infection on glial and neural cell genesis within cognitive regions of the brain. *Brain Behavior and Immunity, 24*, 329–338.

Boku, S., Toda, H., Nakagawa, S., Kato, A., Inoue, T., Koyama, T., et al. (2014). Neonatal maternal separation alters the capacity of adult neural precursor cells to differentiate into neurons via methylation of retinoic acid receptor gene promoter. *Biological Psychiatry, 77*(4), 335–344.

Boldrini, M., Hen, R., Underwood, M. D., Rosoklija, G. B., Dwork, A. J., Mann, J. J., et al. (2012). Hippocampal angiogenesis and progenitor cell proliferation are increased with antidepressant use in major depression. *Biological Psychiatry, 72*, 562–571.

Boldrini, M., Underwood, M. D., Hen, R., Rosoklija, G. B., Dwork, A. J., John Mann, J., et al. (2009). Antidepressants increase neural progenitor cells in the human hippocampus. *Neuropsychopharmacology, 34*, 2376–2389.

Bourdeau, I., Bard, C., Noël, B., Leclerc, I., Cordeau, M.-P., Bélair, M., et al. (2002). Loss of brain volume in endogenous Cushing's syndrome and its reversibility after correction of hypercortisolism. *Journal of Clinical Endocrinology and Metabolism, 87*, 1949–1954.

Braun, S. M., & Jessberger, S. (May 2014). Adult neurogenesis: mechanisms and functional significance. *Development, 141*(10), 1983–1986.

Bruel-Jungerman, E., Lucassen, P. J., & Francis, F. (2011). Cholinergic influences on cortical development and adult neurogenesis. *Behavioural Brain Research, 221*, 379–388.

Brunson, K. L., Kramár, E., Lin, B., Chen, Y., Colgin, L. L., Yanagihara, T. K., et al. (2005). Mechanisms of late-onset cognitive decline after early-life stress. *Journal of Neuroscience, 25*, 9328–9338.

Butovsky, O., Ziv, Y., Schwartz, A., Landa, G., Talpalar, A. E., Pluchino, S., et al. (2006). Microglia activated by IL-4 or IFN-gamma differentially induce neurogenesis and oligodendrogenesis from adult stem/progenitor cells. *Molecular and Cellular Neuroscience, 31*, 149–160.

Cavegn, N., van Dijk, R. M., Menges, D., Brettschneider, H., Phalanndwa, M., Chimimba, C. T., et al. (2013). Habitat-specific shaping of proliferation and neuronal differentiation in adult hippocampal neurogenesis of wild rodents. *Frontiers in Neuroscience, 7*, 59.

Chetty, S., Friedman, A. R., Taravosh-Lahn, K., Kirby, E. D., Mirescu, C., Guo, F., et al. (2014). Stress and glucocorticoids promote oligodendrogenesis in the adult hippocampus. *Molecular Psychiatry, 19*(12), 1275–1283.

Cho, K. O., Lybrand, Z. R., Ito, N., Brulet, R., Tafacory, F., Zhang, L., et al. (2015). Aberrant hippocampal neurogenesis contributes to epilepsy and associated cognitive decline. *Nature Communications, 6*, 6606.

Clelland, C. D., Choi, M., Romberg, C., Clemenson, G. D., Jr., Fragniere, A., Tyers, P., et al. (2009). A functional role for adult hippocampal neurogenesis in spatial pattern separation. *Science, 325*(5937), 210–213.

Cobb, J. A., Simpson, J., Mahajan, G. J., Overholser, J. C., Jurjus, G. J., Dieter, L., et al. (2013). Hippocampal volume and total cell numbers in major depressive disorder. *Journal of Psychiatric Research, 47*, 299–306.

Coe, C. L., Kramer, M., Czéh, B., Gould, E., Reeves, A. J., Kirschbaum, C., et al. (2003). Prenatal stress diminishes neurogenesis in the dentate gyrus of juvenile rhesus monkeys. *Biological Psychiatry, 54*, 1025–1034.

Couillard-Despres, S., Wuertinger, C., Kandasamy, M., Caioni, M., Stadler, K., Aigner, R., et al. (2009). Ageing abolishes the effects of fluoxetine on neurogenesis. *Molecular Psychiatry, 14*, 856–864.

Cowen, D. S., Takase, L. F., Fornal, C. A., & Jacobs, B. L. (2008). Age-dependent decline in hippocampal neurogenesis is not altered by chronic treatment with fluoxetine. *Brain Research, 1228*, 14–19.

Czéh, B., Fuchs, E., Wiborg, O., & Simon, M. (April 17, 2015). Animal models of major depression and their clinical implications. *Progress in Neuro-Psychopharmacology and Biological Psychiatry, 64*, 293–310.

Czéh, B., & Lucassen, P. J. (2007). What causes the hippocampal volume decrease in depression? Are neurogenesis, glial changes and apoptosis implicated? *European Archives of Psychiatry and Clinical Neuroscience, 257*(5), 250–260.

Czéh, B., Michaelis, T., Watanabe, T., Frahm, J., de Biurrun, G., van Kampen, M., et al. (2001). Stress-induced changes in cerebral metabolites, hippocampal volume, and cell proliferation are prevented by antidepressant treatment with tianeptine. *Proceedings of the National Academy of Sciences of the United States of America, 98*, 12796–12801.

Czéh, B., Pudovkina, O., van der Hart, M. G. C., Simon, M., Heilbronner, U., Michaelis, T., et al. (2005). Examining SLV-323, a novel NK1 receptor antagonist, in a chronic psychosocial stress model for depression. *Psychopharmacology (Berlin), 180*, 548–557.

Czéh, B., Simon, M., Schmelting, B., Hiemke, C., & Fuchs, E. (2006). Astroglial plasticity in the hippocampus is affected by chronic psychosocial stress and concomitant fluoxetine treatment. *Neuropsychopharmacology, 31*, 1616–1626.

Czéh, B., Welt, T., Fischer, A. K., Erhardt, A., Schmitt, W., Müller, M. B., et al. (2002). Chronic psychosocial stress and concomitant repetitive transcranial magnetic stimulation: effects on stress hormone levels and adult hippocampal neurogenesis. *Biological Psychiatry, 52*, 1057–1065.

Dagyte, G., Van der Zee, E. A., Postema, F., Luiten, P. G., Den Boer, J. A., Trentani, A., et al. (2009). Chronic but not acute foot-shock stress leads to temporary suppression of cell proliferation in rat hippocampus. *Neuroscience, 162*(4), 904–913.

Datson, N. A., Speksnijder, N., Mayer, J. L., Steenbergen, P. J., Korobko, O., Goeman, J., et al. (2012). The transcriptional response to chronic stress and glucocorticoid receptor blockade in the hippocampal dentate gyrus. *Hippocampus, 22*, 359–371.

David, D. J., Samuels, B. A., Rainer, Q., Wang, J. W., Marsteller, D., Mendez, I., et al. (2009). Neurogenesis-dependent and -independent effects of fluoxetine in an animal model of anxiety/depression. *Neuron, 62*(4), 479–493.

Déry, N., Goldstein, A., & Becker, S. (June 15, 2015). A role for adult hippocampal neurogenesis at multiple time scales: a study of recent and remote memory in humans. *Behavioral Neuroscience, 129*, 435–449.

Djavadian, R. L. (2004). Serotonin and neurogenesis in the hippocampal dentate gyrus of adult mammals. *Acta Neurobiologiae Experimentalis (Warsaw), 64*, 189–200.

Domínguez-Escribà, L., Hernández-Rabaza, V., Soriano-Navarro, M., Barcia, J. A., Romero, F. J., García-Verdugo, J. M., et al. (2006). Chronic cocaine exposure impairs progenitor proliferation but spares survival and maturation of neural precursors in adult rat dentate gyrus. *European Journal of Neuroscience, 24*(2), 586–594.

Drevets, W. C., Price, J. L., & Furey, M. L. (2008). Brain structural and functional abnormalities in mood disorders: implications for neurocircuitry models of depression. *Brain Structure and Function, 213*, 93–118.

Droste, S. K., Gesing, A., Ulbricht, S., Müller, M. B., Linthorst, A. C. E., & Reul, J. M. H.M. (2003). Effects of long-term voluntary exercise on the mouse hypothalamic-pituitary-adrenocortical axis. *Endocrinology, 144*, 3012–3023.

Duman, R. S. (2004). Depression: a case of neuronal life and death? *Biological Psychiatry, 56*, 140–145.

Ehninger, D., & Kempermann, G. (2006). Paradoxical effects of learning the Morris water maze on adult hippocampal neurogenesis in mice may be explained by a combination of stress and physical activity. *Genes, Brain and Behavior, 5*, 29–39.

Eisch, A. J., & Petrik, D. (2012). Depression and hippocampal neurogenesis: a road to remission? *Science, 338*(6103), 72–75.

Ekdahl, C. T., Kokaia, Z., & Lindvall, O. (2009). Brain inflammation and adult neurogenesis: the dual role of microglia. *Neuroscience, 158*, 1021–1029.

Encinas, J. M., Vaahtokari, A., & Enikolopov, G. (May 23, 2006). Fluoxetine targets early progenitor cells in the adult brain. *Proceedings of the National Academy of Sciences of the United States of America, 103*(21), 8233–8238.

Erdmann, G., Berger, S., & Schütz, G. (2008). Genetic dissection of glucocorticoid receptor function in the mouse brain. *Journal of Neuroendocrinology, 20*, 655–659.

Felice, D., O'Leary, O. F., Cryan, J. F., Dinan, T. G., Gardier, A. M., Sánchez, C., et al. (2015). When ageing meets the blues: Are current antidepressants effective in depressed aged patients? *Neuroscience and Biobehavioral Reviews, 55*, 478–497.

Fitzsimons, C. P., van Bodegraven, E., Schouten, M., Lardenoije, R., Kompotis, K., Kenis, G., et al. (2014). Epigenetic regulation of adult neural stem cells: implications for Alzheimer's disease. *Molecular Neurodegeneration, 9*, 25.

Fitzsimons, C. P., van Hooijdonk, L. W. A., Schouten, M., Zalachoras, I., Brinks, V., Zheng, T., et al. (2013). Knockdown of the glucocorticoid receptor alters functional integration of newborn neurons in the adult hippocampus and impairs fear-motivated behavior. *Molecular Psychiatry, 18*, 993–1005.

Galea, L. A. (2008). Gonadal hormone modulation of neurogenesis in the dentate gyrus of adult male and female rodents. *Brain Research Reviews, 57*, 332–341.

Ge, S., Yang, C.-H., Hsu, K.-S., Ming, G.-L., & Song, H. (2007). A critical period for enhanced synaptic plasticity in newly generated neurons of the adult brain. *Neuron, 54*, 559–566.

Gould, E. (2007). How widespread is adult neurogenesis in mammals? *Nature Reviews Neuroscience, 8*(6), 481–488.

Gould, E., McEwen, B. S., Tanapat, P., Galea, L. A., & Fuchs, E. (1997). Neurogenesis in the dentate gyrus of the adult tree shrew is regulated by psychosocial stress and NMDA receptor activation. *Journal of Neuroscience, 17*, 2492–2498.

Guadagno, J., Xu, X., Karajgikar, M., Brown, A., & Cregan, S. P. (March 14, 2013). Microglia-derived TNFα induces apoptosis in neural precursor cells via transcriptional activation of the Bcl-2 family member Puma. *Cell Death and Disease, 4*, e538.

Harris, A. P., Holmes, M. C., de Kloet, E. R., Chapman, K. E., & Seckl, J. R. (2013). Mineralocorticoid and glucocorticoid receptor balance in control of HPA axis and behaviour. *Psychoneuroendocrinology, 38*, 648–658.

Heim, C., Newport, D., Mletzko, T., Miller, A., & Nemeroff, C. (2008). The link between childhood trauma and depression: insights from HPA axis studies in humans. *Psychoneuroendocrinology, 33*, 693–710.

Heine, V. M., Maslam, S., Joëls, M., & Lucassen, P. J. (2004). Increased P27KIP1 protein expression in the dentate gyrus of chronically stressed rats indicates G1 arrest involvement. *Neuroscience, 129*, 593–601.

Heine, V. M., Maslam, S., Zareno, J., Joels, M., & Lucassen, P. J. (2004). Suppressed proliferation and apoptotic changes in the rat dentate gyrus after acute and chronic stress are reversible. *European Journal of Neuroscience, 19*, 131–144.

Heine, V. M., Zareno, J., Maslam, S., Joels, M., & Lucassen, P. J. (2005). Chronic stress in the adult dentate gyrus reduces cell proliferation near the vasculature and VEGF and Flk-1 protein expression. *European Journal of Neuroscience, 21*, 1304–1314.

Henn, F. A., & Vollmayr, B. (2004). Neurogenesis and depression: etiology or epiphenomenon? *Biological Psychiatry, 56*(3), 146–150.

Hinwood, M., Morandini, J., Day, T. A., & Walker, F. R. (2012). Evidence that microglia mediate the neurobiological effects of chronic psychological stress on the medial prefrontal cortex. *Cerebral Cortex, 22*, 1442–1454.

Hodes, G. E., Hill-Smith, T. E., Suckow, R. F., Cooper, T. B., & Lucki, I. (2009). Sex specific effects of chronic fluoxetine treatment on neuroplasticity and pharmacokinetics in mice. *Journal of Pharmacology and Experimental Therapeutics, 332*(1), 266–273.

Hodes, G. E., Yang, L., Van Kooy, J., Santollo, J., & Shors, T. J. (2009). Prozac during puberty: distinctive effects on neurogenesis as a function of age and sex. *Neuroscience, 163*(2), 609–617.

Holick, K. A., Lee, D. C., Hen, R., & Dulawa, S. C. (2008). Behavioral effects of chronic fluoxetine in Balb/cJ mice do not require adult hippocampal neurogenesis or the serotonin 1A receptor. *Neuropsychopharmacology, 33*, 406–417.

Holmes, M. M., Galea, L. A. M., Mistlberger, R. E., & Kempermann, G. (2004). Adult hippocampal neurogenesis and voluntary running activity: circadian and dose-dependent effects. *Journal of Neuroscience Research, 76*, 216–222.

Hu, P., Oomen, C., van Dam, A. M., Wester, J., Zhou, J. N., Joëls, M., et al. (2012). A single-day treatment with mifepristone is sufficient to normalize chronic glucocorticoid induced suppression of hippocampal cell proliferation. *PLoS One, 7*, e46224.

Huang, Y., Coupland, N. J., Lebel, R. M., Carter, R., Seres, P., Wilman, A. H., et al. (2013). Structural changes in hippocampal subfields in major depressive disorder: a high-field magnetic resonance imaging study. *Biological Psychiatry, 74*, 62–68.

Huang, G. J., Bannerman, D., & Flint, J. (2008). Chronic fluoxetine treatment alters behavior, but not adult hippocampal neurogenesis, in BALB/cJ mice. *Molecular Psychiatry, 13*, 119–121.

Hulshof, H. J., Novati, A., Luiten, P. G., den Boer, J. A., & Meerlo, P. (2012). Despite higher glucocorticoid levels and stress responses in female rats, both sexes exhibit similar stress-induced changes in hippocampal neurogenesis. *Behavioural Brain Research, 234*(2), 357–364.

Ivy, A. S., Rex, C. S., Chen, Y., Dubé, C., Maras, P. M., Grigoriadis, D. E., et al. (2010). Hippocampal dysfunction and cognitive impairments provoked by chronic early-life stress involve excessive activation of CRH receptors. *Journal of Neuroscience, 30*, 13005–13015.

Jacobs, B. L., van Praag, H., & Gage, F. H. (2000). Adult brain neurogenesis and psychiatry: a novel theory of depression. *MolecularPsychiatry, 5*(3), 262–269.

Jakubs, K., Bonde, S., Iosif, R. E., Ekdahl, C. T., Kokaia, Z., Kokaia, M., et al. (2008). Inflammation regulates functional integration of neurons born in adult brain. *Journal of Neuroscience, 28*, 12477–12488.

Jessberger, S., Clark, R. E., Broadbent, N. J., Clemenson, G. D., Jr., Consiglio, A., Lie, D. C., et al. (January 29, 2009). Dentate gyrus-specific knockdown of adult neurogenesis impairs spatial and object recognition memory in adult rats. *Learning and Memory, 16*(2), 147–154.

Jessberger, S., & Gage, F. H. (2014). Adult neurogenesis: bridging the gap between mice and humans. *Trends in Cell Biology, 24*(10), 558–563.

Joels, M., & Baram, T. Z. (2009). The neuro-symphony of stress. *Nature Reviews Neuroscience, 10*, 459–466.

Joels, M., Karst, H., Krugers, H. J., & Lucassen, P. J. (2007). Chronic stress: implications for neuronal morphology, function and neurogenesis. *Frontiers in Neuroendocrinology, 28*, 72–96.

Joels, M., Sarabdjitsingh, R. A., & Karst, H. (2012). Unraveling the time domains of corticosteroid hormone influences on brain activity: rapid, slow, and chronic modes. *Pharmacological Reviews, 64*, 901–938.

Johansson, S., Price, J., & Modo, M. (2008). Effect of inflammatory cytokines on major histocompatibility complex expression and differentiation of human neural stem/progenitor cells. *Stem Cells, 26*, 2444–2454.

Kannangara, T. S., Lucero, M. J., Gil-Mohapel, J., Drapala, R. J., Simpson, J. M., Christie, B. R., et al. (2011). Running reduces stress and enhances cell genesis in aged mice. *Neurobiology of Aging, 32*(12), 2279–2286.

Karg, K., Burmeister, M., Shedden, K., & Sen, S. (May 2011). The serotonin transporter promoter variant (5-HTTLPR), stress, and depression meta-analysis revisited: evidence of genetic moderation. *Archives of General Psychiatry, 68*(5), 444–454.

Kempermann, G. (2008). The neurogenic reserve hypothesis: what is adult hippocampal neurogenesis good for? *Trends in Neurosciences, 31*(4), 163–169.

Kempermann, G. (2012). New neurons for 'survival of the fittest'. *Nature Reviews Neuroscience, 13*, 727–736.

Kempermann, G., Fabel, K., Ehninger, D., Babu, H., Leal-Galicia, P., Garthe, A., et al. (2010). Why and how physical activity promotes experience-induced brain plasticity. *Frontiers in Neuroscience, 4*, 189.

Kempermann, G., Krebs, J., & Fabel, K. (2008). The contribution of failing adult hippocampal neurogenesis to psychiatric disorders. *Current Opinion in Psychiatry, 21*, 290–295.

Kempermann, G., & Kronenberg, G. (2003). Depressed new neurons–adult hippocampal neurogenesis and a cellular plasticity hypothesis of major depression. *Biological Psychiatry, 54*(5), 499–503.

Kempton, M. J., Salvador, Z., Munafò, M. R., Geddes, J. R., Simmons, A., Frangou, S., et al. (2011). Structural neuroimaging studies in major depressive disorder. Meta-analysis and comparison with bipolar disorder. *Archives of General Psychiatry, 68*, 675–690.

Kheirbek, M. A., & Hen, R. (January 2011). Dorsal vs ventral hippocampal neurogenesis: implications for cognition and mood. *Neuropsychopharmacology*, *36*(1), 373–374.

Kheirbek, M. A., Klemenhagen, K. C., Sahay, A., & Hen, R. (December 2012). Neurogenesis and generalization: a new approach to stratify and treat anxiety disorders. *Nature Neuroscience*, *15*(12), 1613–1620.

Kim, J. I., Lee, J. W., Lee, Y. A., Lee, D. H., Han, N. S., Choi, Y. K., et al. (2013). Sexual activity counteracts the suppressive effects of chronic stress on adult hippocampal neurogenesis and recognition memory. *Brain Research*, *1538*, 26–40.

Kirby, E. D., Muroy, S. E., Sun, W. G., Covarrubias, D., Leong, M. J., Barchas, L. A., et al. (April 16, 2013). Acute stress enhances adult rat hippocampal neurogenesis and activation of newborn neurons via secreted astrocytic FGF2. *Elife*, *2*, e00362.

de Kloet, E. R., Joëls, M., & Holsboer, F. (2005). Stress and the brain: from adaptation to disease. *Nature Reviews Neuroscience*, *6*, 463–475.

Klok, M. D., Alt, S. R., Irurzun Lafitte, A. J., Turner, J. D., Lakke, E. A., Huitinga, I., et al. (July 2011). Decreased expression of mineralocorticoid receptor mRNA and its splice variants in postmortem brain regions of patients with major depressive disorder. *Journal of Psychiatric Research*, *45*(7), 871–878.

Klok, M. D., Giltay, E. J., Van der Does, A. J., Geleijnse, J. M., Antypa, N., Penninx, B. W., et al. (2011). A common and functional mineralocorticoid receptor haplotype enhances optimism and protects against depression in females. *Translational Psychiatry*, *1*, e62.

Klomp, A., Václavů, L., Meerhoff, G. F., Reneman, L., & Lucassen, P. J. (2014). Effects of chronic fluoxetine treatment on neurogenesis and tryptophan hydroxylase expression in adolescent and adult rats. *PLoS One*, *9*(5) e97603.

Kobilo, T., Liu, Q. R., Gandhi, K., Mughal, M., Shaham, Y., & van Praag, H. (2011). Running is the neurogenic and neurotrophic stimulus in environmental enrichment. *Learning and Memory*, *18*(9), 605–609.

Koehl, M., van der Veen, R., Gonzales, D., Piazza, P. V., & Abrous, D. N. (2012). Interplay of maternal care and genetic influences in programming adult hippocampal neurogenesis. *Biological Psychiatry*, *72*(4), 282–289.

Koo, J. W., & Duman, R. S. (2008). IL-1beta is an essential mediator of the antineurogenic and anhedonic effects of stress. *Proceedings of the National Academy of Sciences of the United States of America*, *105*, 751–756.

Koolhaas, J. M., Bartolomucci, A., Buwalda, B., de Boer, S. F., Flügge, G., Korte, S. M., et al. (2011). Stress revisited: a critical evaluation of the stress concept. *Neuroscience and Biobehavioral Reviews*, *35*, 1291–1301.

Korosi, A., Naninck, E. F. G., Oomen, C. A., Schouten, M., Krugers, H., Fitzsimons, C., et al. (2012). Early-life stress mediated modulation of adult neurogenesis and behavior. *Behavioural Brain Research*, *227*, 400–409.

Kreisel, T., Frank, M. G., Licht, T., Reshef, R., Ben-Menachem-Zidon, O., Baratta, M. V., et al. (2014). Dynamic microglial alterations underlie stress-induced depressive-like behavior and suppressed neurogenesis. *Molecular Psychiatry*, *19*, 699–709.

Kretz, O., Reichardt, H. M., Schütz, G., & Bock, R. (1999). Corticotropin-releasing hormone expression is the major target for glucocorticoid feedback-control at the hypothalamic level. *Brain Research*, *818*, 488–491.

Krzisch, M., Temprana, S. G., Mongiat, L. A., Armida, J., Schmutz, V., Virtanen, M. A., et al. (July 2015). Pre-existing astrocytes form functional perisynaptic processes on neurons generated in the adult hippocampus. *Brain Structure and Function*, *220*(4), 2027–2042.

Lehmann, M. L., Brachman, R. A., Martinowich, K., Schloesser, R. J., & Herkenham, M. (2013). Glucocorticoids orchestrate divergent effects on mood through adult neurogenesis. *Journal of Neuroscience*, *33*, 2961–2972.

Lemaire, V., Koehl, M., Le Moal, M., & Abrous, D. N. (2000). Prenatal stress produces learning deficits associated with an inhibition of neurogenesis in the hippocampus. *Proceedings of the National Academy of Sciences of the United States of America*, *97*, 11032–11037.

Leuner, B. (2010). Parenting and plasticity. *Trends in Neurosciences, 33*, 465–473.

Liston, C., Cichon, J. M., Jeanneteau, F., Jia, Z., Chao, M. V., & Gan, W.-B. (2013). Circadian glucocorticoid oscillations promote learning-dependent synapse formation and maintenance. *Nature Publishing Group, 16*, 698–705.

Loi, M., Koricka, S., Lucassen, P. J., & Joëls, M. (2014). Age- and sex-dependent effects of early life stress on hippocampal neurogenesis. *Frontiers in Endocrinology (Lausanne), 5*, 13.

Loman, M. M., Gunnar, M. R., & Early Experience, Stress, and Neurobehavioral Development Center (2010). Early experience and the development of stress reactivity and regulation in children. *Neuroscience and Biobehavioral Reviews, 34*, 867–876.

Lorenzetti, V., Allen, N. B., Fornito, A., & Yücel, A. (2009). Structural brain abnormalities in major depressive disorder: a selective review of recent MRI studies. *Journal of Affective Disorders, 117*, 1–17.

Lucassen, P. J., Bosch, O. J., Jousma, E., Krömer, S. A., Andrew, R., Seckl, J. R., et al. (2009). Prenatal stress reduces postnatal neurogenesis in rats selectively bred for high, but not low, anxiety: possible key role of placental 11β-hydroxysteroid dehydrogenase type 2. *European Journal of Neuroscience, 29*, 97–103.

Lucassen, P. J., Fitzsimons, C. P., Korosi, A., Joëls, M., Belzung, C., & Abrous, D. N. (2013). Stressing new neurons into depression? *Molecular Psychiatry, 18*, 396–397.

Lucassen, P. J., Meerlo, P., Naylor, A. S., van Dam, A. M., Dayer, A. G., Fuchs, E., et al. (2010). Regulation of adult neurogenesis by stress, sleep disruption, exercise and inflammation: Implications for depression and antidepressant action. *European Neuropsychopharmacology, 20*, 1–17.

Lucassen, P. J., Naninck, E. F., van Goudoever, J. B., Fitzsimons, C., Joels, M., & Korosi, A. (2013). Perinatal programming of adult hippocampal structure and function; emerging roles of stress, nutrition and epigenetics. *Trends in Neurosciences, 36*, 621–631.

Lucassen, P. J., Pruessner, J., Sousa, N., Almeida, O. F., Van Dam, A. M., Rajkowska, G., et al. (2014). Neuropathology of stress. *Acta Neuropathologica, 127*, 109–135.

Lucassen, P. J., Stumpel, M. W., Wang, Q., & Aronica, E. (2010). Decreased numbers of progenitor cells but no response to antidepressant drugs in the hippocampus of elderly depressed patients. *Neuropharmacology, 58*, 940–949.

MacQueen, G. M., Yucel, K., Taylor, V. H., Macdonald, K., & Joffe, R. (2008). Posterior hippocampal volumes are associated with remission rates in patients with major depressive disorder. *Biological Psychiatry, 64*, 880–883.

Malberg, J. E., & Duman, R. S. (2003). Cell proliferation in adult hippocampus is decreased by inescapable stress: reversal by fluoxetine treatment. *Neuropsychopharmacology, 28*, 1562–1571.

Manganas, L. N., Zhang, X., Li, Y., Hazel, R. D., Smith, S. D., Wagshul, M. E., et al. (2007). Magnetic resonance spectroscopy identifies neural progenitor cells in the live human brain. *Science, 318*, 980–985.

Marlatt, M. W., Lucassen, P. J., & van Praag, H. (2010). Comparison of neurogenic effects of fluoxetine, duloxetine and running in mice. *Brain Research, 1341*, 93–99.

Marlatt, M. W., Potter, M. C., Lucassen, P. J., & van Praag, H. (2012). Running throughout middle-age improves memory function, hippocampal neurogenesis, and BDNF levels in female C57BL/6J mice. *Developmental Neurobiology, 72*(6), 943–952.

Maselko, J., Kubzansky, L., Lipsitt, L., & Buka, S. L. (2011). Mother's affection at 8 months predicts emotional distress in adulthood. *Journal of Epidemiology and Community Health, 65*, 621–625.

Maya Vetencourt, J. F., Sale, A., Viegi, A., Baroncelli, L., De Pasquale, R., O'Leary, O. F., et al. (2008). The antidepressant fluoxetine restores plasticity in the adult visual cortex. *Science, 320*(5874), 385–388. http://dx.doi.org/10.1126/science.1150516.

Mayer, J. L., Klumpers, L., Maslam, S., de Kloet, E. R., Joëls, M., & Lucassen, P. J. (2006). Brief treatment with the glucocorticoid receptor antagonist mifepristone normalises the corticosterone-induced reduction of adult hippocampal neurogenesis. *Journal of Neuroendocrinology, 18*, 629–631.

Medina, A., Seasholtz, A. F., Sharma, V., Burke, S., Bunney, W., Myers, R. M., et al. (2013). Glucocorticoid and mineralocorticoid receptor expression in the human hippocampus in major depressive disorder. *Journal of Psychiatric Research*, *47*, 307–314.

Mendez-David, I., David, D. J., Darcet, F., Wu, M. V., Kerdine-Römer, S., Gardier, A. M., et al. (2014). Rapid anxiolytic effects of a 5-HT$_4$ receptor agonist are mediated by a neurogenesis-independent mechanism. *Neuropsychopharmacology*, *39*(6), 1366–1378.

Miller, J. A., Nathanson, J., Franjic, D., Shim, S., Dalley, R. A., Shapouri, S., et al. (2013). Conserved molecular signatures of neurogenesis in the hippocampal subgranular zone of rodents and primates. *Development*, *140*, 4633–4644.

Mirescu, C., & Gould, E. (2006). Stress and adult neurogenesis. *Hippocampus*, *16*, 233–238.

Mirescu, C., Peters, J. D., & Gould, E. (2004). Early life experience alters response of adult neurogenesis to stress. *Nature Neuroscience*, *7*, 841–846.

Montaron, M. F., Piazza, P. V., Aurousseau, C., Urani, A., Le Moal, M., & Abrous, D. N. (2003). Implication of corticosteroid receptors in the regulation of hippocampal structural plasticity. *European Journal of Neuroscience*, *18*, 3105–3111.

Morris, G. P., Clark, I. A., Zinn, R., & Vissel, B. (2013). Microglia: a new frontier for synaptic plasticity, learning and memory, and neurodegenerative disease research. *Neurobiology of Learning and Memory*, *105*, 40–53.

Nacher, J., & McEwen, B. S. (2006). The role of N-methyl-D-asparate receptors in neurogenesis. *Hippocampus*, *16*, 267–270.

Naninck, E. F. G., Hoeijmakers, L., Kakava-Georgiadou, N., Meesters, A., Lazic, S. E., Lucassen, P. J., et al. (2015). Chronic early-life stress alters developmental and adult neurogenesis and impairs cognitive function in mice. *Hippocampus*, *25*(3), 309–328.

Navailles, S., Hof, P. R., & Schmauss, C. (2008). Antidepressant drug-induced stimulation of mouse hippocampal neurogenesis is age-dependent and altered by early life stress. *Journal of Comparative Neurology*, *509*(4), 372–381.

Oomen, C. A., Bekinschtein, P., Kent, B. A., Saksida, L. M., & Bussey, T. J. (2014). Adult hippocampal neurogenesis and its role in cognition. *WIREs Cognitive Science*, *5*, 573–587. http://dx.doi.org/10.1002/wcs.1304.

Oomen, C. A., Girardi, C. E. N., Cahyadi, R., Verbeek, E. C., Krugers, H., Joels, M., et al. (2009). Opposite effects of early maternal deprivation on neurogenesis in male versus female rats. *PLoS One*, *4*, e3675.

Oomen, C. A., Hvoslef-Eide, M., Heath, C. J., Mar, A. C., Horner, A. E., Bussey, T. J., et al. (2013). The touchscreen operant platform for testing working memory and pattern separation in rats and mice. *Nature Protocols*, *8*, 2006–2021.

Oomen, C. A., Mayer, J. L., de Kloet, E. R., Joels, M., & Lucassen, P. J. (2007). Brief treatment with the glucocorticoid receptor antagonist mifepristone normalizes the reduction in neurogenesis after chronic stress. *European Journal of Neuroscience*, *26*, 3395–3401.

Oomen, C. A., Soeters, H., Audureau, N., Vermunt, L., van Hasselt, F. N., Manders, E. M. M., et al. (2010). Severe early life stress hampers spatial learning and neurogenesis, but improves hippocampal synaptic plasticity and emotional learning under high-stress conditions in adulthood. *Journal of Neuroscience*, *30*, 6635–6645.

Opendak, M., & Gould, E. (October 4, 2011). New neurons maintain efficient stress recovery. *Cell Stem Cell*, *9*(4), 287–288.

Opendak, M., & Gould, E. (2015). Adult neurogenesis: a substrate for experience-dependent change. *Trends in Cognitive Sciences*, *19*(3), 151–161.

Parihar, V. K., Hattiangady, B., Kuruba, R., Shuai, B., & Shetty, A. K. (2011). Predictable chronic mild stress improves mood, hippocampal neurogenesis and memory. *Molecular Psychiatry*, *16*, 171–183.

Perera, T. D., Dwork, A. J., Keegan, K. A., Thirumangalakudi, L., Lipira, C. M., Joyce, N., et al. (2011). Necessity of hippocampal neurogenesis for the therapeutic action of antidepressants in adult nonhuman primates. *PLoS One*, *6*, e17600.

Petrik, D., Lagace, D. C., & Eisch, A. J. (January 2012). The neurogenesis hypothesis of affective and anxiety disorders: are we mistaking the scaffolding for the building? *Neuropharmacology*, *62*(1), 21–34.

Pham, K., Nacher, J., Hof, P. R., & McEwen, B. S. (2003). Repeated restraint stress suppresses neurogenesis and induces biphasic PSA-NCAM expression in the adult rat dentate gyrus. *European Journal of Neuroscience*, *17*, 879–886.

Pruessner, J. C., Dedovic, K., Pruessner, M., Lord, C., Buss, C., Collins, L., et al. (2010). Stress regulation in the central nervous system: evidence from structural and functional neuroimaging studies in human populations – 2008 Curt Richter Award Winner. *Psychoneuroendocrinology*, *35*, 179–191.

Qi, X.-R., Kamphuis, W., Wang, S., Wang, Q., Lucassen, P. J., Zhou, J.-N., et al. (2013). Aberrant stress hormone receptor balance in the human prefrontal cortex and hypothalamic paraventricular nucleus of depressed patients. *Psychoneuroendocrinology*, *38*, 863–870.

Qian, X., Droste, S. K., Lightman, S. L., Reul, J. M., & Linthorst, A. C. E. (2012). Circadian and ultradian rhythms of free glucocorticoid hormone are highly synchronized between the blood, the subcutaneous tissue, and the brain. *Endocrinology*, *153*, 4346–4353.

Raadsheer, F. C., van Heerikhuize, J. J., Lucassen, P. J., Hoogendijk, W. J., Tilders, F. J., & Swaab, D. F. (September 1995). Corticotropin-releasing hormone mRNA levels in the paraventricular nucleus of patients with Alzheimer's disease and depression. *American Journal of Psychiatry*, *152*(9), 1372–1376.

Reif, A., Fritzen, S., Finger, M., Strobel, A., & Lauer, M. (2006). Neural stem cell proliferation is decreased in schizophrenia, but not in depression. *Molecular Psychiatry*, *11*, 514–522.

Revest, J. M., Dupret, D., Koehl, M., Funk-Reiter, C., Grosjean, N., Piazza, P. V., et al. (2009). Adult hippocampal neurogenesis is involved in anxiety related behaviors. *Molecular Psychiatry*, *14*, 959–967.

Richetin, K., Leclerc, C., Toni, N., Gallopin, T., Pech, S., Roybon, L., et al. (2015). Genetic manipulation of adult-born hippocampal neurons rescues memory in a mouse model of Alzheimer's disease. *Brain*, *138*(Pt 2), 440–455.

Ridder, S., Chourbaji, S., Hellweg, R., Urani, A., Zacher, C., Schmid, W., et al. (2005). Mice with genetically altered glucocorticoid receptor expression show altered sensitivity for stress-induced depressive reactions. *Journal of Neuroscience*, *25*, 6243–6250.

Risch, N., Herrell, R., Lehner, T., Liang, K. Y., Eaves, L., Hoh, J., et al. (2009). Interaction between the serotonin transporter gene (5-HTTLPR), stressful life events, and risk of depression: a meta-analysis. *Journal of the American Medical Association*, *301*(23), 2462–2471.

Sahay, A., & Hen, R. (2007). Adult hippocampal neurogenesis in depression. *Nature Neuroscience*, *10*, 1110–1115.

Sahay, A., Scobie, K. N., Hill, A. S., O'Carroll, C. M., Kheirbek, M. A., Burghardt, N. S., et al. (2011). Increasing adult hippocampal neurogenesis is sufficient to improve pattern separation. *Nature*, *472*(7344), 466–470.

Santarelli, L., Saxe, M., Gross, C., Surget, A., Battaglia, F., Dulawa, S., et al. (2003). Requirement of hippocampal neurogenesis for the behavioral effects of antidepressants. *Science*, *301*, 805–809.

Sapolsky, R. M. (October 2000). Glucocorticoids and hippocampal atrophy in neuropsychiatric disorders. *Archives of General Psychiatry*, *57*(10), 925–935.

Schmidt, H. D., & Duman, R. S. (2007). The role of neurotrophic factors in adult hippocampal neurogenesis, antidepressant treatments and animal models of depressive-like behavior. *Behavioural Pharmacology*, *18*, 391–418.

Schoenfeld, T. J., & Gould, E. (2013). Differential effects of stress and glucocorticoids on adult neurogenesis. *Current Topics in Behavioral Neurosciences*, *15*, 139–164.

Schouten, M., Buijink, M. R., Lucassen, P. J., & Fitzsimons, C. P. (2012). New neurons in aging brains: molecular control by small non-coding RNAs. *Frontiers in Neuroscience, 6*, 25. http://dx.doi.org/10.3389/fnins.2012.00025.

Schouten, M., Fratantoni, S. A., Hubens, C. J., Piersma, S. R., Pham, T. V., Bielefeld, P., et al. (2015). MicroRNA-124 and -137 cooperativity controls caspase-3 activity through BCL2L13 in hippocampal neural stem cells. *Scientific Reports, 5*, 12448. http://dx.doi.org/10.1038/srep12448.

Sierra, A., Encinas, J. M., Deudero, J. J. P., Chancey, J. H., Enikolopov, G., Overstreet-Wadiche, L. S., et al. (2010). Microglia shape adult hippocampal neurogenesis through apoptosis-coupled phagocytosis. *Cell Stem Cell, 7*, 483–495.

Sierra, A., Martín-Suárez, S., Valcárcel-Martín, R., Pascual-Brazo, J., Aelvoet, S. A., Abiega, O., et al. (2015). Neuronal hyperactivity accelerates depletion of neural stem cells and impairs hippocampal neurogenesis. *Cell Stem Cell, 16*(5), 488–503.

Sinclair, D., Tsai, S. Y., Woon, H. G., & Weickert, C. S. (2011). Abnormal glucocorticoid receptor mRNA and protein isoform expression in the prefrontal cortex in psychiatric illness. *Neuropsychopharmacology, 36*(13), 2698–2709.

Snyder, J. S., Choe, J. S., Clifford, M. A., Jeurling, S. I., Hurley, P., Brown, A., et al. (2009). Adult-born hippocampal neurons are more numerous, faster maturing, and more involved in behavior in rats than in mice. *Journal of Neuroscience., 29*(46), 14484–14495.

Snyder, J. S., Soumier, A., Brewer, M., Pickel, J., & Cameron, H. A. (2011). Adult hippocampal neurogenesis buffers stress responses and depressive behaviour. *Nature, 476*, 458–461.

Song, J., Zhong, C., Bonaguidi, M. A., Sun, G. J., Hsu, D., Gu, Y., et al. (September 6, 2012). Neuronal circuitry mechanism regulating adult quiescent neural stem-cell fate decision. *Nature, 489*(7414), 150–154.

Sousa, N., Cerqueira, J. J., & Almeida, O. (2008). Corticosteroid receptors and neuroplasticity. *Brain Research Reviews, 57*, 561–570.

Spijker, A. T., Giltay, E. J., van Rossum, E. F., Manenschijn, L., DeRijk, R. H., Haffmans, J., et al. (2011). Glucocorticoid and mineralocorticoid receptor polymorphisms and clinical characteristics in bipolar disorder patients. *Psychoneuroendocrinology, 36*(10), 1460–1469.

Spulber, S., Oprica, M., Bartfai, T., Winblad, B., & Schultzberg, M. (2008). Blunted neurogenesis and gliosis due to transgenic overexpression of human soluble IL-1ra in the mouse. *European Journal of Neuroscience, 27*, 549–558.

Starkman, M. N., Giordani, B., Gebarski, S. S., Berent, S., Schork, M. A., & Schteingart, D. E. (1999). Decrease in cortisol reverses human hippocampal atrophy following treatment of Cushing's disease. *Biological Psychiatry, 46*(12), 1595–1602.

Stockmeier, C. A., Mahajan, G. J., Konick, L. C., Overholser, J. C., Jurjus, G. J., Meltzer, H. Y., et al. (2004). Cellular changes in the postmortem hippocampus in major depression. *Biological Psychiatry, 56*, 640–650.

Surget, A., Saxe, M., Leman, S., Ibarguen-Vargas, Y., Chalon, S., Griebel, G., et al. (2008). Drug-dependent requirement of hippocampal neurogenesis in a model of depression and of antidepressant reversal. *Biological Psychiatry, 64*, 293–301.

Surget, A., Tanti, A., Leonardo, E. D., Laugeray, A., Rainer, Q., Touma, C., et al. (2011). Antidepressants recruit new neurons to improve stress response regulation. *Molecular Psychiatry, 16*, 1177–1188.

Swaab, D., Bao, A., & Lucassen, P. (2005). The stress system in the human brain in depression and neurodegeneration. *Ageing Research Reviews, 4*, 141–194.

Takamura, N., Nakagawa, S., Masuda, T., Boku, S., Kato, A., Song, N., et al. (2014). The effect of dopamine on adult hippocampal neurogenesis. *Progress in Neuropsychopharmacology and Biological Psychiatry, 50*, 116–124. http://dx.doi.org/10.1016/j.pnpbp.2013.12.011. Epub 2013 Dec 26.

Tauber, S. C., Bunkowski, S., Schlumbohm, C., Rühlmann, M., Fuchs, E., Nau, R., et al. (2008). No long-term effect two years after intrauterine exposure to dexamethasone on dentate gyrus volume, neuronal proliferation and differentiation in common marmoset monkeys. *Brain Pathology, 18*(4), 497–503.

Teicher, M. H., Anderson, C. M., & Polcari, A. (2012). Childhood maltreatment is associated with reduced volume in the hippocampal subfields CA3, dentate gyrus, and subiculum. *Proceedings of the National Academy of Sciences of the United States of America, 109*, 563–572.

Tronel, S., Charrier, V., Sage, C., Maitre, M., Leste-Lasserre, T., & Abrous, D. N. (April 25, 2015). Adult-born dentate neurons are recruited in both spatial memory encoding and retrieval. *Hippocampus*. http://dx.doi.org/10.1002/hipo.22468.

Ursin, H., & Eriksen, H. R. (2004). The cognitive activation theory of stress. *Psychoneuroendocrinology, 29*, 567–592.

Vallières, L., Campbell, I. L., Gage, F. H., & Sawchenko, P. E. (2002). Reduced hippocampal neurogenesis in adult transgenic mice with chronic astrocytic production of interleukin-6. *Journal of Neuroscience, 22*, 486–492.

Van Bokhoven, P., Oomen, C. A., Hoogendijk, W. J., Smit, A. B., Lucassen, P. J., & Spijker, S. (2011). Reduction in hippocampal neurogenesis after social defeat is long-lasting and responsive to late antidepressant treatment. *European Journal of Neuroscience, 33*, 1833–1840.

Vinkers, C. H., Joëls, M., Milaneschi, Y., Kahn, R. S., Penninx, B. W., & Boks, M. P. (2014). Stress exposure across the life span cumulatively increases depression risk and is moderated by neuroticism. *Depression and Anxiety, 31*(9), 737–745.

Vivar, C., Potter, M. C., & van Praag, H. (2013). All about running: synaptic plasticity, growth factors and adult hippocampal neurogenesis. *Current Topics in Behavioral Neurosciences, 15*, 189–210.

Vollmayr, B., Simonis, C., Weber, S., Gass, P., & Henn, F. (2003). Reduced cell proliferation in the dentate gyrus is not correlated with the development of learned helplessness. *Biological Psychiatry, 54*, 1035–1040.

Voss, M. W., Vivar, C., Kramer, A. F., & van Praag, H. (October 2013). Bridging animal and human models of exercise-induced brain plasticity. *Trends in Cognitive Sciences, 17*(10), 525–544.

Walker, F. R., Nilsson, M., & Jones, K. (October 2013). Acute and chronic stress-induced disturbances of microglial plasticity, phenotype and function. *Current Drug Targets, 14*(11), 1262–1276.

Wang, Q., Joëls, M., Swaab, D. F., & Lucassen, P. J. (2012). Hippocampal GR expression is increased in elderly depressed females. *Neuropharmacology, 62*, 527–533.

Wang, Q., Van Heerikhuize, J., Aronica, E., Kawata, M., Seress, L., Joels, M., et al. (2013). Glucocorticoid receptor protein expression in human hippocampus; stability with age. *Neurobiology of Aging, 34*, 1662–1673.

Wei, Q., Hebda-Bauer, E. K., Pletsch, A., Luo, J., Hoversten, M. T., Osetek, A. J., et al. (2007). Overexpressing the glucocorticoid receptor in forebrain causes an aging-like neuroendocrine phenotype and mild cognitive dysfunction. *Journal of Neuroscience, 27*, 8836–8844.

Wilson, C. B. (2014). Predator exposure/psychosocial stress animal model of post-traumatic stress disorder modulates neurotransmitters in the rat hippocampus and prefrontal cortex. *PLoS One, 9*, e89104.

Wong, E. Y. H., & Herbert, J. (2004). The corticoid environment: a determining factor for neural progenitors' survival in the adult hippocampus. *European Journal of Neuroscience, 20*, 2491–2498.

Wong, E. Y. H., & Herbert, J. (2005). Roles of mineralocorticoid and glucocorticoid receptors in the regulation of progenitor proliferation in the adult hippocampus. *European Journal of Neuroscience, 22*, 785–792.

Wu, X., & Castrén, E. (2009). Co-treatment with diazepam prevents the effects of fluoxetine on the proliferation and survival of hippocampal dentate granule cells. *Biological Psychiatry, 66*, 5–8.

Wu, M. V., Shamy, J. L., Bedi, G., Choi, C. W., Wall, M. M., Arango, V., et al. (2014). Impact of social status and antidepressant treatment on neurogenesis in the baboon hippocampus. *Neuropsychopharmacology, 39*, 1861–1871.

Zalachoras, I., Houtman, R., Atucha, E., Devos, R., Tijssen, A. M., Hu, P., et al. (May 7, 2013). Differential targeting of brain stress circuits with a selective glucocorticoid receptor modulator. *Proceedings of the National Academy of Sciences of the United States of America, 110*(19), 7910–7915.
Zhao, C., Deng, W., & Gage, F. H. (2008). Mechanisms and functional implications of adult neurogenesis. *Cell, 132*(4), 645–660.

Further Reading

Battista, D., Ferrari, C. C., Gage, F. H., & Pitossi, F. J. (2006). Neurogenic niche modulation by activated microglia: transforming growth factor beta increases neurogenesis in the adult dentate gyrus. *European Journal of Neuroscience, 23*, 83–93.
Brené, S., Bjørnebekk, A., Åberg, E., Mathé, A. A., Olson, L., & Werme, M. (2007). Running is rewarding and antidepressive. *Physiology and Behavior, 92*, 136–140.
Dantzer, R., O'Connor, J. C., Freund, G. G., Johnson, R. W., & Kelley, K. W. (2008). From inflammation to sickness and depression: when the immune system subjugates the brain. *Nature Reviews Neuroscience, 9*, 46–56.
Kronenberg, G., Bick-Sander, A., Bunk, E., Wolf, C., Ehninger, D., & Kempermann, G. (2006). Physical exercise prevents age-related decline in precursor cell activity in the mouse dentate gyrus. *Neurobiology of Aging, 27*(10), 1505–1513.
Monje, M. L., Toda, H., & Palmer, T. D. (2003). Inflammatory blockade restores adult hippocampal neurogenesis. *Science, 302*, 1760–1765.
Morrens, J., Van Den Broeck, W., & Kempermann, G. (2012). Glial cells in adult neurogenesis. *Glia, 60*, 159–174.
Naylor, A. S., Bull, C., Nilsson, M. K. L., Zhu, C., Bjork-Eriksson, T., Eriksson, P. S., et al. (2008). Voluntary running rescues adult hippocampal neurogenesis after irradiation of the young mouse brain. *Proceedings of the National Academy of Sciences of the United States of America, 105*, 14632–14637.
Naylor, A. S., Persson, A. I., Eriksson, P. S., Jonsdottir, I. H., & Thorlin, T. (2005). Extended voluntary running inhibits exercise-induced adult hippocampal progenitor proliferation in the spontaneously hypertensive rat. *Journal of Neurophysiology, 93*, 2406–2414.
Suri, D., Veenit, V., Sarkar, A., Thiagarajan, D., Kumar, A., Nestler, E. J., et al. (2013). Early stress evokes age-dependent biphasic changes in hippocampal neurogenesis, BDNF expression, and cognition. *Biological Psychiatry, 73*, 658–666.
Wachs, F.-P., Winner, B., Couillard-Despres, S., Schiller, T., Aigner, R., Winkler, J., et al. (2006). Transforming growth factor-beta1 is a negative modulator of adult neurogenesis. *Journal of Neuropathology and Experimental Neurology, 65*, 358–370.
Weaver, I. C. G., Grant, R. J., & Meaney, M. J. (2002). Maternal behavior regulates long-term hippocampal expression of BAX and apoptosis in the offspring. *Journal of Neurochemistry, 82*, 998–1002.
Zhu, C., Gao, J., Karlsson, N., Li, Q., & Zhang, Y. (2010). Isoflurane anesthesia induced persistent, progressive memory impairment, caused a loss of neural stem cells, and reduced neurogenesis in young, but not adult rodents. *Journal of Cerebral Blood Flow and Metabolism, 30*, 1017–1030.
Ziv, Y., Avidan, H., Pluchino, S., Martino, G., & Schwartz, M. (2006). Synergy between immune cells and adult neural stem/progenitor cells promotes functional recovery from spinal cord injury. *Proceedings of the National Academy of Sciences of the United States of America, 103*, 13174–13179.
Zunszain, P. A., Anacker, C., Cattaneo, A., Choudhury, S., Musaelyan, K., Myint, A. M., et al. (2012). Interleukin-1β: a new regulator of the kynurenine pathway affecting human hippocampal neurogenesis. *Neuropsychopharmacology, 37*, 939–949.

CHAPTER

9

Acute Stress and Anxiety

L. Culig, C. Belzung

Francois Rabelais University, Tours, France

INTRODUCTION

In an influential book published in 1982, Gray and McNaughton (2000), pioneers in the field, proposed that anxiety could involve the hippocampus (as well as the septum). Their basic idea was in fact very simple and elegant: the septum is, together with the hippocampal formation, able to detect competition between two different goals (for eg, escaping an aversive location and searching for food in the same area). This goal competition is inducing a kind of uncertainty that the system will try to disambiguate by inhibiting certain behaviors, which will enable the subject to gain further information about the situation before generating a response. This will translate into avoidance behaviors in response to threat and punishment. This septo-hippocampal system can thus be described as a "behavioral inhibition system."

In humans, two scales have been designed to measure behavioral inhibition: the "Behavioral-inhibition, behavioral activation, and affective responses to impending reward and punishment scale" (Carver & White, 1994) and the "Sensitivity to Punishment (StP) scale of the Sensitivity to Punishment and Sensitivity to Reward questionnaire" (Torrubia, Ávila, Moltó, & Caseras, 2001). These scales capture anxiety in normal human subjects and are able to predict anxiety-related pathological states, such as generalized anxiety disorder (Maack, Tull, & Gratz, 2012). Anxiety and behavioral inhibition are frequently dissociated from fear, a response that occurs when a subject is facing a real and imminent threat with no level of uncertainty. If fear is a rapid and phasic response, anxiety is characterized by long-lasting apprehension sometimes also termed as sustained fear (Davis, Walker, Miles, & Grillon, 2010). In any case, situations triggering anxiety are always stressful; thus it has also been proposed that

Adult Neurogenesis in the Hippocampus
http://dx.doi.org/10.1016/B978-0-12-801977-1.00009-X

anxiety can be considered as a reaction to an acute stressor. Likewise, in rodents some situations have been designed to assess anxiety behavior. They involve both unconditioned and conditioned protocols. In unconditioned assays, to asses anxiety behaviors, rodents are often forced into a novel arena, which induces two conflicting motivations: on the one hand, the animal displays a tendency to move around the whole arena, to find a way to escape from it, and on the other hand, it prefers avoiding the most stressful parts of the device such as the arms of a labyrinth that are open over an elevated space (elevated plus maze), highly illuminated places (light/dark boxes), or the center of a big arena (open field). In conditioned assays, for example, fear conditioning, a specific cue or context is associated with an electric shock. In this case, the rodent is experiencing fear (where the conditioned stimulus is highly predictive) or anxiety (where the moment of the delivery is less certain) when the conditioned stimulus (cue or context) is presented. Therefore, in this protocol, anxiety corresponds to behavior displayed by the rodent once the learning has been achieved: it does not concern the acquisition or the extinction of the conditioning, but only its expression.

Interestingly, the theoretical framework proposed by J. Gray has been confirmed by experimental evidence from human as well as animal studies. In humans, support came from structural as well as functional neuroimaging. A first approach consisted in associating high scores on the behavioral inhibition scales with function or volume of the hippocampus. Interestingly, higher scores on these scales are associated with higher functional connectivity between the hippocampus and the amygdala (Hahn et al., 2010) and to enhanced hippocampal volume (Barrós-Loscertales et al., 2006; Cherbuin et al., 2008; Levita et al., 2014). Other functional neuroimaging studies focused on contextual fear conditioning, using a paradigm in which a subject has been placed in a room previously associated with an aversive stimulus, without being able to predict the exact moment at which the painful stimulus will be delivered. This situation elicits a strong anxiety, under real world conditions as well as under virtual reality conditions (Alvarez, Chen, Bodurka, Kaplan, & Grillon, 2011; Grillon, Baas, Lissek, Smith, & Milstein, 2004; Huff et al., 2011). This approach confirmed the involvement of the hippocampus, as this brain area is activated when subjects are anticipating the delivery of an aversive stimulus (ie, when they are placed in a context that has been previously associated with the aversive stimulus) (Alvarez et al., 2011; Marschner et al., 2008). Further, larger hippocampal volume is associated with a significantly stronger cortisol increase in response to a social stress test and to a significantly greater cortisol response to awakening (Pruessner, Pruessner, Hellhammer, Bruce Pike, & Lupien, 2007). Studies undertaken in rodents confirmed this view, as hippocampal lesions have been shown to produce anxiolytic effects

(Deacon, Bannerman, & Rawlins, 2002) and as stress-induced hippocampal volume changes have been associated with resilience or susceptibility to stress in mice (Tse et al., 2014).

The hippocampus is not a homogeneous brain area: not only does it run along a posterior to anterior axis in human subjects (which corresponds to a dorsal/septal to a ventral/temporal axis in rodents), but it can also be subdivided in to several subfields, such as the dentate gyrus (DG), the Cornu Ammonis (CA, including the CA3 and CA1 subfields), and the subiculum. Unfortunately, these multiple boundaries are poorly delineated on neuroimaging scans, which renders assessing the involvement of these different subparts in anxiety in human studies difficult. Nevertheless, recent studies in humans showed that childhood maltreatment, which elicits increased anxiety in the subjects once adults, also induces a decreased volume specifically of the left DG, CA3, and subiculum (Teicher, Anderson, & Polcari, 2012) while in rodents stress provokes cell death particularly in the granule cell layer of the DG (Jayatissa, Bisgaard, West, & Wiborg, 2008). However, additional studies are necessary to more precisely elucidate causal relationships between a determined subfield and an anxiety, and this requires invasive methods that can be used only in animals. Furthermore, regarding the dorso/ventral segregation, it has been proposed that this involvement of the hippocampus on anxiety concerned its more anterior (in humans) or ventral (in rodents) portions (Bannerman et al., 2004; Fanselow & Dong, 2010). Indeed, in rodents, the ventral/temporal part is densely connected with the amygdala and with hypothalamic nuclei involved in the control of glucocorticoid release, which are associated with stress processing and anxiety (Canteras & Swanson, 1992; Risold & Swanson, 1996; Van Groen & Wyss, 1990). In primates, this corresponds to the most anterior (uncal) CA1 and prosubiculum (Fudge, deCampo, & Becoats, 2012).

Interestingly, this functional segregation has received support from experimental investigations that have been undertaken in rodents. The first evidence came from lesion studies in rodents showing that lesion of the ventral but not of the dorsal hippocampus was able to decrease anxiety behavior in an elevated plus maze, which is typically a situation enabling detection of anxiety behaviors (Kjelstrup et al., 2002). Recent studies, particularly using optogenetics (Belzung, Turiault, & Griebel, 2014) showed that this also concerned specific subfields. Indeed, optogenetic excitation of ventral DG granule cells specifically suppresses innate anxiety behaviors while the same stimulation in the dorsal DG interferes with learning of conditioned fear (Kheirbek et al., 2013). Further, lesion of the ventral DG in rats decreased anxiety behavior in rats subjected to an elevated plus maze or an open field test (Weeden, Roberts, Kamm, & Kesner, 2015).

Interestingly, the DG is one of the few brain areas in the adult brain where neural progenitor cells continue to proliferate and to differentiate into neurons (Ming & Song, 2011) and so the question of the functional implication of adult hippocampal neurogenesis in anxiety behavior has particular relevance. This review will explore the relationship between acute stress/anxiety and adult hippocampal neurogenesis answering two questions: (1) Does neurogenesis impact anxiety? (2) Do anxiogenic situations have an impact on neurogenesis? We will mainly focus on normal anxiogenic situations, and not on pathological conditions such as posttraumatic stress or other anxiety disorders, even if we briefly address this issue.

DOES HIPPOCAMPAL NEUROGENESIS IMPACT ON ANXIETY BEHAVIOR?

Different strategies can be used to assess the impact of neurogenesis on anxiety behavior. The first strategy is rather indirect and mainly relies on a correlational approach. It consists in assessing whether experimental manipulations that increase neurogenesis (environmental enrichment, physical exercise, chronic antidepressant treatment) also alter anxiety behavior. For example, long-lasting environmental enrichment decreases anxiety-like behavior (Chapillon, Manneché, Belzung, & Caston, 1999) and increases hippocampal cell proliferation, survival, and immature newborn neurons (Tanti et al., 2013; Tanti, Rainer, Minier, Surget, & Belzung, 2012). Even if this approach suggests a correlation between both phenomena, it does not really provide evidence to establish a causal relationship (Chapillon et al., 1999). A second strategy consists in looking at whether adult newborn neurons are activated when a subject is faced with an acute stressor that evokes anxiety. This strategy already enabled Denny et al. (2014) to show that mice exposed to a situation triggering anxiety (reexposure to a context previously associated with an aversive situation) had a higher percentage of reactivated cells within the DG and CA3 than mice exposed to a less anxiogenic novel context. In the same vein, Liu et al. (2012) showed that stimulation of neurons that have been activated during acquisition of fear conditioning ("the hippocampal conditioned fear engram") induces the expression of anxiety behavior. However, this does not indicate whether adult-born DG cells were specifically involved in anxiety behavior. This was investigated by Surget et al. (2011) who showed that forced confrontation of mice with an open field induced fos immunoreactivity (which is an indirect marker of neuronal activity) in adult-generated neurons. However, the proportion of cells activated by confrontation with the open field was the same in mature granule cells from the DG and in adult-generated neurons aged 4 weeks (around 1% of the neurons were activated). Nevertheless, this recruitment of the newborn neurons could be an epiphenomenon and this study does not establish that adult newborn neurons are necessary to evoke anxiety behavior. To assess a causal relationship, the sole valid approach consists in suppressing adult neurogenesis. To this end, different methods can be used, including focal irradiation of the hippocampus, genetic strategies (consisting in inducing death of cells undergoing proliferation such as Nestin-expressing cells), or peripheral injection of an antimitotic agent such as MAM. Usually, these methods induce the death of cells undergoing proliferation, and after some time, the number of newborn cells is diminished totally or partially. Only after focal hippocampal irradiation or by using Nestin-rtTA/TRE-Bax mice it was possible to ablate neurogenesis, specifically adult hippocampal newborn cells, because it spares adult neurogenesis in the other brain regions in which neurogenesis takes place, such as the olfactory bulbs (Dupret et al., 2008) or the hypothalamus. Other methods with more specificity for a subregion have been designed, such as novel knock-in DCX^{DTR} mice (Vukovic et al., 2013) or intersectional genetic strategies using Cre and Flp (Imayoshi, Sakamoto, & Kageyama, 2011), but to our knowledge they have not been used yet to investigate the involvement of hippocampal neurogenesis in anxiety behavior or stress sensitivity.

Results of studies of experimentally induced loss of adult neurogenesis have already been reviewed (reviewed in Petrik, Lagace, & Eisch, 2012, and in Tanti & Belzung, 2013). Concerning the effect of ablation of neurogenesis on anxiety, the picture much depends on the assay that has been used to elicit anxiety. For example, using the novelty suppression of feeding test, which relies on two competing drives (venturing in the center of an unfamiliar area to obtain food or avoid the most dangerous part of the device), a large number of studies found that neurogenesis did not have an impact on anxiety behavior. Indeed, no effect was seen in this test using focal irradiation in mice (David et al., 2007; Meshi et al., 2006; Santarelli et al., 2003; Surget et al., 2008; Wang, David, Monckton, Battaglia, & Hen, 2008) or in rats (Zhu et al., 2010), or using genetic strategies such as Nestin-bax mice (Revest et al., 2009) or hGFAPtk mice (Snyder, Soumier, Brewer, Pickel, & Cameron, 2011). However, the picture is not monolithic and effects of ablation of neurogenesis could be seen in this device using an antimitotic agent to suppress neurogenesis (Bessa et al., 2009; Mateus-Pinheiro et al., 2013) or when combining ablation of neurogenesis with acute stress (Snyder et al., 2011). Using the open field test, no effects of ablation of neurogenesis per se could be detected in mice genetically modified to suppress neurogenesis (hGFAPtk) (Schloesser, Lehmann, Martinowich, Manji, & Herkenham, 2010; Snyder et al., 2011) or using focal irradiation alone (Fuss et al., 2010) or combined with chronic corticosterone (CORT) (David et al., 2009). In the elevated plus maze, most studies did not find an effect of ablation of neurogenesis when using MAM (an antimitotic agent) to suppress newborn neurons (Shors, Townsend, Zhao, Kozorovitskiy, & Gould, 2002) or using focal hippocampal irradiation (Saxe et al., 2006) or a genetic strategy (hGFAPtk) (Snyder et al., 2011). The sole exception concerns the Nestin-bax mice, which displayed increased anxiety behavior (Revest et al., 2009). The zero maze is a variant of the elevated plus maze: it detected no effect (Fuss et al., 2010) or decreased anxiety (Onksen, Brown, & Blendy, 2011). In the light/dark boxes, most studies found no effect (Lehmann, Brachman, Martinowich, Schloesser, & Herkenham, 2013; Saxe et al., 2006; Schloesser et al., 2010) while others found increased anxiety (Fuss et al., 2010; Revest et al., 2009). Only one study investigated predator avoidance and found that decreased adult neurogenesis increased avoidance (Revest et al., 2009). Finally, also only one study investigated marble burying and found decreased anxiety after suppression of neurogenesis (Onksen et al., 2011), which is to our knowledge the sole study that showed anxiogenesis after increase in newborn neurons. Interestingly, all studies that have found an effect of the ablation of newborn neurons on anxiety relied on mitotic and genetic strategies, which destroyed neurogenesis in the hippocampus and also in extrahippocampal areas such as the olfactive bulbs or the hypothalamus, which are brain areas that are also involved in anxiety (Belzung et al., 2014). Therefore, it is difficult to establish the part of the behavioral effect related to absence of the hippocampal neurogenesis from the part related to the contribution of neurogenesis within the extrahippocampal structures. Of interest is the fact that none of the studies using focal hippocampal irradiation found alterations in anxiety, suggesting that the observed effects using genetic approaches or antimitotic agents were not related to the hippocampus. Further, in all these studies, irradiation included the dorsal as well as the ventral subpart of the DG. To address the specific contribution of ventral and/or dorsal hippocampal neurogenesis, Wu & Hen (2014) destroyed specific subareas using focal low dose irradiation. They found that adult newborn neurons from the dorsal DG were required for rapid acquisition of contextual fear discrimination, while specific suppression of newborn neurons from the ventral DG did not produce any effect per se in anxiety tests (open field and NSF test). To summarize, our analysis of the literature reveals that using the novelty suppression of feeding test, nine studies found no effects of ablation of neurogenesis, while three found heightened anxiety behavior. The same proportions were found in the elevated plus maze and light/dark tests (in each case three studies with no effect, and one with an increased anxiety). In the open field, none of the five studies found an effect of suppression of neurogenesis. Some exceptions to this rule suggesting poor involvement of neurogenesis in anxiety can be found, but they generally involve behavioral assays used in few experiments (two experiments only used the zero maze, one predator avoidance and one marble burying) and should therefore receive further empirical support. Of interest is the fact that almost all of these studies relied (except the study by Revest et al., 2009, which used predator stimuli) on situations based on forced confrontation of rodents with novelty to induced anxiogenesis and therefore these finding must be confirmed with other kinds of behavioral assays.

So, effects of an ablation of adult neurogenesis on anxiety are still rather controversial, even if a majority of studies did not find that hippocampal neurogenesis was necessary for anxiety behavior and even if an effect was never observed using focal irradiation. However, could neurogenesis be sufficient to modify anxiety? This can be investigated by inducing a specific increase of neurogenesis, which was done by Sahay et al. (2011) using mice with a genetic deletion of the proapoptotic *Bax* gene in Nestin+ cells. Overall, they found that increased neurogenesis had no effect in the open field test or in the novelty suppression of feeding test when mice were reared under normal conditions. However, the mice that were transferred to cages with running wheels, which further increases neurogenesis, displayed reduced anxiety in both situations, when compared to mice that were just subjected to running wheels with no neurogenesis increase. Thus, increased neurogenesis could dampen anxiety but this requires an increase of neurogenesis of high magnitude. In a study where the same strain of mice had been exposed to a 4-week-long chronic CORT treatment 6 weeks after the administration of tamoxifen, an effect of increased neurogenesis on stress resilience was observed (Hill, Sahay, & Hen, 2015). More precisely, mice with increased levels of newborn neurons in the DG counteracted the anxiogenic effects of CORT, which manifested itself as decreased time spent in the open arms of the elevated plus maze, as well as fewer open arm entries. Additionally, CORT also decreased mobility in the tail suspension test, and tamoxifen reversed this effect, showing that increased neurogenesis is sufficient for rescuing certain deficits in depressive- and anxiety-like behavior. However, not all behaviors were affected by CORT administration, and the effects of tamoxifen were evident only on the behaviors affected by CORT, which is consistent with the lack of effect of increased neurogenesis in baseline (no stress) conditions. Interestingly, there was no effect on the HPA axis regulation, suggesting some other mechanism through which increased neurogenesis can affect behavior, possibly through connections with the amygdala or nucleus accumbens, structures that receive ventral hippocampal inputs. It would be of interest to study the effects of increased neurogenesis on these structures, as well as on the functional network containing other stress-related structures. Additionally, CORT administration does not model all the effects of stress, so it would be of interest to use a model that more closely resembles a depressive behavioral phenotype. Lastly, while this study has indeed shown that increased neurogenesis can protect against the effects of stress, as TAM administration occurred 4 weeks before the CORT administration, it would be of interest to find out if increased neurogenesis can rescue behavioral deficits after an insult (TAM after CORT). Antidepressants are usually not prescribed before the onset of depression, so in the clinical sense it would be of interest to investigate the paradigm where neurogenesis is increased after stress, which could lead to new approaches in treating depression.

An open question is related to the fact that neurogenesis could be involved in some form of anxiety or stress-related disorders, even if not specifically involved in normal anxiety. Even if the effects of experimental manipulations of neurogenesis have not yet been investigated in animal models of pathological anxiety, some indirect or theoretical evidence suggests such a relationship, particularly in relation to posttraumatic stress disorder (PTSD). PTSD is a form of pathological anxiety that is triggered by confrontation with a traumatic event; that is, with a situation involving threat to the physical integrity of oneself or others. It has been shown since the middle of the 1990s that PTSD is associated with a decreased volume of the hippocampus, which is, according to the authors, either conceived as a consequence of the traumatic exposure (the stressful traumatic event is inducing reduced hippocampal volume) or seen as a predisposition factor (a small hippocampus renders one vulnerable to the effects of trauma, thus increasing the percent of people developing PTSD after the trauma) (see Pitman et al., 2012, for a review). Recently, using sMRI, it was shown that this reduction of the hippocampal volume particularly concerned the CA3 and DG subfields (Wang et al., 2010), and not the other hippocampal subfields. This suggests that neurogenesis within the DG could be of particular relevance to PTSD. Within this framework, two (not mutually exclusive) views are possible: (1) reduced initial adult DG neurogenesis could increase the vulnerability for the development of PTSD symptoms when a subject has been confronted with a traumatic event; (2) traumatic stress could have an impact on adult neurogenesis. The second of these views will be extensively discussed in the next paragraph focusing on the effects of acute stress on hippocampal neurogenesis, and here we will focus on the first of these proposals.

Not all subjects who are confronted with a traumatic event develop PTSD symptoms, and neuroimaging data have already shown that a small hippocampal volume before the trauma can be considered as a risk factor for the onset of a PTSD (Gilbertson et al., 2002). Even if this study did not investigate the specific contribution of the DG, because high-resolution MRI scans were not available at that time, and even if reduced hippocampal volume can be related to many factors, including mature cell death and reduced dendritic branching, the Gilbertson et al. (2002) study is coherent with the view that adult hippocampal neurogenesis could be involved in vulnerability to psychological trauma. Even if such involvement has not yet been directly investigated empirically, this suggests that hippocampal newborn neurons could have a specific function crucial to the onset of PTSD symptoms. Interestingly, some theories proposed such a view, particularly highlighting that neurogenesis could be relevant for pattern separation, a process enabling disambiguating overlapping sensorial representations, for example, risk-related situations from similar ones that are not associated with any danger. It has repeatedly been proposed that people with maladaptive anxiety could exhibit overgeneralization, as they generalize from cues associated with the trauma to other unrelated cues, which could be related to a defect in pattern separation (Kheirbek, Klemenhagen, Sahay, & Hen, 2012). Interestingly, neuroimaging findings in humans and in vivo electrophysiology studies in rodents supported the view that the DG is crucially involved in pattern separation (Bakker, Kirwan, Miller, & Stark, 2008; Leutgeb, Leutgeb, Moser, & Moser, 2007). So what about the contribution of newborn neurons from the DG in pattern separation? Several studies reported evidence of a pattern separation deficit when neurogenesis had been decreased and conversely, an increased pattern separation after an increased adult neurogenesis. This applies to situations in which subjects must discriminate fear-related as well as fear-unrelated stimuli. For example, after hippocampal X-irradiation, mice display decreased nonmatching-to-place in a radial arm maze with poor spatial separation between two arms (Clelland et al., 2009). Similar findings have been obtained using more stressful protocols such as contextual fear discrimination in which mice learn over repeated trials to distinguish very similar contexts associated either with a single foot shock (shock context; A) or with none (safe context; B). Indeed, when adult newborn hippocampal neurons have been decreased using either a genetic strategy (Nakashiba et al., 2012; Tronel et al., 2012) or a hippocampal focal irradiation, pattern separation is decreased, while the opposite is found in the case of increased neurogenesis (Sahay et al., 2011). However, it is important to consider these findings with some caution, as pattern separation is an expression that has initially been used to refer to computational orthogonalization, so that its extension to the processes investigated in a behavioral context can be considered an extrapolation. For example, Santoro (2013) proposes to replace it by the term "behavioral context discrimination."

Taken together, these data suggest that hippocampal neurogenesis could be poorly involved in normal anxiety, and that it could underlie a function crucial to the development of some forms of pathological anxiety such as pattern separation. However, further experiments are necessary using animal models of PTSD.

Finally, many studies investigated not a direct impact of ablation of neurogenesis on anxiety behavior, but an effect of ablation of adult newborn neurons on the anxiolytic effects of chronic antidepressants such as assessed in the novelty suppression of feeding, the open field, the light/dark boxes tests, or in the zero maze. Most findings showed that the anxiolytic effects of chronic treatment with a selective serotonin reuptake inhibitor such as fluoxetine were suppressed in the novelty-induced suppression of feeding test (Bessa et al., 2009;David et al., 2009; Santarelli et al., 2003; Surget et al., 2008; Wang et al., 2008). However, the picture is not monolithic and it must be noted that when using another test to detect anxiolytic effects of chronic antidepressants, the same results are not found: indeed, focal hippocampal irradiation of the DG does not suppress the anxiolytic effects of chronic fluoxetine in the open field (David et al., 2009). Further, when using another antidepressant such as the tricyclic imipramine, its anxiolytic effects measured in the novelty suppression of feeding test are suppressed by irradiation in some (Bessa et al., 2009; Santarelli et al., 2003) but not all studies (Surget et al., 2008). Additionally, the anxiolytic effects of environmental enrichment are suppressed by irradiation in the open field and in the light/dark boxes in one study (Schloesser et al., 2010) but not in the novelty-induced suppression of feeding test (Meshi et al., 2006). Finally, the anxiolytic effects of running are not altered by disruption of neurogenesis in the open field, the light/dark boxes, or the zero maze (Fuss et al., 2010). In most of these studies reporting the effects of chronic antidepressants, the anxiolytic effects of chronic antidepressants were assessed in animals that had been subjected to experimental manipulations such as the unpredictable chronic mild stress or chronic CORT, which induce sustained anxiety behavior that is sometimes associated with pathological anxiety.

Taken together, the data reported in this section suggest or support an involvement of neurogenesis in anxiety disorders such as PTSD, or in remission after pathological anxiety, rather than in normal anxiety situations.

DOES ACUTE STRESS IMPACT ON NEUROGENESIS?

The DG of the hippocampus is a dynamic structure that not only has a role in regulating the stress response and anxiety but is also particularly susceptible to stress (Herman, Ostrander, Mueller, & Figueiredo, 2005). This notable susceptibility is mainly due to the exceptionally high level of corticosteroid receptors, which on binding stress hormones result in an impairment in the function of the hippocampus (Joëls, Krugers, & Karst, 2008). Many aspects are affected, ranging from synaptic plasticity and neural morphology to adult neurogenesis, the last of these being the focus of this chapter. In humans, childhood stress represents one of the major factors for the development of mood disorders such as depression and bipolar disorder, and these changes can result in the reduction of the volume of the hippocampus (Brietzke et al., 2012). A proposed explanation of hippocampal volume reduction in depressed patients is stress-induced reduction in neurogenesis (Sheline, 2000). In juvenile rodents, acute stress reduces the number of tagged granule cells in the fascia dentata, and there are many studies showing that early life trauma in rodents is also a contributing factor leading to heightened anxiety- and depressive-like symptoms in adult animals (Kikusui & Mori, 2009; King et al., 2004; Schmidt, Wang, & Meijer, 2011). Thus, it is suggested that the stress-induced decrease in neurogenesis acts as a contributor to the pathology of depression, a link which is supported by the fact that chronic treatment with antidepressants has an opposite effect, increasing the rate of adult neurogenesis in the DG (Czéh et al., 2001; Dranovsky & Hen, 2006; Malberg, Eisch, Nestler, & Duman, 2000).

Adult neurogenesis itself can be separated into several main stages: cell proliferation, neuronal differentiation, and cell survival, with some studies suggesting at least six different stages of neurogenesis (Christie & Cameron, 2006; Kempermann, Jessberger, Steiner, & Kronenberg, 2004). Briefly, cell proliferation involves the division of progenitor cells located in the subgranular zone of the DG. Neuronal differentiation is the stage referring to the selection of the neuronal fate of newborn daughter cells—depending on the species, animal age, and other variables, between 80% and 95% differentiate into neurons, and around 10% into glia (Cameron, Woolley, McEwen, & Gould, 1993; Snyder et al., 2009; Steiner et al., 2004). During maturation, newborn neurons go through morphological and electrophysiological changes. As a result, one notable change is enhanced synaptic plasticity, and also the shift in the reaction to GABA, the main inhibitory neurotransmitter (Snyder, Kee, & Wojtowicz, 2001). Curiously, immature newborn neurons respond to GABA with excitation, due to their high cytoplasmic chloride ion content, and as they mature they respond with inhibition, a phenomenon that is related to the predominant expression of NKCC1 (Na–K–Cl cotransporter) in immature neurons and an upregulation of KCC2 (neuron-specific K^+/Cl^- cotransporter) in late neuronal developmental stages (Ge et al., 2006; Pontes, Zhang, & Hu, 2013). This lack of GABA-related inhibition along with enhanced synaptic plasticity during maturation leads to newborn neurons being an ideal substrate for influencing hippocampal function (Schoenfeld & Gould, 2012). Lastly, cell survival involves long-lasting incorporation of these newborn cells into the hippocampal circuitry. A large proportion of these cells, around 50% in the rat DG, do not survive longer than a few weeks, but some can survive up to a year or longer (Dayer, Ford, Cleaver, Yassaee, & Cameron, 2003). Although all of the noted stages can be influenced by stress and other environmental factors, the bulk of data suggests that the effects of stress alter the cell proliferation stage.

In general, acute stress exposure tends to reduce the levels of adult neurogenesis by decreasing cell proliferation, with few studies reporting an effect on cell survival (Table 9.1). Stressors used in the study of

TABLE 9.1 Effects of Exposure to Stressors at Different Stages of Neurogenesis

Stressor	Animal model	Cell proliferation	Differentiation	Survival	References
Natural stressors Psychosocial stress	Tree shew	–	n/a	n/a	Gould, McEwen, Tanapat, Galea, and Fuchs (1997)
	Marmoset	–	n/a	n/a	Gould et al. (1998)
	Mouse	– (after 10 episodes)	n/a	n/a	Yap et al. (2006)
		– (transient)	n/a	n/a	Lagace et al. (2010)
	Rat	0	n/a	–	Thomas et al. (2007)
Predator odor	Rat	–	n/a	n/a	Hill et al. (2006), Kambo and Galea (2006), Mirescu et al. (2004), Tanapat et al. (2001b)
	Rat (female)	0	n/a	n/a	Falconer and Galea (2003)
	Rat	0	n/a	n/a	Thomas et al. (2006)
Laboratory stressors	Rat	–	n/a	n/a	
Physical restraint	Mouse	+	n/a	n/a	Bain et al. (2004)
	Rat	–	n/a	n/a	Vollmayr et al. (2003)
	Rat	0	n/a	n/a	Duric and McCarson (2006), Hanson et al. (2011), Pham et al. (2003), Rosenbrock et al. (2005)
Restraint/swimming	Rat	– (transient)	n/a	n/a	Heine et al. (2004)
Electric shock	Rat	–	n/a	n/a	Malberg and Duman (2003)
	Rat	0	n/a	0	Vollmayr et al. (2003)
	Rat	– (after 7 days)	n/a	n/a	Fornal et al. (2007)
	Rat	–	n/a	–	Bland et al. (2006)

n/a, no available data; 0, no effect; -, decrease; +, increase.

acute stress consequences on neurogenesis range from "natural stressors," such as social defeat and exposure to predator odor, to "laboratory stressors," such as electric foot shock. It is considered that natural stressors could more closely mimic the onset of stress-related pathology in humans and provide a socially more relevant model (Blanchard, McKittrick, & Blanchard, 2001; Thomas, Hotsenpiller, & Peterson, 2007). In a study utilizing exposure to acute psychosocial stress in tree shews and marmosets researchers reported a decrease in cell proliferation in the DG (Gould, McEwen, Tanapat, Galea, & Fuchs, 1997; Gould, Tanapat, McEwen, Flügge, & Fuchs, 1998). The picture is not so clear in mice, with one study reporting a nonsignificant decrease in neurogenesis after a single defeat episode and a significant decrease after 10 defeat episodes (Yap et al., 2006), and one study reporting a significant (but only transient) decrease in the number of cells labeled with the S-phase marker BrdU, also after 10 days of social defeat (Lagace et al., 2010). One study conducted with rats reported no difference in cell proliferation, but instead a reduction in the survival of newborn neurons (Thomas et al., 2007). However, this could be due to the experimental design, in which the cells had been labeled before or during exposure to stress, and not after it, which thus does not make it a direct test of effects of stress on proliferating cells (Schoenfeld & Gould, 2012).

Another stressor considered to belong in the family of "natural stressors" is exposure to trimethylthiazoline (TMT), a component in fox feces, which is designed to mimic exposure to the odor of natural predators. However, it should be noted that certain criticisms have been made, bringing into question if TMT is a true predator odor. Cat odor is suggested as an alternative, and it is easily distinguished from TMT both on the behavioral and the neural level. This could be due to the fact that TMT does not have the "pheromone-like" quality of cat odor, and ethological relevance of mammalian chemosignaling is known to be dependent on odor complexity, which in turn is absent from synthetic compounds such as TMT (Staples, McGregor, Apfelbach, & Hunt, 2008). Another explanation could involve the odor context, meaning that predator feces odor is durable and not predictive of the areas where the predators are located, thus not providing adequate focus as danger cues, especially having in mind that there is evidence which suggests that predators tend to defecate in areas where they do not hunt (Blanchard, Yang, Li, Gervacio, & Blanchard, 2001; Kats & Dill, 1998). Despite these shortcomings, this does not rule out the usefulness of TMT as a stimulus, but should be kept in mind when drawing conclusions from the behavioral data. Also the behavioral data are mostly in accord: exposure to TMT decreases cell proliferation in the DG of rats (Hill, Kambo, Sun, Gorzalka, & Galea, 2006; Kambo & Galea, 2006; Mirescu, Peters, & Gould, 2004; Tanapat, Hastings, Rydel, Galea, & Gould, 2001b), but a sex difference has been observed, with a decrease detected in male rats and no difference in cell proliferation in female rats after acute exposure to TMT (Falconer & Galea, 2003). One study observed no decrease in cell proliferation after acute exposure to TMT, despite physiological and behavioral evidence of stress. As with the group's previously noted study (Thomas et al., 2007), this could be due to the fact that new cells were labeled during, and not after, the stress, therefore not making it a direct test of effects of stress on proliferating cells (Thomas, Urban, & Peterson, 2006).

When examining the consequences of laboratory stressors such as acute physical restraint on neurogenesis, one finds that the results are mixed. One study found that after 3h of restraint stress in plastic tubes (for mice) or glass restrainers (for rats) resulted in the observation of decreased cell proliferation in adult rats, but also of increased cell proliferation in adult mice (Bain, Dwyer, & Rusak, 2004). A study found a decrease in cell proliferation after 45min of restraint stress in rats (Vollmayr, Simonis, Weber, Gass, & Henn, 2003), while several studies found no significant effect of 45min to 6h of restraint stress on cell proliferation in the DG (Duric & McCarson, 2006; Hanson, Owens, Boss-Williams, Weiss, & Nemeroff, 2011; Pham, Nacher, Hof, & McEwen, 2003; Rosenbrock, Koros, Bloching, Podhorna, & Borsini, 2005). A different study reported a transient decrease in cell proliferation in rats after one episode of cold immobilization and forced swimming (Heine et al., 2004).

Electric foot (or tail) shock also belongs to the family of laboratory stressors, which in rodents usually activate the HPA axis and induce anxiety-like behavior. One study found a decrease in cell proliferation in adult male rats after exposure to inescapable foot shock (Malberg & Duman, 2003), another found a decrease in both cell proliferation and survival (Bland, Schmid, Greenwood, Watkins, & Maier, 2006), while another study also utilizing adult male rats found no difference in cell proliferation after uncontrollable foot shock (Vollmayr et al., 2003). The results are further complicated by a study where a decrease in cell proliferation has been observed, but at 7days after the tail shock, a time when plasma CORT levels have reverted back to normal (Fornal et al., 2007). This suggests that there could be a temporal disconnect between increased CORT levels and suppression of cell proliferation in the DG and/or that the physiological response to a stressor may include release of a counteracting or protective factor to the CORT effect. This is disparate from the studies that show a direct relationship between the acute elevations of CORT and the suppression of cell proliferation (Cameron & Gould, 1994; Cameron, Tanapat, & Gould, 1998; Mirescu & Gould, 2006; Tanapat, Hastings, Rydel, Galea, & Gould, 2001a), hinting at a more complicated mechanism between electric shock and cell proliferation, which is yet to be uncovered.

CONCLUSIONS

Taken together, these studies support the idea that heightened adult hippocampal neurogenesis does not influence anxiety behavior in normal situations in which an organism has not been challenged by stressful events. However, in the case where the subject is faced with an acute stressor, this could induce a decrease in cell proliferation and neuronal survival, which could cause later on some symptoms related to pathological anxiety such as PTSD, inducing a deficit in pattern separation and thus the psychiatric symptomatology. Additionally, as PTSD as well as other forms of pathological anxiety includes not only emotional but also cognitive symptomatology, for example, decreased semantic memory which is also associated with decreased neurogenesis, the loss of adult DG newborn neurons caused by the acute stressor could contribute not only to the heightened anxiety seen in PTSD but also to the cognitive symptomatology. Interestingly, this can be prevented using chronic antidepressants, as clinical evidence established that these compounds are effective not only in alleviating depression-related symptoms but also in inducing remission of several forms of pathological anxiety. However, this proposal is purely speculative and further empirical studies are necessary to investigate the relationship of adult hippocampal neurogenesis with PTSD as well as other pathological situations.

References

Alvarez, R. P., Chen, G., Bodurka, J., Kaplan, R., & Grillon, C. (2011). Phasic and sustained fear in humans elicits distinct patterns of brain activity. *NeuroImage*, *55*(1), 389–400. http://doi.org/10.1016/j.neuroimage.2010.11.057.

Bain, M. J., Dwyer, S. M., & Rusak, B. (2004). Restraint stress affects hippocampal cell proliferation differently in rats and mice. *Neuroscience Letters*, *368*(1), 7–10. http://doi.org/10.1016/j.neulet.2004.04.096.

Bakker, A., Kirwan, C. B., Miller, M., & Stark, C. E. L. (2008). Pattern separation in the human hippocampal CA3 and dentate gyrus. *Science (New York, N.Y.)*, *319*(5870), 1640–1642. http://doi.org/10.1126/science.1152882.

Bannerman, D. M., Rawlins, J. N. P., McHugh, S. B., Deacon, R. M. J., Yee, B. K., Bast, T., et al. (2004). Regional dissociations within the hippocampus - memory and anxiety. *Neuroscience and Biobehavioral Reviews*, *28*, 273–283. http://doi.org/10.1016/j.neubiorev.2004.03.004.

Barrós-Loscertales, A., Meseguer, V., Sanjuán, A., Belloch, V., Parcet, M. A., Torrubia, R., et al. (2006). Behavioral inhibition system activity is associated with increased amygdala and hippocampal gray matter: a voxel-based morphometry study. *NeuroImage*, *33*, 1011–1015. http://doi.org/10.1016/j.neuroimage.2006.07.025.

Belzung, C., Turiault, M., & Griebel, G. (2014). Optogenetics to study the circuits of fear- and depression-like behaviors: a critical analysis. *Pharmacology, Biochemistry, and Behavior*, *122*, 144–157. http://doi.org/10.1016/j.pbb.2014.04.002.

Bessa, J. M., Ferreira, D., Melo, I., Marques, F., Cerqueira, J. J., Palha, J. A., et al. (2009). The mood-improving actions of antidepressants do not depend on neurogenesis but are associated with neuronal remodeling. *Molecular Psychiatry*, *14*(8), 764–773. 739 http://doi.org/10.1038/mp.2008.119.

Blanchard, R. J., McKittrick, C. R., & Blanchard, D. C. (2001). Animal models of social stress: effects on behavior and brain neurochemical systems. *Physiology & Behavior*, *73*(3), 261–271. http://doi.org/10.1016/S0031-9384(01)00449-8.

Blanchard, R. J., Yang, M., Li, C.-I., Gervacio, A., & Blanchard, D. C. C. (2001). Cue and context conditioning of defensive behaviors to cat odor stimuli. *Neuroscience & Biobehavioral Reviews*, *25*(7–8), 587–595. http://doi.org/10.1016/S0149-7634(01)00043-4.

Bland, S. T., Schmid, M. J., Greenwood, B. N., Watkins, L. R., & Maier, S. F. (2006). Behavioral control of the stressor modulates stress-induced changes in neurogenesis and fibroblast growth factor-2. *Neuroreport*, *17*(6), 593–597. Retrieved from: http://www.ncbi.nlm.nih.gov/pubmed/16603918.

Brietzke, E., Sant'anna, M. K., Jackowski, A., Grassi-Oliveira, R., Bucker, J., Zugman, A., & Bressan, R. A., et al. (2012). Impact of childhood stress on psychopathology. *Revista Brasileira de Psiquiatria*, *34*(4), 480–488. http://doi.org/10.1016/j.rbp.2012.04.009.

Cameron, H. A., & Gould, E. (1994). Adult neurogenesis is regulated by adrenal steroids in the dentate gyrus. *Neuroscience*, *61*(2), 203–209. Retrieved from: http://www.ncbi.nlm.nih.gov/pubmed/7969902.

Cameron, H. A., Tanapat, P., & Gould, E. (1998). Adrenal steroids and N-methyl-D-aspartate receptor activation regulate neurogenesis in the dentate gyrus of adult rats through a common pathway. *Neuroscience*, *82*(2), 349–354. Retrieved from: http://www.ncbi.nlm.nih.gov/pubmed/9466447.

Cameron, H., Woolley, C. S., Mcewen, B. S., & Gould, E. (1993). Differentiation of newly born neurons and glia in the dentate gyrus of the adult rat. *Neuroscience*, *56*(2), 337–344. Retrieved from: http://www.sciencedirect.com/science/article/pii/030645229390335D.

Canteras, N. S., & Swanson, L. W. (1992). Projections of the ventral subiculum to the amygdala, septum, and hypothalamus: a PHAL anterograde tract-tracing study in the rat. *The Journal of Comparative Neurology*, *324*(2), 180–194. http://doi.org/10.1002/cne.903240204.

Carver, C. S., & White, T. L. (1994). Behavioral inhibition, behavioral activation, and affective responses to impending reward and punishment: the BIS/BAS scales. *Journal of Personality and Social Psychology*, *67*, 319–333. http://doi.org/http://dx.doi.org/10.1037/0022-3514.67.2.319.

Chapillon, P., Manneché, C., Belzung, C., & Caston, J. (1999). Rearing environmental enrichment in two inbred strains of mice: 1. Effects on emotional reactivity. *Behavior Genetics*, *29*(1), 41–46. http://doi.org/10.1023/A:1021437905913.

Cherbuin, N., Windsor, T. D., Anstey, K. J., Maller, J. J., Meslin, C., & Sachdev, P. S. (2008). Hippocampal volume is positively associated with behavioural inhibition (BIS) in a large community-based sample of mid-life adults: the PATH through life study. *Social Cognitive and Affective Neuroscience*, *3*, 262–269. http://doi.org/10.1093/scan/nsn018.

Christie, B. R., & Cameron, H. A. (2006). Neurogenesis in the adult hippocampus. *Hippocampus*, *16*(3), 199–207. http://doi.org/10.1002/hipo.20151.

Clelland, C. D., Choi, M., Romberg, C., Clemenson, G. D., Fragniere, A., Tyers, P., et al. (2009). A functional role for adult hippocampal neurogenesis in spatial pattern separation. *Science (New York, N.Y.)*, *325*(5937), 210–213. http://doi.org/10.1126/science.1173215.

Czéh, B., Michaelis, T., Watanabe, T., Frahm, J., de Biurrun, G., van Kampen, M., et al. (2001). Stress-induced changes in cerebral metabolites, hippocampal volume, and cell proliferation are prevented by antidepressant treatment with tianeptine. *Proceedings of the National Academy of Sciences of the United States of America*, *98*(22), 12796–12801. http://doi.org/10.1073/pnas.211427898.

David, D. J., Klemenhagen, K. C., Holick, K. A., Saxe, M. D., Mendez, I., Santarelli, L., et al. (2007). Efficacy of the MCHR1 antagonist N-[3-(1-{[4-(3,4-difluorophenoxy)phenyl] methyl}(4-piperidyl))-4-methylphenyl]-2-methylpropanamide (SNAP 94847) in mouse models of anxiety and depression following acute and chronic administration is independent of hippocampal neurogenesis. *The Journal of Pharmacology and Experimental Therapeutics*, *321*, 237–248. http://doi.org/10.1124/jpet.106.109678.

David, D. J., Samuels, B. A., Rainer, Q., Wang, J. W., Marsteller, D., Mendez, I., et al. (2009). Neurogenesis-dependent and -independent effects of fluoxetine in an animal model of anxiety/depression. *Neuron, 62*(4), 479–493. http://doi.org/10.1016/j.neuron.2009.04.017.

Davis, M., Walker, D. L., Miles, L., & Grillon, C. (2010). Phasic vs sustained fear in rats and humans: role of the extended amygdala in fear vs anxiety. *Neuropsychopharmacology, 35*(1), 105–135. http://doi.org/10.1038/npp.2009.109.

Dayer, A. G., Ford, A. A., Cleaver, K. M., Yassaee, M., & Cameron, H. A. (2003). Short-term and long-term survival of new neurons in the rat dentate gyrus. *The Journal of Comparative Neurology, 460*(4), 563–572. http://doi.org/10.1002/cne.10675.

Deacon, R. M. J., Bannerman, D. M., & Rawlins, J. N. P. (2002). Anxiolytic effects of cytotoxic hippocampal lesions in rats. *Behavioral Neuroscience, 116*(3), 494–497. Retrieved from: http://www.ncbi.nlm.nih.gov/pubmed/12049331.

Denny, C. A., Kheirbek, M. A., Alba, E. L., Tanaka, K. F., Brachman, R. A., Laughman, K. B., et al. (2014). Hippocampal memory traces are differentially modulated by experience, time, and adult neurogenesis. *Neuron, 83*(1), 189–201. http://doi.org/10.1016/j.neuron.2014.05.018.

Dranovsky, A., & Hen, R. (2006). Hippocampal neurogenesis: regulation by stress and antidepressants. *Biological Psychiatry, 59*(12), 1136–1143. http://doi.org/10.1016/j.biopsych.2006.03.082.

Dupret, D., Revest, J.-M., Koehl, M., Ichas, F., De Giorgi, F., Costet, P., et al. (2008). Spatial relational memory requires hippocampal adult neurogenesis. *PLoS One, 3*(4), e1959. http://doi.org/10.1371/journal.pone.0001959.

Duric, V., & McCarson, K. E. (2006). Persistent pain produces stress-like alterations in hippocampal neurogenesis and gene expression. *The Journal of Pain, 7*(8), 544–555. http://doi.org/10.1016/j.jpain.2006.01.458.

Falconer, E., & Galea, L. (2003). Sex differences in cell proliferation, cell death and defensive behavior following acute predator odor stress in adult rats. *Brain Research, 975*, 22–36. Retrieved from: http://www.sciencedirect.com/science/article/pii/S0006899303025423.

Fanselow, M. S., & Dong, H.-W. (2010). Are the dorsal and ventral hippocampus functionally distinct structures? *Neuron, 65*(1), 7–19. http://doi.org/10.1016/j.neuron.2009.11.031.

Fornal, C. A., Stevens, J., Barson, J. R., Blakley, G. G., Patterson-Buckendahl, P., & Jacobs, B. L. (2007). Delayed suppression of hippocampal cell proliferation in rats following inescapable shocks. *Brain Research, 1130*(1), 48–53. http://doi.org/10.1016/j.brainres.2006.10.081.

Fudge, J. L., deCampo, D. M., & Becoats, K. T. (2012). Revisiting the hippocampal-amygdala pathway in primates: association with immature-appearing neurons. *Neuroscience, 212*, 104–119. http://doi.org/10.1016/j.neuroscience.2012.03.040.

Fuss, J., Ben Abdallah, N. M. B., Hensley, F. W., Weber, K. J., Hellweg, R., & Gass, P. (2010). Deletion of running-induced hippocampal neurogenesis by irradiation prevents development of an anxious phenotype in mice. *PLoS One, 5*(9), 1–9. http://doi.org/10.1371/journal.pone.0012769.

Ge, S., Goh, E. L. K., Sailor, K. A., Kitabatake, Y., Ming, G., & Song, H. (2006). GABA regulates synaptic integration of newly generated neurons in the adult brain. *Nature, 439*(7076), 589–593. http://doi.org/10.1038/nature04404.

Gilbertson, M. W., Shenton, M. E., Ciszewski, A., Kasai, K., Lasko, N. B., Orr, S. P., et al. (2002). Smaller hippocampal volume predicts pathologic vulnerability to psychological trauma. *Nature Neuroscience, 5*(11), 1242–1247. http://doi.org/10.1038/nn958.

Gould, E., McEwen, B. S., Tanapat, P., Galea, L. A., & Fuchs, E. (1997). Neurogenesis in the dentate gyrus of the adult tree shrew is regulated by psychosocial stress and NMDA receptor activation. *The Journal of Neuroscience, 17*(7), 2492–2498. Retrieved from: http://www.ncbi.nlm.nih.gov/pubmed/9065509.

Gould, E., Tanapat, P., McEwen, B. S., Flügge, G., & Fuchs, E. (1998). Proliferation of granule cell precursors in the dentate gyrus of adult monkeys is diminished by stress. *Proceedings of the National Academy of Sciences of the United States of America, 95*(6), 3168–3171. Retrieved from: http://www.pubmedcentral.nih.gov/articlerender.fcgi?artid=19713&tool=pmcentrez&rendertype=abstract.

Gray, J. A., & McNaughton, N. (2000). *The neuropsychology of anxiety: An enquiry into the functions of the septo-hippocampal system*. Oxford University Press. 1982.

Grillon, C., Baas, J. P., Lissek, S., Smith, K., & Milstein, J. (2004). Anxious responses to predictable and unpredictable aversive events. *Behavioral Neuroscience, 118*(5), 916–924. http://doi.org/10.1037/0735-7044.118.5.916.

Hahn, T., Dresler, T., Plichta, M. M., Ehlis, A.-C. C., Ernst, L. H., Markulin, F., et al. (2010). Functional amygdala-hippocampus connectivity during anticipation of aversive events is associated with Gray's trait "sensitivity to punishment". *Biological Psychiatry, 68*(5), 459–464. http://doi.org/10.1016/j.biopsych.2010.04.033.

Hanson, N. D., Owens, M. J., Boss-Williams, K. A., Weiss, J. M., & Nemeroff, C. B. (2011). Several stressors fail to reduce adult hippocampal neurogenesis. *Psychoneuroendocrinology, 36*(10), 1520–1529. http://doi.org/10.1016/j.psyneuen.2011.04.006.

Heine, V. M., Maslam, S., Zareno, J., Joels, M., Lucassen, P. J., & Joe, M. (2004). Suppressed proliferation and apoptotic changes in the rat dentate gyrus after acute and chronic stress are reversible. *European Journal of Neuroscience, 19*(1), 131–144. http://doi.org/10.1046/j.1460-9568.2003.03100.x.

Herman, J. P., Ostrander, M. M., Mueller, N. K., & Figueiredo, H. (2005). Limbic system mechanisms of stress regulation: hypothalamo-pituitary-adrenocortical axis. *Progress in Neuro-Psychopharmacology & Biological Psychiatry, 29*(8), 1201–1213. http://doi.org/10.1016/j.pnpbp.2005.08.006.

Hill, M. N., Kambo, J. S., Sun, J. C., Gorzalka, B. B., & Galea, L. A. M. (2006). Endocannabinoids modulate stress-induced suppression of hippocampal cell proliferation and activation of defensive behaviours. *The European Journal of Neuroscience, 24*(7), 1845–1849. http://doi.org/10.1111/j.1460-9568.2006.05061.x.

Hill, A. S., Sahay, A., & Hen, R. (2015). Increasing adult hippocampal neurogenesis is sufficient to reduce anxiety and depression-like behaviors. *Neuropsychopharmacology*, 1–11. http://doi.org/10.1038/npp.2015.85.

Huff, N. C., Hernandez, J. A., Fecteau, M. E., Zielinski, D. J., Brady, R., & Labar, K. S. (November 2011). Revealing context-specific conditioned fear memories with full immersion virtual reality. *Frontiers in Behavioral Neuroscience, 5*, 1–8. http://doi.org/10.3389/fnbeh.2011.00075.

Imayoshi, I., Sakamoto, M., & Kageyama, R. (May 2011). Genetic methods to identify and manipulate newly born neurons in the adult brain. *Frontiers in Neuroscience, 5*, 64. http://doi.org/10.3389/fnins.2011.00064.

Jayatissa, M. N., Bisgaard, C. F., West, M. J., & Wiborg, O. (2008). The number of granule cells in rat hippocampus is reduced after chronic mild stress and re-established after chronic escitalopram treatment. *Neuropharmacology, 54*(3), 530–541. http://doi.org/10.1016/j.neuropharm.2007.11.009.

Joëls, M., Krugers, H., & Karst, H. (2008). Stress-induced changes in hippocampal function. *Progress in Brain Research, 167*, 3–15. http://doi.org/10.1016/S0079-6123(07)67001-0.

Kambo, J. S., & Galea, L. A. M. (2006). Activational levels of androgens influence risk assessment behaviour but do not influence stress-induced suppression in hippocampal cell proliferation in adult male rats. *Behavioural Brain Research, 175*(2), 263–270. http://doi.org/10.1016/j.bbr.2006.08.032.

Kats, L., & Dill, L. (1998). The scent of death: chemosensory assessment of predation risk by prey animals. *Ecoscience, 5*(3), 361–394. Retrieved from: http://scholar.google.com/scholar?hl=hr&q=The+scent+of+death:+Chemosensory+assessment+of+predation+risk+by+prey+animals.+&btnG=#0.

Kempermann, G., Jessberger, S., Steiner, B., & Kronenberg, G. (2004). Milestones of neuronal development in the adult hippocampus. *Trends in Neurosciences, 27*(8), 447–452. http://doi.org/10.1016/j.tins.2004.05.013.

Kheirbek, M. A., Drew, L. J., Burghardt, N. S., Costantini, D. O., Tannenholz, L., Ahmari, S. E., et al. (2013). Differential control of learning and anxiety along the dorsoventral axis of the dentate gyrus. *Neuron, 77*(5), 955–968. http://doi.org/10.1016/j.neuron.2012.12.038.

Kheirbek, M. A., Klemenhagen, K. C., Sahay, A., & Hen, R. (2012). Neurogenesis and generalization: a new approach to stratify and treat anxiety disorders. *Nature Neuroscience*, *15*(12), 1613–1620. http://doi.org/10.1038/nn.3262.

Kikusui, T., & Mori, Y. (2009). Behavioural and neurochemical consequences of early weaning in rodents. *Journal of Neuroendocrinology*, *21*(4), 427–431. http://doi.org/10.1111/j.1365-2826.2009.01837.x.

King, R. S., DeBassio, W. A., Kemper, T. L., Rosene, D. L., Tonkiss, J., Galler, J. R., et al. (2004). Effects of prenatal protein malnutrition and acute postnatal stress on granule cell genesis in the fascia dentata of neonatal and juvenile rats. *Brain Research. Developmental Brain Research*, *150*(1), 9–15. http://doi.org/10.1016/j.devbrainres.2004.02.002.

Kjelstrup, K. G., Tuvnes, F. A., Steffenach, H.-A., Murison, R., Moser, E. I., & Moser, M.-B. (2002). Reduced fear expression after lesions of the ventral hippocampus. *Proceedings of the National Academy of Sciences of the United States of America*, *99*, 10825–10830. http://doi.org/10.1073/pnas.152112399.

Lagace, D. C., Donovan, M. H., DeCarolis, N. A., Farnbauch, L. A., Malhotra, S., Berton, O., et al. (2010). Adult hippocampal neurogenesis is functionally important for stress-induced social avoidance. *Proceedings of the National Academy of Sciences of the United States of America*, *107*(9), 4436–4441. http://doi.org/10.1073/pnas.0910072107.

Lehmann, M. L., Brachman, R. A., Martinowich, K., Schloesser, R. J., & Herkenham, M. (2013). Glucocorticoids orchestrate divergent effects on mood through adult neurogenesis. *The Journal of Neuroscience*, *33*(7), 2961–2972. http://doi.org/10.1523/JNEUROSCI.3878-12.2013.

Leutgeb, J. K., Leutgeb, S., Moser, M.-B., & Moser, E. I. (2007). Pattern separation in the dentate gyrus and CA3 of the hippocampus. *Science (New York, N.Y.)*, *315*(5814), 961–966. http://doi.org/10.1126/science.1135801.

Levita, L., Bois, C., Healey, A., Smyllie, E., Papakonstantinou, E., Hartley, T., et al. (2014). The behavioural inhibition system, anxiety and hippocampal volume in a non-clinical population. *Biology of Mood & Anxiety Disorders*, *4*(1), 4. http://doi.org/10.1186/2045-5380-4-4.

Liu, X., Ramirez, S., Pang, P. T., Puryear, C. B., Govindarajan, A., Deisseroth, K., et al. (2012). Optogenetic stimulation of a hippocampal engram activates fear memory recall. *Nature*, *484*(7394), 381–385. http://doi.org/10.1038/nature11028.

Maack, D. J., Tull, M. T., & Gratz, K. L. (2012). Examining the incremental contribution of behavioral inhibition to generalized anxiety disorder relative to other Axis I disorders and cognitive-emotional vulnerabilities. *Journal of Anxiety Disorders*, *26*(6), 689–695. http://doi.org/10.1016/j.janxdis.2012.05.005.

Malberg, J. E., & Duman, R. S. (2003). Cell proliferation in adult hippocampus is decreased by inescapable stress: reversal by fluoxetine treatment. *Neuropsychopharmacology*, *28*(9), 1562–1571. http://doi.org/10.1038/sj.npp.1300234.

Malberg, J. E., Eisch, A. J., Nestler, E. J., & Duman, R. S. (2000). Chronic antidepressant treatment increases neurogenesis in adult rat hippocampus. *The Journal of Neuroscience*, *20*(24), 9104–9110.

Marschner, A., Kalisch, R., Vervliet, B., Vansteenwegen, D., Büchel, C., & Bu, C. (2008). Dissociable roles for the hippocampus and the amygdala in human cued versus context fear conditioning. *The Journal of Neuroscience*, *28*(36), 9030–9036. http://doi.org/10.1523/JNEUROSCI.1651-08.2008.

Mateus-Pinheiro, A., Pinto, L., Bessa, J. M., Morais, M., Alves, N. D., Monteiro, S., et al. (2013). Sustained remission from depressive-like behavior depends on hippocampal neurogenesis. *Translational Psychiatry*, *3*(October 2012), e210. http://doi.org/10.1038/tp.2012.141.

Meshi, D., Drew, M. R., Saxe, M., Ansorge, M. S., David, D., Santarelli, L., et al. (2006). Hippocampal neurogenesis is not required for behavioral effects of environmental enrichment. *Nature Neuroscience*, *9*(6), 729–731. http://doi.org/10.1038/nn1696.

Ming, G.-L., & Song, H. (2011). Adult neurogenesis in the mammalian brain: significant answers and significant questions. *Neuron*, *70*(4), 687–702. http://doi.org/10.1016/j.neuron.2011.05.001.

Mirescu, C., & Gould, E. (2006). Stress and adult neurogenesis. *Hippocampus, 16*(3), 233–238. http://doi.org/10.1002/hipo.20155.

Mirescu, C., Peters, J. D., & Gould, E. (2004). Early life experience alters response of adult neurogenesis to stress. *Nature Neuroscience, 7*(8), 841–846. http://doi.org/10.1038/nn1290.

Nakashiba, T., Cushman, J. D., Pelkey, K. A., Renaudineau, S., Buhl, D. L., McHugh, T. J., et al. (2012). Young dentate granule cells mediate pattern separation, whereas old granule cells facilitate pattern completion. *Cell, 149*(1), 188–201. http://doi.org/10.1016/j.cell.2012.01.046.

Onksen, J. L., Brown, E. J., & Blendy, J. A. (2011). Selective deletion of a cell cycle checkpoint kinase (ATR) reduces neurogenesis and alters responses in rodent models of behavioral affect. *Neuropsychopharmacology, 36*(5), 960–969. http://doi.org/10.1038/npp.2010.234.

Petrik, D., Lagace, D. D. C., & Eisch, A. A. J. (2012). The neurogenesis hypothesis of affective and anxiety disorders: are we mistaking the scaffolding for the building? *Neuropharmacology, 62*(1), 21–34. http://doi.org/10.1016/j.neuropharm.2011.09.003.

Pham, K., Nacher, J., Hof, P. R., & McEwen, B. S. (2003). Repeated restraint stress suppresses neurogenesis and induces biphasic PSA-NCAM expression in the adult rat dentate gyrus. *European Journal of Neuroscience, 17*(4), 879–886. http://doi.org/10.1046/j.1460-9568.2003.02513.x.

Pitman, R. K., Rasmusson, A. M., Koenen, K. C., Shin, L. M., Orr, S. P., Gilbertson, M. W., et al (2012). Biological studies of post-traumatic stress disorder. *Nature Reviews Neuroscience, 13*(11), 769–787. http://doi.org/10.1038/nrn3339.

Pontes, A., Zhang, Y., & Hu, W. (2013). Novel functions of GABA signaling in adult neurogenesis. *Frontiers in Biology, 8*(5), 496–507. http://doi.org/10.1007/s11515-013-1270-2.

Pruessner, M., Pruessner, J. C., Hellhammer, D. H., Bruce Pike, G., & Lupien, S. J. (2007). The associations among hippocampal volume, cortisol reactivity, and memory performance in healthy young men. *Psychiatry Research, 155*(1), 1–10.

Revest, J.-M., Dupret, D., Koehl, M., Funk-Reiter, C., Grosjean, N., Piazza, P.-V., et al. (2009). Adult hippocampal neurogenesis is involved in anxiety-related behaviors. *Molecular Psychiatry, 14*(10), 959–967. http://doi.org/10.1038/mp.2009.15.

Risold, P. Y., & Swanson, L. W. (1996). Structural evidence for functional domains in the rat hippocampus. *Science, 272*(5267), 1484–1486. http://doi.org/10.1126/science.272.5267.1484.

Rosenbrock, H., Koros, E., Bloching, A., Podhorna, J., & Borsini, F. (2005). Effect of chronic intermittent restraint stress on hippocampal expression of marker proteins for synaptic plasticity and progenitor cell proliferation in rats. *Brain Research, 1040*(1–2), 55–63. http://doi.org/10.1016/j.brainres.2005.01.065.

Sahay, A., Scobie, K. N., Hill, A. S., O'Carroll, C. M., Kheirbek, M. A., Burghardt, N. S., et al. (2011). Increasing adult hippocampal neurogenesis is sufficient to improve pattern separation. *Nature, 472*(7344), 466–470. http://doi.org/10.1038/nature09817.

Santarelli, L., Saxe, M., Gross, C., Surget, A., Battaglia, F., Dulawa, S., et al. (2003). Requirement of hippocampal neurogenesis for the behavioral effects of antidepressants. *Science (New York, N.Y.), 301*(5634), 805–809. http://doi.org/10.1126/science.1083328.

Santoro, A. (2013). Reassessing pattern separation in the dentate gyrus. *Frontiers in Behavioral Neuroscience, 7*, 96. http://doi.org/10.3389/fnbeh.2013.00096.

Saxe, M. D., Battaglia, F., Wang, J., Malleret, G., David, D. J., Monckton, J. E., et al. (2006). Ablation of hippocampal neurogenesis impairs contextual fear conditioning and synaptic plasticity in the dentate gyrus. *Proceedings of the National Academy of Sciences, 103*(46), 17501–17506.

Schloesser, R. J., Lehmann, M., Martinowich, K., Manji, H. K., & Herkenham, M. (2010). Environmental enrichment requires adult neurogenesis to facilitate the recovery from psychosocial stress. *Molecular Psychiatry, 15*(12), 1152–1163. http://doi.org/10.1038/mp.2010.34.

Schmidt, M. V., Wang, X.-D., & Meijer, O. C. (2011). Early life stress paradigms in rodents: potential animal models of depression? *Psychopharmacology, 214*(1), 131–140. http://doi.org/10.1007/s00213-010-2096-0.

Schoenfeld, T. J., & Gould, E. (2012). Stress, stress hormones, and adult neurogenesis. *Experimental Neurology*, *233*(1), 12–21. http://doi.org/10.1016/j.expneurol.2011.01.008.

Sheline, Y. I. (2000). 3D MRI studies of neuroanatomic changes in unipolar major depression: the role of stress and medical comorbidity. *Biological Psychiatry*, *48*(8), 791–800. http://doi.org/10.1016/S0006-3223(00)00994-X.

Shors, T. J., Townsend, D. A., Zhao, M., Kozorovitskiy, Y., & Gould, E. (2002). Neurogenesis may relate to some but not all types of hippocampal-dependent learning. *Hippocampus*, *12*, 578–584. http://doi.org/10.1002/hipo.10103.

Snyder, J. S., Choe, J. S., Clifford, M. A., Jeurling, S. I., Hurley, P., Brown, A., et al. (2009). Adult-born hippocampal neurons are more numerous, faster maturing, and more involved in behavior in rats than in mice. *The Journal of Neuroscience*, *29*(46), 14484–14495. http://doi.org/10.1523/JNEUROSCI.1768-09.2009.

Snyder, J. S., Kee, N., & Wojtowicz, J. M. (2001). Effects of adult neurogenesis on synaptic plasticity in the rat dentate gyrus. *Journal of Neurophysiology*, *85*(6), 2423–2431. Retrieved from: http://jn.physiology.org/content/85/6/2423.

Snyder, J. S., Soumier, A., Brewer, M., Pickel, J., & Cameron, H. A. (2011). Adult hippocampal neurogenesis buffers stress responses and depressive behaviour. *Nature*, *476*(7361), 458–461. http://doi.org/10.1038/nature10287.

Staples, L. G., McGregor, I. S., Apfelbach, R., & Hunt, G. E. (2008). Cat odor, but not trimethylthiazoline (fox odor), activates accessory olfactory and defense-related brain regions in rats. *Neuroscience*, *151*(4), 937–947. http://doi.org/10.1016/j.neuroscience.2007.11.039.

Steiner, B., Kronenberg, G., Jessberger, S., Brandt, M. D., Reuter, K., Kempermann, G., et al. (2004). Differential regulation of gliogenesis in the context of adult hippocampal neurogenesis in mice. *Glia*, *46*(1), 41–52. http://doi.org/10.1002/glia.10337.

Surget, A., Saxe, M., Leman, S., Ibarguen-Vargas, Y., Chalon, S., Griebel, G., et al. (2008). Drug-dependent requirement of hippocampal neurogenesis in a model of depression and of antidepressant reversal. *Biological Psychiatry*, *64*(4), 293–301. http://doi.org/10.1016/j.biopsych.2008.02.022.

Surget, A., Tanti, A., Leonardo, E. D., Laugeray, A., Rainer, Q., Touma, C., et al. (2011). Antidepressants recruit new neurons to improve stress response regulation. *Molecular Psychiatry*, *16*(12), 1177–1188. http://doi.org/10.1038/mp.2011.48.

Tanapat, P., Hastings, N. B., Rydel, T. A., Galea, L. A. M.M., & Gould, E. (2001b). Exposure to fox odor inhibits cell proliferation in the hippocampus of adult rats via an adrenal hormone-dependent mechanism. *The Journal of Comparative Neurology*, *437*(4), 496–504. http://doi.org/10.1002/cne.1297.

Tanti, A., & Belzung, C. (2013). Hippocampal neurogenesis: a biomarker for depression or antidepressant effects? Methodological considerations and perspectives for future research. *Cell and Tissue Research*, *354*(1), 203–219. http://doi.org/10.1007/s00441-013-1612-z.

Tanti, A., Rainer, Q., Minier, F., Surget, A., & Belzung, C. (2012). Differential environmental regulation of neurogenesis along the septo-temporal axis of the hippocampus. *Neuropharmacology*, *63*(3), 374–384. http://doi.org/10.1016/j.neuropharm.2012.04.022.

Tanti, A., Westphal, W.-P., Girault, V., Brizard, B., Devers, S., Leguisquet, A.-M., et al. (2013). Region-dependent and stage-specific effects of stress, environmental enrichment, and antidepressant treatment on hippocampal neurogenesis. *Hippocampus*, *23*(9), 797–811. http://doi.org/10.1002/hipo.22134.

Teicher, M. H., Anderson, C. M., & Polcari, A. (2012). PNAS Plus: childhood maltreatment is associated with reduced volume in the hippocampal subfields CA3, dentate gyrus, and subiculum. *Proceedings of the National Academy of Sciences*, *109*(35), E563–E572. http://doi.org/10.1073/pnas.1115396109.

Thomas, R. M., Hotsenpiller, G., & Peterson, D. A. (2007). Acute psychosocial stress reduces cell survival in adult hippocampal neurogenesis without altering proliferation. *The Journal of Neuroscience*, *27*(11), 2734–2743. http://doi.org/10.1523/JNEUROSCI.3849-06.2007.

Thomas, R. M., Urban, J. H., & Peterson, D. A. (2006). Acute exposure to predator odor elicits a robust increase in corticosterone and a decrease in activity without altering proliferation in the adult rat hippocampus. *Experimental Neurology*, *201*(2), 308–315. http://doi.org/10.1016/j.expneurol.2006.04.010.

Torrubia, R., Ávila, C., Moltó, J., & Caseras, X. (2001). The sensitivity to punishment and sensitivity to reward questionnaire (SPSRQ) as a measure of Gray's anxiety and impulsivity dimensions. *Personality and Individual Differences*, *31*, 837–862. http://doi.org/10.1016/S0191-8869(00)00183-5.

Tronel, S., Belnoue, L., Grosjean, N., Revest, J.-M., Piazza, P.-V., Koehl, M., et al. (2012). Adult-born neurons are necessary for extended contextual discrimination. *Hippocampus*, *22*(2), 292–298. http://doi.org/10.1002/hipo.20895.

Tse, Y. C., Montoya, I., Wong, A. S., Mathieu, A., Lissemore, J., Lagace, D. C., et al. (2014). A longitudinal study of stress-induced hippocampal volume changes in mice that are susceptible or resilient to chronic social defeat. *Hippocampus*, *24*(9), 1120–1128. http://doi: 10.1002/hipo.22296.

Van Groen, T., & Wyss, J. M. (1990). Extrinsic projections from area CA1 of the rat hippocampus: olfactory, cortical, subcortical, and bilateral hippocampal formation projections. *The Journal of Comparative Neurology*, *302*(3), 515–528. http://doi.org/10.1002/cne.903020308.

Vollmayr, B., Simonis, C., Weber, S., Gass, P., & Henn, F. (2003). Reduced cell proliferation in the dentate gyrusis not correlated with the development of learned helplessness. *Biological Psychiatry*, *54*(10), 1035–1040. http://doi.org/10.1016/S0006-3223(03)00527-4.

Vukovic, J., Borlikova, G. G., Ruitenberg, M. J., Robinson, G. J., Sullivan, R. K. P., Walker, T. L., et al. (2013). Immature doublecortin-positive hippocampal neurons are important for learning but not for remembering. *The Journal of Neuroscience*, *33*(15), 6603–6613. http://doi.org/10.1523/JNEUROSCI.3064-12.2013.

Wang, J.-W., David, D. J., Monckton, J. E., Battaglia, F., & Hen, R. (2008). Chronic fluoxetine stimulates maturation and synaptic plasticity of adult-born hippocampal granule cells. *The Journal of Neuroscience*, *28*(6), 1374–1384. http://doi.org/10.1523/JNEUROSCI.3632-07.2008.

Wang, Z., Neylan, T. C., Mueller, S. G., Lenoci, M., Truran, D., Marmar, C. R., et al. (2010). Magnetic resonance imaging of hippocampal subfields in posttraumatic stress disorder. *Archives of General Psychiatry*, *67*(3), 296–303. http://doi.org/10.1001/archgenpsychiatry.2009.205.

Weeden, C. S. S., Roberts, J. M., Kamm, A. M., & Kesner, R. P. (2015). Neurobiology of learning and memory the role of the ventral dentate gyrus in anxiety-based behaviors. *Neurobiology of Learning and Memory*, *118*, 143–149. http://doi.org/10.1016/j.nlm.2014.12.002.

Wu, M. V., & Hen, R. (2014). Functional dissociation of adult-born neurons along the dorsoventral axis of the dentate gyrus. *Hippocampus*, *24*, 751–761. http://doi.org/10.1002/hipo.22265.

Yap, J. J., Takase, L. F., Kochman, L. J., Fornal, C. A., Miczek, K. A., & Jacobs, B. L. (2006). Repeated brief social defeat episodes in mice: effects on cell proliferation in the dentate gyrus. *Behavioural Brain Research*, *172*(2), 344–350. http://doi.org/10.1016/j.bbr.2006.05.027.

Zhu, X.-H., Yan, H.-C., Zhang, J., Qu, H.-D., Qiu, X.-S., Chen, L., et al. (2010). Intermittent hypoxia promotes hippocampal neurogenesis and produces antidepressant-like effects in adult rats. *The Journal of Neuroscience*, *30*(38), 12653–12663. http://doi.org/10.1523/JNEUROSCI.6414-09.2010.

CHAPTER

10

Addiction

J.J. Canales

University of Leicester, Leicester, United Kingdom

INTRODUCTION

Addiction is a multifaceted disorder characterized by persistent episodes of excessive and compulsive drug taking, preoccupation over drug availability, and recurrent relapses. It is a disorder that takes a very significant toll on the people affected, particularly in terms of quality of life and health status, and on the health system and the society as a whole. Many advances have been made in recent years in identifying some of the risk factors that may predispose to addiction and in delineating the brain mechanisms and adaptations linked to drug-related behaviors. The Diagnostic and Statistical Manual of Mental Disorders, Fifth Edition (DSM-V), recognizes that not all individuals have the same degree of vulnerability to develop an addiction if exposed to drugs (American Psychiatric Association, 2013). Thus one of the key objectives of current addiction research is to understand the biological and psychological underpinnings of such differential predisposition to problematic drug use. The DSM-V further recognizes that activation of the so-called brain reward systems is a landmark that is common across addictive substances. Research in the last 30 years into the neurobiology of drug addiction has been mainly focused on the basal ganglia and their affiliated cortical networks as key components of the "addiction circuitry." Studies conducted in the 1980s and 1990s established that the nucleus accumbens (NAcb) and its dopamine (DA) innervation played a critical role in the positive reinforcing and motivational effects of addictive drugs, including cocaine, alcohol, and opiates (Koob & Weiss, 1992; White & Kalivas, 1998; Wise, 1987). Furthermore, strong evidence indicated that chronic drug exposure was associated with lasting neuroadaptations in the DA pathways that originate in the ventral tegmental area (VTA) and terminate in the ventral striatum (Feltenstein & See, 2008; Self, 2004; Wolf, Sun, Mangiavacchi, & Chao, 2004).

Adult Neurogenesis in the Hippocampus
http://dx.doi.org/10.1016/B978-0-12-801977-1.00010-6

Recently, the research focus has shifted toward understanding the neurobiological mechanisms that mediate the transition from sporadic drug use to full-blown addiction. On the one hand, addiction is thought to be associated with the emergence of a negative emotional state that is driven by hedonic dysregulation and appears when access to the drug is prevented (Koob & Le Moal, 2008). According to this notion, escalation of drug taking and progression to compulsion are facilitated by negative reinforcement and dysregulation of the reward and stress systems, which are under the control of DA and endogenous opioids in the NAcb, and of corticotropin-releasing factor in the extended amygdala (AMY) network (Koob, 2009). On the other hand, such transition appears to be paralleled by a shift from ventral to dorsal striatum (DSt) in the control of drug-associated behaviors (Belin, Jonkman, Dickinson, Robbins, & Everitt, 2009; Everitt & Robbins, 2013; Robbins, Cador, Taylor, & Everitt, 1989). Drugs of abuse possess, as do natural reinforcers such as food and sexual contact, the ability to strengthen stimulus–response associations and form pervasive drug-related habits, which are thought to be controlled by the DSt (Belin & Everitt, 2008). This latter notion integrates concepts of the learning theory, most notably Pavlovian and instrumental conditioning, into the defining algorithm of addiction and provides a biological framework that pictures addiction as a staging process that evolves from occasional, reward-driven drug use to compulsive abuse triggered by drug-related cues. Although the process seems to depend on serial connectivity and information transfer between the midbrain DA neurons and the striatum (Belin & Everitt, 2008; Haber, Fudge, & McFarland, 2000), the contribution of cortical and allocortical areas projecting to the ventral and dorsal striatum to the progression of changes associated with addiction has not yet been fully elucidated.

In this context, one of the principal allocortical areas projecting to the striatum is the hippocampus. Traditionally the hippocampus has not been considered to be part of the main network of connections implicated in the habit-forming effects of drugs; however, it is widely interconnected with the systems that mediate drug reward and reinforcement, as well as with brain regions that are involved in endocrine regulation, associative learning, and decision making, including the hypothalamus, the AMY, and the prefrontal cortex (PFC) (Fig. 10.1). The precommissural branch of the fornix innervates the ventral striatum, and the orbital and the anterior cingulate cortices, which are generally involved in drug reinforcement and decision making, respectively. Moreover, glutamatergic afferents from the hippocampus to the NAcb strongly regulate the firing rate of DA neurons in the VTA (Floresco, Todd, & Grace, 2001) and a transsynaptic CA3–VTA loop has been demonstrated (Luo, Tahsili-Fahadan, Wise, Lupica, & Aston-Jones, 2011), suggesting that hippocampal activity can influence drug-stimulated effects on DA function. Since drug-seeking behaviors are critically linked to specific contexts, the presence of this

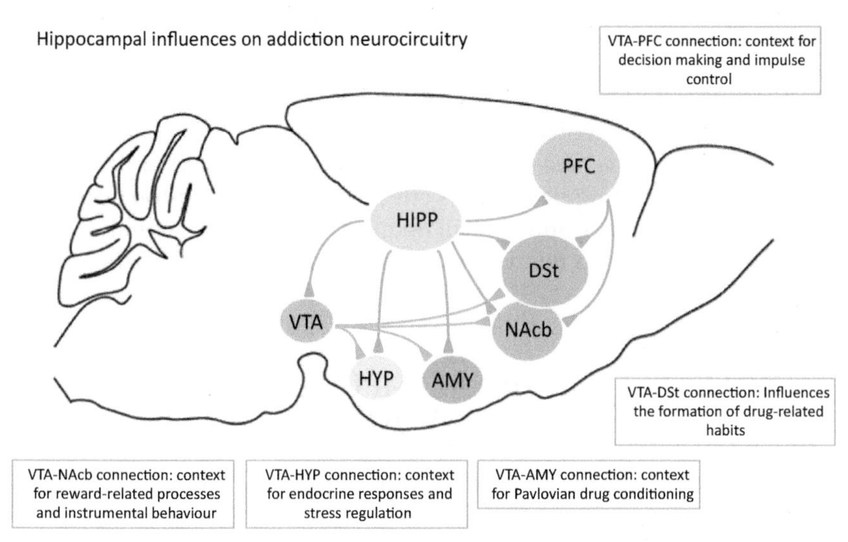

FIGURE 10.1 Connections of the hippocampus with functionally relevant brain regions mediating drug-related behaviors. The discovery of neurogenesis in the adult hippocampus, coupled with the conclusive accumulated evidence that addictive drugs impair the neurogenic process, has led to the expansion of the addiction neurocircuitry. The hippocampus exerts strong influences on midbrain dopamine neurons in the ventral tegmental area (VTA) that mediate drug reward and habit learning by way of projections to the nucleus accumbens (NAcb) and dorsal striatum (DSt) (note that for simplicity connections between the substantia nigra and the DSt are omitted). Hippocampal projections to the hypothalamus (HYP) provide a context for, and regulate, stress responsiveness, while reciprocating connections with the amygdala network (AMY) may influence drug conditioning and emotional responses to drugs. Through direct connections with the prefrontal cortex (PFC), the hippocampus may also provide a context for decision making and impulse control, with impaired neurogenesis resulting in cognitive inflexibility.

hippocampal–NAcb–VTA transneuronal connection provides a scaffold for the hippocampus to engage in drug-mediated behaviors. It is also important to consider that midbrain DA neurons project to the hippocampus and surrounding areas in the medial temporal lobe (Gasbarri, Sulli, & Packard, 1997), thus addictive drugs may have direct effects on hippocampal function through stimulation of dopaminergic afferents.

The remarkable ability of the hippocampal dentate gyrus (DG) to sustain postembryonic neurogenesis not only has added a new piece to the puzzle of addiction but has also created a new paradigm for understanding the neurobiology of this intricate disorder. Undoubtedly, adult neurogenesis represents one of the most intriguing and interesting forms of brain plasticity. The discovery that addictive drugs alter neurogenesis in the adult hippocampus (Canales, 2007; Eisch & Harburg, 2006), together with mounting evidence of a significant functional role of the adult-born hippocampal neurons, has consolidated the position of the hippocampal network as part of a wider circuitry dynamically engaged in influencing addictive behavior. To provide a framework for understanding the relationship between adult neurogenesis and addiction, in this chapter we will first discuss the involvement of the hippocampus in the phenomenology of drug addiction. We will then examine accumulated experimental evidence demonstrating deleterious effects of abused drugs on adult hippocampal neurogenesis, and we will finally consider the potential involvement of adult-generated neurons in the hippocampus in the pathogenesis of addiction.

ROLE OF THE HIPPOCAMPUS IN ADDICTIVE BEHAVIOR

As indicated previously, the hippocampus has not traditionally been regarded as an integral part of the neurocircuitry involved in drug reward and reinforcement. However, multiple lines of evidence suggest that the hippocampus critically contributes to shaping the context of drug-related memories. Drug abuse can generate strong and persistent associations between drug subjective experiences and specific contexts, broadly defined here as a constellation of both exteroceptive and interoceptive cues linked to drug taking. Such contexts can evoke intense desires to take drugs, thus acting as a powerful driving force for relapse. The hippocampus has been involved in these processes and also in more basic behaviors linked to repeated drug exposure in experimental animals, such as sensitization and place preference.

Repeated drug exposure is known to produce long-lasting behavioral adaptations, including both tolerance and sensitization, which to a large extent are behavior specific; that is, the same drug may produce tolerance to certain behavioral effects and sensitization to others. For example, repeated treatment with psychomotor stimulants such as nicotine or amphetamines induces progressively increased locomotor behavior, which is associated with neuroadaptations in the mesolimbic DA system and changes in glutamatergic transmission in the NAcb (Kalivas, 1995; Wolf et al., 2004). Sensitized activation of the DA system by conditioned cues may underlie the increased subjective salience of drug-related stimuli and the enhanced motivation to seek and take drugs (Robinson & Berridge, 2008; Leyton, 2007). Previous studies have shown that reversible inactivation of the dorsal, but not the ventral hippocampus, prevented the expression of behavioral sensitization to amphetamines (Degoulet, Rouillon, Rostain, David, & Abraini, 2008), suggesting that the hippocampus plays a role in maintaining a representation of the drug-associated context and the cues that powerfully influence drug-related behaviors. Moreover, electrophysiological evidence indicated that sensitization to amphetamine was related to both increased theta coherence between the hippocampus and the PFC and suppression of cross-correlations between theta and gamma frequencies in both regions (Lapish, Chiang, Wang, & Phillips, 2012), which indicates that timing and synchronization of communication between those two regions might be disrupted after drug sensitization. Similarly, long-term potentiation (LTP), a key neurophysiological mechanism implicated in hippocampal plasticity and memory, appears to be altered by chronic drug exposure. Furthermore, repeated cocaine treatment reduced the threshold for induction of LTP in the rat DG, which could in turn facilitate the retention of drug-related contextual memories (Perez et al., 2010).

A behavioral paradigm that has been extensively used to characterize the role of the hippocampus in forming contextual representations of the drug environment is the conditioned place preference (CPP) paradigm. In this Pavlovian conditioning task, which is typically performed in a two-compartment apparatus, rats learn the association between a drug-paired and a non-drug-paired environment, such that a preference is exhibited for the drug-paired compartment after drug conditioning, thereby indicating that the drug has rewarding properties (Tzschentke, 1998). The CPP and the contextual fear paradigms have been instrumental in establishing that the integrity of the DG of the hippocampus is essential for maintaining a coherent representation of the context to which emotional experiences, either hedonic or aversive, are bound. The establishment of cocaine CPP was blocked by colchicine-induced lesions of the DG (Hernandez-Rabaza et al., 2008). In agreement with these findings, inactivation of the dorsal hippocampus with the GABAa receptor agonist muscimol blocked the expression of cocaine CPP (Raybuck & Lattal, 2014). Further suggesting that activation of the hippocampus is necessary for the retrieval of the drug-associated context, recent observations revealed that such retrieval is accompanied by strong induction of the early gene, cFos, a marker of cellular activity, in granular neurons of the DG (Rivera et al., 2015). Moreover, reexposure to the drug-associated environment led to increased phosphorylation of cAMP-response element binding protein (CREB), extracellular signal-regulated kinase (ERK), and GluR1 (Tropea, Kosofsky, & Rajadhyaksha, 2008). In addition to activation of the ERK/CREB pathway, signal transduction mediated through calcium/calmodulin protein kinase II (CaMKII) is essential for hippocampal-dependent learning and memory processes, including CPP. Microinjections of the CaMKII inhibitor, KN-62, into the hippocampus reduced naloxone-precipitated morphine withdrawal symptoms and suppressed the formation and reactivation of morphine CPP (Lu, Zeng, Liu, & Ceng, 2000). Recent data have further demonstrated a role for NMDA receptor-mediated transmission in the hippocampus in morphine CPP, with increased synaptic expression of NR1 and NR2b NMDA receptor subunits and impaired LTP (Portugal et al., 2014), as previously shown in sensitized rats (Perez et al., 2010). Collectively, these data clearly indicate that the formation of contextual drug reward associations involves recruitment of hippocampal circuits, which are likely to act in concert with other brain regions implicated in the encoding and retrieval of drug-associated memories, including AMY, NAcb, and PFC.

The animal model of addiction with the greatest face validity is the self-administration model, which allows assessment of voluntary drug intake as well as relapse behaviors. Relapse can be elicited after abstinence and reexposure to the self-administration context or after extinction and reinstatement. Extinguished drug-seeking behaviors in experimental animals can be reinstated by conditioned cues, reexposure to the self-administered drug, and stressors, as indeed occurs in humans (Marchant, Li, & Shaham, 2013). Evidence accrued so far shows clearly that the hippocampus plays a role in drug taking, in drug withdrawal, and especially in relapse to drug seeking. Lidocaine inactivation of the dorsal subiculum significantly reduced cocaine taking during the drug maintenance phase (Black, Green-Jordan, Eichenbaum, & Kantak, 2004). Magnetic resonance imaging studies in rats following chronic, extended access to cocaine self-administration showed reduced cerebral blood volume, a measure of brain metabolism, in fronto-cortical areas, NAcb and ventral hippocampus (Gozzi et al., 2011). Moreover, withdrawal from chronic cocaine self-administration is linked to alterations in LTP, including an increased magnitude in the first 3 days after cocaine withdrawal and a persistent decrease 100 days postwithdrawal (Thompson, Swant, Gosnell, & Wagner, 2004).

Reinstatement of drug-seeking behavior is also affected by different types of hippocampal manipulations. Bilateral microinjection of baclofen, a GABA receptor agonist, into the ventral hippocampus attenuated both cue-induced and cocaine-primed reinstatement of cocaine seeking without altering food-maintained responding (Rogers & See, 2007), suggesting a participation of the ventral hippocampus in forms of drug reinstatement not depending exclusively on processing of contextual cues. On the other hand, tetrodotoxin inactivation abolished contextual, but failed to alter explicit conditioned cue- or cocaine-induced, reinstatement of cocaine-seeking behavior (Fuchs et al., 2005; Martin-Fardon, Ciccocioppo, Aujla, & Weiss, 2008). Conversely, theta burst stimulation of glutamatergic fibers in the ventral subiculum effectively reinstated cocaine seeking (Vorel, Liu, Hayes, Spector, & Gardner, 2001). Similarly, focal electrical stimulation of the ventral subiculum (20 Hz) reinstated amphetamine seeking, which was accompanied by DA release in the NAcb and was prevented by DA receptor antagonists (Taepavarapruk, Butts, & Phillips, 2014).

Human studies are consistent with the notion that the hippocampus plays a role in mediating some aspects of drug craving and relapse. A positron emission tomography (PET) study showed that internally generated drug craving through the stimulation of mental imaginery of personalized drug use led to activation of the hippocampus in crack cocaine-dependent subjects (Kilts et al., 2001). Moreover, in cocaine-dependent individuals the presentation of cocaine-related cues elicited DA release in the AMY and the hippocampus, the release being markedly higher in high-craving responders as determined by PET (Fotros et al., 2013). Using arterial spin-labeled perfusion magnetic resonance imaging, Wang et al. (2007) showed that abstinence-induced cravings to smoke elicited increased cerebral blood flow in the orbitofrontal, cingulate, and PFCs, striatum, AMY, insula, and hippocampus, confirming previous observations (Zubieta et al., 2005). Taken together, these findings indicate that drug craving is associated with the activation of the hippocampus, and coincide with the general notion that the preoccupation/anticipation stage in addiction involves recruitment of a network of limbic, striatal, and paralimbic regions, including the orbitofrontal/prefrontal cortex, dorsal striatum, AMY, insula, and hippocampus (Koob & Volkow, 2010).

TOXIC EFFECTS OF DRUGS ON ADULT HIPPOCAMPAL NEUROGENESIS

Understanding how adult-generated neurons in the adult hippocampus influence the mental processes and behaviors associated with addiction remains a major challenge. The discovery of the existence of adult-generated progenitors that eventually mature and integrate in the hippocampal circuitry has spurred a wealth of research into two main aspects: (1) the characterization of the toxic effects of drugs on neurogenic processes and (2) the relevance of such effects for understanding drug-induced cognitive dysfunction and addictive behavior. Many of the early studies in this field focused on assessing the neurotoxic effects of various drug classes on the multiple steps involved in the neurogenic process, including proliferation of neural progenitors and survival, growth, maturation, and integration of adult-born neurons. In this section we will briefly revise the main findings of such experiments, which by and large demonstrate deleterious effects of addictive drugs on hippocampal neurogenesis. Some of these findings have been previously reviewed (Canales, 2007, 2010). We will consider the principal drug classes, including alcohol, psychomotor stimulants, opiates, and hallucinogenic drugs (Table 10.1).

The effects of alcohol on adult hippocampal neurogenesis have been extensively investigated in different animal models and at different stages of brain development, ranging from prenatal to adult exposure. Initial observations indicated that chronic exposure to alcohol reduced the survival of newly generated neurons in the rat hippocampus, an effect that was prevented by antioxidant treatment (Herrera et al., 2003). Similarly, binge-like or chronic alcohol exposure for 1, 2, or 4 weeks in rats decreased cell proliferation in the DG, measured with proliferating cell nuclear antigen

TABLE 10.1 Studies on the Principal Drug Classes and Their Effects

Drug	Subjects	Treatment	Cytogenesis	Maturation	Survival	References
Amphetamine	Adult rats	Chronic 2-week treatment + withdrawal	↓			Barr et al. (2010)
Alcohol	Adult mice	Two-bottle free choice	↑			Aberg et al. (2015)
Alcohol	Adult rats	2-week free access	↓			Anderson et al. (2012)
Alcohol	Adolescent rats	3-week intragastric	=		↓	Broadwater et al. (2014)
Alcohol	Adult mice	Free access	↓			Crews et al. (2004)
Alcohol	Rats	Prenatal, chronic	=		=	Gil-Mohapel et al. (2011)
Alcohol	Rats	Early postnatal, binge-like			↓	Hamilton et al. (2011)
Alcohol	Adult rats	Chronic	↓	↓	↓	He et al. (2005)
Alcohol	Adult rats	6-week alcohol feeding			↓	Herrera et al. (2003)
Alcohol	Rats	Early postnatal, single treatment	↓			Ieraci and Herrera (2007)
Alcohol	Rats	Early postnatal, binge-like			↓	Klintsova et al. (2007)
Alcohol	Adolescent rats	Binge-like	↓	↓	↓	Morris et al. (2010)
Alcohol	Adult rats	Binge-like	↓		↓	Nixon and Crews (2002)
Alcohol	Adult rats	Prenatal, chronic	↓		↓	Redila et al. (2006)
Alcohol + MDMA	Rats	Prenatal, binge-like	↓	↓		Canales and Ferrer-Donato (2014)

Alcohol + MDMA	Adolescent rats	Binge-like		=	↓	Hernandez-Rabaza et al. (2010)
Cocaine	Adult rats	Chronic 8-day and 24-day treatment	↓	=	=	Dominguez-Escriba et al. (2006)
Cocaine	Adult rats	Self-administration	↓			Noonan et al. (2008)
Heroin	Adult rats	Self-administration	↓			Eisch et al, (2000)
MDMA	Adult rats	Chronic	↑		↓	Catlow et al. (2010)
MDMA	Adult rats	Prenatal exposure	↓		↓	Cho et al. (2008)
MDMA	Adult rats	Binge-like exposure	=	=	↓	Hernandez-Rabaza et al. (2006)
Methamphetamine	Adult rats	Self-administration	↓			Recinto et al. (2012)
Methamphetamine	Gerbil	Single treatment	↓			Teuchert-Noodt et al. (2000)
Morphine	Adult rats	Chronic 5-day treatment	↓	=	↓	Eisch et al. (2000)
Morphine	Mouse	48-h treatment	↓			Arguello et al. (2008)

expression or 5-bromodeoxyuridine (BrdU) incorporation, impaired the growth and maturation of doublecortin-positive neurons, and decreased survival rates and neurogenesis several weeks after treatment (He, Nixon, Shetty, & Crews, 2005; Nixon & Crews, 2002). Likewise, moderate consumption of alcohol in adult rats (ie, free exposure to a 4% ethanol replacement diet for 2 weeks) produced a 40% decrease in proliferation rates in the DG (Anderson, Nokia, Govindaraju, & Shors, 2012), albeit longer exposure accompanied by lower blood ethanol concentrations led to an increase in proliferation and neurogenesis (Aberg, Hofstetter, Olson, & Brene, 2005). Interestingly, inhibition of neurogenesis after alcohol exposure is typically followed by compensatory bursts in proliferation during abstinence (Nixon & Crews, 2004), suggesting that neurogenesis may contribute to recovery of function during abstinence from drug abuse. Moreover, the negative effects of alcohol exposure on cell proliferation in the DG were completely reversed by exercise in mice (Crews, Nixon, & Wilkie, 2004), which may have implications for functional recovery.

Exposure to addictive drugs during periods of developmental vulnerability (fetal through adolescent stages) can have multiple adverse health consequences. Initiating drug taking during adolescence can also increase the likelihood that a drug use disorder will develop later in life. Using a rat model of fetal alcohol syndrome with intermittent treatments, Gil-Mohapel et al. (2011) found no significant enduring alterations in cell proliferation and survival of adult-generated hippocampal cells, measuring these parameters at different stages of development, including adolescence, young adulthood, and adulthood. However, others have reported deficits in proliferation and survival in adult rats exposed to alcohol throughout the embryonic period (Redila et al., 2006). In addition, early postnatal, binge-like ethanol treatment impaired in adult rats the ability of new neurons to incorporate into the DG network and caused deficits in contextual fear conditioning (Hamilton et al., 2011), a finding that has been replicated (Klintsova et al., 2007). Even a single episode of alcohol exposure shortly after birth appears to produce persistent toxic effects on the pool of neural stem/progenitor cells in the DG (Ieraci & Herrera, 2007).

Evidence has shown that chronic intermittent exposure to ethanol during adolescence in rats (postnatal day 28–48) induced long-term decreases in doublecortin (DCX) immunoreactivity, which were associated with elevated cell death of immature neurons (Broadwater, Liu, Crews, & Spear, 2014). Similarly, a binge protocol of alcohol treatment reduced the long-term survival of BrdU-labeled cells in the DG, albeit this treatment did not alter the length of dendritic arbors of DCX-positive neurons or the number of DCX gaps in the same region (Hernandez-Rabaza et al., 2010). It appears that binge alcohol treatment during adolescence impairs neurogenesis by inhibiting neural stem cell proliferation and decreasing the survival of newly generated neurons (Morris, Eaves, Smith, & Nixon, 2010).

Stimulant effects on neurogenesis have also been widely investigated. A thorough investigation of the effects of cocaine on neurogenesis showed that chronic short-term (8-day) and long-term (24-day) exposure significantly reduce proliferation of neural progenitors in the DG without altering the maturation or the survival of newly generated neurons (Dominguez-Escriba et al., 2006). Significant decreases in proliferation rates were also observed following voluntary cocaine self-administration in rats, an effect that was reversed after extended withdrawal (Noonan, Choi, Self, & Eisch, 2008). Also, in the gerbil, a single dose of methamphetamine produced a decrease in cytogenesis in the DG (Teuchert-Noodt, Dawirs, & Hildebrandt, 2000), as did chronic methamphetamine self-administration in rats (Recinto et al., 2012). However, more persistent deficits in neurogenesis were observed after chronic amphetamine treatment, characterized by reduced cytogenesis several weeks after withdrawal, which was paralleled by increased anxiety-like behavior (Barr, Renner, & Forster, 2010). Taken together, these observations clearly indicate that psychomotor stimulants impair hippocampal neurogenesis, mainly through alterations in proliferation dynamics.

MDMA (3,4-methylenedioxymethamphetamine), popularly known as ecstasy, is an abused synthetic, psychoactive drug that has similarities to both the stimulant amphetamine and the hallucinogen mescaline. The more delayed course of MDMA psychoactive actions (ie, effects typically last 3–6h) implies that the pattern of MDMA abuse in humans is quite distinct from classical psychostimulants, such as cocaine, with binge administration being common. Hernandez-Rabaza et al. (2006) were first to show that binge-like exposure to MDMA in adult rats did not affect cytogenesis but reduced by c. 50% the number of new cells incorporated to the DG 2weeks after the treatments. These authors evidenced that decreased survival was not accompanied by impairments in neuronal maturation, as assessed by DCX immunohistochemistry. Decreased survival rates of adult-generated cells have also been observed in adolescent rats treated chronically with MDMA (Catlow et al., 2010; Hernandez-Rabaza et al., 2010). Moreover, prenatal exposure to MDMA resulted in impaired proliferation and survival of newborn cells in the DG of adult rats (Cho, Rhee, Kwack, Chung, & Kim, 2008), highlighting the risks associated with fetal exposure to MDMA.

One of the first demonstrations of the negative consequences of drug exposure on neurogenesis was provided by Eisch, Barrot, Schad, Self, and Nestler (2000), investigating the effects of morphine and heroin. Both these opiates reduced neurogenesis in the DG when administered chronically. Chronic but not acute morphine decreased proliferation rates and long-term survival of neural precursors (Eisch et al., 2000). Similarly, the same authors showed that chronic self-administration of heroin reduced proliferation of neural progenitors in the DG. Further experiments revealed that chronic morphine treatment exerted a disproportionate effect on S-phase cells compared to other cycling cells (Arguello et al., 2008). These data demonstrate that the opiate-induced decrease in hippocampal neurogenesis could be one of the mechanisms underlying opiate addiction symptomatology and the cognitive deficits associated with opiate addiction.

A large proportion of drug abusers are users of many psychoactive substances. A few experiments have been conducted to explore the effects of polydrug exposure on adult neurogenesis. Treatment with combinations of alcohol and MDMA during the embryonic development was shown to induce impairments in exploratory activity and working memory function and reductions in proliferation and neurogenesis that expanded into adulthood (Canales & Ferrer-Donato, 2014). Similarly, in adolescent rats, the combination of alcohol and MDMA produced severe long-term memory deficits that correlated with increased microgliosis, reduced survival of neural precursors, and granule cell depletion in the DG (Hernandez-Rabaza et al., 2010).

NEUROGENESIS, COGNITION, AND PERSISTENT DRUG SEEKING

The adverse health outcomes associated with chronic drug exposure are not limited to addiction but encompass a wide range of neurological and behavioral alterations as well as a broad spectrum of cognitive impairments (Devlin & Henry, 2008; Chen & Lin, 2009). This evidence is significant because a rapidly growing body of literature indicates that adult neurogenesis in the hippocampus plays a key role in learning and memory functions, thereby supporting the view that at least some of the cognitive deficits induced by chronic drug exposure may derive from decreased neurogenesis. Although the tenet that hippocampal neurogenesis participates in cognitive processing was initially met with skepticism, recent experiments using a variety of neurogenesis-altering approaches, including pharmacological treatments, irradiation, selective genetic knockdown techniques, and environmental enrichment, have made it increasingly clear that such participation is substantial (Aimone et al., 2014). The chronic abuse of drugs in humans has a negative impact on a variety of neuropsychological functions, the severity of which reflects in part the length of exposure and the intrinsic toxicity of the drug subject of abuse. A large number of studies in drug users has documented deficits in problem solving, decision making, attentional control, episodic memory, working memory, and cognitive flexibility, among other alterations (Canales, 2010; Novier, Diaz-Granados, & Matthews, 2015; Rogers & Robbins, 2001; Scheurich, 2005). It is well known that at least some of these functions require the integrity of the hippocampus. However, understanding the extent to which drug-induced impairments in the neurogenic processes underlie the deleterious effects of chronic drug exposure on cognition requires delineating first the functional role of the newborn hippocampal neurons themselves. Several hypotheses and functions have been put forward, which should not be regarded as mutually exclusive given the unique nature and functional capabilities of the adult-born neurons at different stages of their development. Adult-generated hippocampal neurons exhibit age-dependent plasticity, such as enhanced synaptic excitability and lower induction threshold for LTP at early stages of their development compared to mature granular neurons (Chancey et al., 2013; Snyder, Kee, & Wojtowicz, 2001), suggesting that their role may be diverse and dependent on maturational parameters.

One cardinal function of the hippocampus is to generate contextual representations of the environment, critically combining the "what," "when," and "where" elements into a coherent percept (Howard & Eichenbaum, 2013; Smith & Mizumori, 2006). The context in which emotional experiences occur is a key feature linked to the encoding of memory episodes, a function traditionally ascribed to the temporal lobe (Eichenbaum, Sauvage, Fortin, Komorowski, & Lipton, 2012; Mayes & Roberts, 2001). Contextual memory for aversive and hedonic experiences, such as foot shock or cocaine exposure, respectively, is known to be disrupted by selective lesions of the DG (Hernandez-Rabaza et al., 2008). Newborn hippocampal neurons play a role in such processes because depletion of hippocampal neurogenesis by means of ionizing radiation interferes with contextual fear conditioning (Hernandez-Rabaza et al., 2009; Saxe et al., 2006), albeit not significantly so when extended training is provided (Drew, Denny, & Hen, 2010). By contrast, the involvement of adult-generated hippocampal neurons in drug conditioning appears to be more complex. The acquisition of cocaine-induced place preference was not blocked by irradiation (Brown et al., 2010), which may be accounted for by an alternative recruitment of hedonic hotspots in the medial PFC, NAcb, and paraventricular hypothalamic nucleus instead of the hippocampus (Castilla-Ortega et al., 2015). Interestingly, depletion of hippocampal neurogenesis, while not preventing the acquisition of place preference, caused delayed extinction of conditioned responses and enhanced reinstatement of drug-seeking behavior (Castilla-Ortega et al., 2015), in agreement with previous observations using the self-administration paradigm (Noonan, Bulin, Fuller, & Eisch, 2010). These authors similarly showed that ablation of hippocampal neurogenesis was associated with increased cocaine intake, and increased motivation to seek and take cocaine, suggesting that drug-induced impairments of adult hippocampal neurogenesis could increase the vulnerability to engage in drug taking and develop an addiction (Noonan et al., 2010).

Such deficits, which could manifest themselves as an inability to shift from, reverse, or extinguish previously learned behavioral strategies, denote not only cognitive inflexibility, which may ultimately predispose to drug abuse, but also inability to discriminate changes in familiar contexts to flexibly adapt behavior. While the primary mechanisms underlying such maladaptive behavior are still unknown, one intriguing possibility is that chronic drug exposure impairs adult neurogenesis and interferes with a principal process referred to as pattern separation (Aimone, Deng, & Gage, 2011; Clelland et al., 2009). Such process allows disambiguating and reducing interference in situations that generate similar, though not identical, input firing patterns, such as two separate, but similar, contexts (eg, drug availability vs. no drug availability in a drug self-administration extinction paradigm). Contexts in which we experience events or perform actions can be highly similar (eg, parking our car daily in the same busy parking lot) and only subtle details allow us to keep memories distinct and resistant to interference. In the context of drug abuse, chronic impairment of adult neurogenesis could lead to deficits in pattern separation which could, in turn, elicit inflexible and generalized responses (eg, craving and drug seeking). Indeed, this mechanism could be invoked to explain not only the expression of uncontrollable craving in drug and non-drug-related environments but also situations where generalized fear or anxiety is elicited in patients suffering from posttraumatic stress disorders and chronic anxiety-spectrum disorders (Besnard & Sahay, 2016; Kheirbek, Klemenhagen, Sahay, & Hen, 2012). Moreover, decreased pattern separation and behavioral and cognitive inflexibility subsequent to chronic drug exposure could also be linked to increased propensity to develop pervasive habits. Chronic social stress in mice has been shown to decrease hippocampal neurogenesis and promote the use of habit-based learning strategies at the expense of spatial (ie, cognitive) strategies (Ferragud et al., 2010). This evidence emphasizes the complex interplay between the hippocampal system and the DA-modulated basal ganglia network (Canales, 2013; Gruber & McDonald, 2012; Packard & Knowlton, 2002), and highlights the decisive influence of stress (Schwabe, 2013), which typically associates with addiction symptomatology, in the modulation of learning. Such a link could, together with the aforementioned primary deficits in pattern separation, in part explain the persistent drug seeking and taking exhibited by addicts in the face of the negative consequences produced by repeated drug exposure.

CONCLUSIONS

In summary, the study of adult hippocampal neurogenesis has added a new dimension to the investigation of the biological basis of addiction. There is now overwhelming evidence indicating that addictive substances such as alcohol, stimulants, and opiates produce deleterious effects on adult neurogenesis at all stages of the ontogenetic development, including the embryonic phase, adolescence, and adulthood. Similarly, evidence has begun to emerge in recent years suggesting that drug-associated deficits in neurogenesis may have functional significance, influencing many of the key processes that characterize the staging process of addiction. The hippocampus is strategically positioned to modulate endocrine activity, associative learning, reward, reinforcement, habit learning, and decision making, providing a spatiotemporal context for such fundamental processes. Neurogenesis may also play a significant role in the recovery process as evidenced by the changes in cell proliferation, maturation, and survival stimulated by abstinence from chronic drug exposure. However, many of these hypotheses require further experimental support and many challenges remain. Critically, there remains some uncertainty regarding the extent to which many of the extensive findings accumulated using animal models can be extrapolated to the human population, albeit there are many reasons to assume that many of the mechanisms that regulate neurogenesis are common among mammals, including humans. Undoubtedly, we can look forward to new and exciting discoveries in this field in the years to come, including a more specific association of adult neurogenic processes in the hippocampus with addictive-like behaviors in animals and clinically relevant behaviors in humans.

References

Aberg, E., Hofstetter, C. P., Olson, L., & Brene, S. (2005). Moderate ethanol consumption increases hippocampal cell proliferation and neurogenesis in the adult mouse. *International Journal of Neuropsychopharmacology, 8*, 557–567.

Aimone, J. B., Deng, W., & Gage, F. H. (2011). Resolving new memories: a critical look at the dentate gyrus, adult neurogenesis, and pattern separation. *Neuron, 70*, 589–596.

Aimone, J. B., Li, Y., Lee, S. W., Clemenson, G. D., Deng, W., & Gage, F. H. (2014). Regulation and function of adult neurogenesis: from genes to cognition. *Physiological Reviews, 94*, 991–1026.

American Psychiatric Association. (2013). *Diagnostic and statistical manual of mental disorders* (5th ed.). Arlington, VA: American Psychiatric Publishing.

Anderson, M. L., Nokia, M. S., Govindaraju, K. P., & Shors, T. J. (2012). Moderate drinking? Alcohol consumption significantly decreases neurogenesis in the adult hippocampus. *Neuroscience, 224*, 202–209.

Arguello, A. A., Harburg, G. C., Schonborn, J. R., Mandyam, C. D., Yamaguchi, M., & Eisch, A. J. (2008). Time course of morphine's effects on adult hippocampal subgranular zone reveals preferential inhibition of cells in S phase of the cell cycle and a subpopulation of immature neurons. *Neuroscience, 157*, 70–79.

Barr, J. L., Renner, K. J., & Forster, G. L. (2010). Withdrawal from chronic amphetamine produces persistent anxiety-like behavior but temporally-limited reductions in monoamines and neurogenesis in the adult rat dentate gyrus. *Neuropharmacology, 59*, 395–405.

Belin, D., & Everitt, B. J. (2008). Cocaine seeking habits depend upon dopamine-dependent serial connectivity linking the ventral with the dorsal striatum. *Neuron, 57*, 432–441.

Belin, D., Jonkman, S., Dickinson, A., Robbins, T. W., & Everitt, B. J. (2009). Parallel and interactive learning processes within the basal ganglia: relevance for the understanding of addiction. *Behavioural Brain Research, 199*, 89–102.

Besnard, A., & Sahay, A. (2016). Adult hippocampal neurogenesis, fear generalization and stress. *Neuropsychopharmacology, 41*(1), 24–44.

Black, Y. D., Green-Jordan, K., Eichenbaum, H. B., & Kantak, K. M. (2004). Hippocampal memory system function and the regulation of cocaine self-administration behavior in rats. *Behavioural Brain Research, 151*, 225–238.

Broadwater, M. A., Liu, W., Crews, F. T., & Spear, L. P. (2014). Persistent loss of hippocampal neurogenesis and increased cell death following adolescent, but not adult, chronic ethanol exposure. *Developmental Neuroscience, 36*, 297–305.

Brown, T. E., Lee, B. R., Ryu, V., Herzog, T., Czaja, K., & Dong, Y. (2010). Reducing hippocampal cell proliferation in the adult rat does not prevent the acquisition of cocaine-induced conditioned place preference. *Neuroscience Letters, 481*, 41–46.

Canales, J. J. (2007). Adult neurogenesis and the memories of drug addiction. *European Archives of Psychiatry and Clinical Neuroscience, 257*, 261–270.

Canales, J. J. (2010). Comparative neuroscience of stimulant-induced memory dysfunction: role for neurogenesis in the adult hippocampus. *Behavioural Pharmacology, 21*, 379–393.

Canales, J. J. (2013). Deficient plasticity in the hippocampus and the spiral of addiction: focus on adult neurogenesis. *Current Topics in Behavioral Neuroscience, 15*, 293–312.

Canales, J. J., & Ferrer-Donato, A. (2014). Prenatal exposure to alcohol and 3,4-methylenedioxymethamphetamine (ecstasy) alters adult hippocampal neurogenesis and causes enduring memory deficits. *Developmental Neuroscience, 36*, 10–17.

Castilla-Ortega, E., Blanco, E., Serrano, A., Ladron, d. G.-M., Pedraz, M., Estivill-Torrus, G., et al. (2015). Pharmacological reduction of adult hippocampal neurogenesis modifies functional brain circuits in mice exposed to a cocaine conditioned place preference paradigm. *Addiction Biology*.

Catlow, B. J., Badanich, K. A., Sponaugle, A. E., Rowe, A. R., Song, S., Rafalovich, I., et al. (2010). Effects of MDMA ("ecstasy") during adolescence on place conditioning and hippocampal neurogenesis. *European Journal of Pharmacology, 628*, 96–103.

Chancey, J. H., Adlaf, E. W., Sapp, M. C., Pugh, P. C., Wadiche, J. I., & Overstreet-Wadiche, L. S. (2013). GABA depolarization is required for experience-dependent synapse unsilencing in adult-born neurons. *Journal of Neuroscience, 33*, 6614–6622.

Chen, C. Y., & Lin, K. M. (2009). Health consequences of illegal drug use. *Current Opinion in Psychiatry, 22*, 287–292.

Cho, K. O., Rhee, G. S., Kwack, S. J., Chung, S. Y., & Kim, S. Y. (2008). Developmental exposure to 3,4-methylenedioxymethamphetamine results in downregulation of neurogenesis in the adult mouse hippocampus. *Neuroscience, 154*, 1034–1041.

Clelland, C. D., Choi, M., Romberg, C., Clemenson, G. D., Jr., Fragniere, A., Tyers, P., et al. (2009). A functional role for adult hippocampal neurogenesis in spatial pattern separation. *Science, 325*, 210–213.

Crews, F. T., Nixon, K., & Wilkie, M. E. (2004). Exercise reverses ethanol inhibition of neural stem cell proliferation. *Alcohol, 33*, 63–71.

Degoulet, M., Rouillon, C., Rostain, J. C., David, H. N., & Abraini, J. H. (2008). Modulation by the dorsal, but not the ventral, hippocampus of the expression of behavioural sensitization to amphetamine. *International Journal of Neuropsychopharmacology, 11*, 497–508.

Devlin, R. J., & Henry, J. A. (2008). Clinical review: major consequences of illicit drug consumption. *Critical Care, 12*, 202.

Dominguez-Escriba, L., Hernandez-Rabaza, V., Soriano-Navarro, M., Barcia, J. A., Romero, F. J., Garcia-Verdugo, J. M., et al. (2006). Chronic cocaine exposure impairs progenitor proliferation but spares survival and maturation of neural precursors in adult rat dentate gyrus. *European Journal of Neuroscience, 24*, 586–594.

Drew, M. R., Denny, C. A., & Hen, R. (2010). Arrest of adult hippocampal neurogenesis in mice impairs single- but not multiple-trial contextual fear conditioning. *Behavioral Neuroscience, 124*, 446–454.

Eichenbaum, H., Sauvage, M., Fortin, N., Komorowski, R., & Lipton, P. (2012). Towards a functional organization of episodic memory in the medial temporal lobe. *Neuroscience & Biobehavioral Reviews, 36*, 1597–1608.

Eisch, A. J., Barrot, M., Schad, C. A., Self, D. W., & Nestler, E. J. (2000). Opiates inhibit neurogenesis in the adult rat hippocampus. *Proceedings of the National Academy of Sciences of the United States of America, 97*, 7579–7584.

Eisch, A. J., & Harburg, G. C. (2006). Opiates, psychostimulants, and adult hippocampal neurogenesis: insights for addiction and stem cell biology. *Hippocampus, 16*, 271–286.

Everitt, B. J., & Robbins, T. W. (2013). From the ventral to the dorsal striatum: devolving views of their roles in drug addiction. *Neuroscience & Biobehavioral Reviews, 37*, 1946–1954.

Feltenstein, M. W., & See, R. E. (2008). The neurocircuitry of addiction: an overview. *British Journal of Pharmacology, 154*, 261–274.

Ferragud, A., Haro, A., Sylvain, A., Velazquez-Sanchez, C., Hernandez-Rabaza, V., & Canales, J. J. (2010). Enhanced habit-based learning and decreased neurogenesis in the adult hippocampus in a murine model of chronic social stress. *Behavioural Brain Research, 210*, 134–139.

Floresco, S. B., Todd, C. L., & Grace, A. A. (2001). Glutamatergic afferents from the hippocampus to the nucleus accumbens regulate activity of ventral tegmental area dopamine neurons. *Journal of Neuroscience, 21*, 4915–4922.

Fotros, A., Casey, K. F., Larcher, K., Verhaeghe, J. A., Cox, S. M., Gravel, P., et al. (2013). Cocaine cue-induced dopamine release in amygdala and hippocampus: a high-resolution PET [(1)(8)F]fallypride study in cocaine dependent participants. *Neuropsychopharmacology, 38*, 1780–1788.

Fuchs, R. A., Evans, K. A., Ledford, C. C., Parker, M. P., Case, J. M., Mehta, R. H., et al. (2005). The role of the dorsomedial prefrontal cortex, basolateral amygdala, and dorsal hippocampus in contextual reinstatement of cocaine seeking in rats. *Neuropsychopharmacology, 30*, 296–309.

Gasbarri, A., Sulli, A., & Packard, M. G. (1997). The dopaminergic mesencephalic projections to the hippocampal formation in the rat. *Progress in Neuropsychopharmacology & Biological Psychiatry, 21*, 1–22.

Gil-Mohapel, J., Boehme, F., Patten, A., Cox, A., Kainer, L., Giles, E., et al. (2011). Altered adult hippocampal neuronal maturation in a rat model of fetal alcohol syndrome. *Brain Research, 1384*, 29–41.

Gozzi, A., Tessari, M., Dacome, L., Agosta, F., Lepore, S., Lanzoni, A., et al. (2011). Neuroimaging evidence of altered fronto-cortical and striatal function after prolonged cocaine self-administration in the rat. *Neuropsychopharmacology, 36*, 2431–2440.

Gruber, A. J., & McDonald, R. J. (2012). Context, emotion, and the strategic pursuit of goals: interactions among multiple brain systems controlling motivated behavior. *Frontiers in Behavioral Neuroscience, 6*, 50.

Haber, S. N., Fudge, J. L., & McFarland, N. R. (2000). Striatonigrostriatal pathways in primates form an ascending spiral from the shell to the dorsolateral striatum. *Journal of Neuroscience, 20*, 2369–2382.

Hamilton, G. F., Murawski, N. J., St Cyr, S. A., Jablonski, S. A., Schiffino, F. L., Stanton, M. E., et al. (2011). Neonatal alcohol exposure disrupts hippocampal neurogenesis and contextual fear conditioning in adult rats. *Brain Research, 1412*, 88–101.

He, J., Nixon, K., Shetty, A. K., & Crews, F. T. (2005). Chronic alcohol exposure reduces hippocampal neurogenesis and dendritic growth of newborn neurons. *European Journal of Neuroscience, 21*, 2711–2720.

Hernandez-Rabaza, V., Dominguez-Escriba, L., Barcia, J. A., Rosel, J. F., Romero, F. J., Garcia-Verdugo, J. M., et al. (2006). Binge administration of 3,4-methylenedioxymethamphetamine ("ecstasy") impairs the survival of neural precursors in adult rat dentate gyrus. *Neuropharmacology, 51*, 967–973.

Hernandez-Rabaza, V., Hontecillas-Prieto, L., Velazquez-Sanchez, C., Ferragud, A., Perez-Villaba, A., Arcusa, A., et al. (2008). The hippocampal dentate gyrus is essential for generating contextual memories of fear and drug-induced reward. *Neurobiology of Learning and Memory, 90*, 553–559.

Hernandez-Rabaza, V., Llorens-Martin, M., Velazquez-Sanchez, C., Ferragud, A., Arcusa, A., Gumus, H. G., et al. (2009). Inhibition of adult hippocampal neurogenesis disrupts contextual learning but spares spatial working memory, long-term conditional rule retention and spatial reversal. *Neuroscience, 159*, 59–68.

Hernandez-Rabaza, V., Navarro-Mora, G., Velazquez-Sanchez, C., Ferragud, A., Marin, M. P., Garcia-Verdugo, J. M., et al. (2010). Neurotoxicity and persistent cognitive deficits induced by combined MDMA and alcohol exposure in adolescent rats. *Addiction Biology, 15*, 413–423.

Herrera, D. G., Yague, A. G., Johnsen-Soriano, S., Bosch-Morell, F., Collado-Morente, L., Muriach, M., et al. (2003). Selective impairment of hippocampal neurogenesis by chronic alcoholism: protective effects of an antioxidant. *Proceedings of the National Academy of Sciences of the United States of America, 100*, 7919–7924.

Howard, M. W., & Eichenbaum, H. (2013). The hippocampus, time, and memory across scales. *Journal of Experimental Psychology.General, 142*, 1211–1230.

Ieraci, A., & Herrera, D. G. (2007). Single alcohol exposure in early life damages hippocampal stem/progenitor cells and reduces adult neurogenesis. *Neurobiology of Disease, 26*, 597–605.

Kalivas, P. W. (1995). Interactions between dopamine and excitatory amino acids in behavioral sensitization to psychostimulants. *Drug and Alcohol Dependence, 37*, 95–100.

Kheirbek, M. A., Klemenhagen, K. C., Sahay, A., & Hen, R. (2012). Neurogenesis and generalization: a new approach to stratify and treat anxiety disorders. *Nature Neuroscience, 15*, 1613–1620.

Kilts, C. D., Schweitzer, J. B., Quinn, C. K., Gross, R. E., Faber, T. L., Muhammad, F., et al. (2001). Neural activity related to drug craving in cocaine addiction. *Archives of General Psychiatry, 58*, 334–341.

Klintsova, A. Y., Helfer, J. L., Calizo, L. H., Dong, W. K., Goodlett, C. R., & Greenough, W. T. (2007). Persistent impairment of hippocampal neurogenesis in young adult rats following early postnatal alcohol exposure. *Alcoholism, Clinical and Experimental Research, 31*, 2073–2082.

Koob, G. F. (2009). Dynamics of neuronal circuits in addiction: reward, antireward, and emotional memory. *Pharmacopsychiatry, 42*(Suppl. 1), S32–S41.

Koob, G. F., & Le Moal, M. (2008). Review. Neurobiological mechanisms for opponent motivational processes in addiction. *Philosophical Transactions of the Royal Society of London. Series B, Biological Sciences, 363*, 3113–3123.

Koob, G. F., & Volkow, N. D. (2010). Neurocircuitry of addiction. *Neuropsychopharmacology, 35*, 217–238.

Koob, G. F., & Weiss, F. (1992). Neuropharmacology of cocaine and ethanol dependence. *Recent Developments in Alcoholism, 10*, 201–233.

Lapish, C. C., Chiang, J., Wang, J. Z., & Phillips, A. G. (2012). Oscillatory power and synchrony in the rat forebrain are altered by a sensitizing regime of D-amphetamine. *Neuroscience, 203*, 108–121.

Leyton, M. (2007). Conditioned and sensitized responses to stimulant drugs in humans. *Progress in Neuropsychopharmacology & Biological Psychiatry, 31*, 1601–1613.

Luo, A. H., Tahsili-Fahadan, P., Wise, R. A., Lupica, C. R., & Aston-Jones, G. (2011). Linking context with reward: a functional circuit from hippocampal CA3 to ventral tegmental area. *Science, 333*, 353–357.

Lu, L., Zeng, S., Liu, D., & Ceng, X. (2000). Inhibition of the amygdala and hippocampal calcium/calmodulin-dependent protein kinase II attenuates the dependence and relapse to morphine differently in rats. *Neuroscience Letters, 291*, 191–195.

Marchant, N. J., Li, X., & Shaham, Y. (2013). Recent developments in animal models of drug relapse. *Current Opinion in Neurobiology, 23*, 675–683.

Martin-Fardon, R., Ciccocioppo, R., Aujla, H., & Weiss, F. (2008). The dorsal subiculum mediates the acquisition of conditioned reinstatement of cocaine-seeking. *Neuropsychopharmacology*, *33*, 1827–1834.

Mayes, A. R., & Roberts, N. (2001). Theories of episodic memory. *Philosophical Transactions of the Royal Society of London. Series B, Biological Sciences*, *356*, 1395–1408.

Morris, S. A., Eaves, D. W., Smith, A. R., & Nixon, K. (2010). Alcohol inhibition of neurogenesis: a mechanism of hippocampal neurodegeneration in an adolescent alcohol abuse model. *Hippocampus*, *20*, 596–607.

Nixon, K., & Crews, F. T. (2002). Binge ethanol exposure decreases neurogenesis in adult rat hippocampus. *Journal of Neurochemistry*, *83*, 1087–1093.

Nixon, K., & Crews, F. T. (2004). Temporally specific burst in cell proliferation increases hippocampal neurogenesis in protracted abstinence from alcohol. *Journal of Neuroscience*, *24*, 9714–9722.

Noonan, M. A., Bulin, S. E., Fuller, D. C., & Eisch, A. J. (2010). Reduction of adult hippocampal neurogenesis confers vulnerability in an animal model of cocaine addiction. *Journal of Neuroscience*, *30*, 304–315.

Noonan, M. A., Choi, K. H., Self, D. W., & Eisch, A. J. (2008). Withdrawal from cocaine self-administration normalizes deficits in proliferation and enhances maturity of adult-generated hippocampal neurons. *Journal of Neuroscience*, *28*, 2516–2526.

Novier, A., Diaz-Granados, J. L., & Matthews, D. B. (2015). Alcohol use across the lifespan: an analysis of adolescent and aged rodents and humans. *Pharmacology Biochemistry and Behavior*, *133*, 65–82.

Packard, M. G., & Knowlton, B. J. (2002). Learning and memory functions of the basal ganglia. *Annual Review of Neuroscience*, *25*, 563–593.

Perez, M. F., Gabach, L. A., Almiron, R. S., Carlini, V. P., De Barioglio, S. R., & Ramirez, O. A. (2010). Different chronic cocaine administration protocols induce changes on dentate gyrus plasticity and hippocampal dependent behavior. *Synapse*, *64*, 742–753.

Portugal, G. S., Al-Hasani, R., Fakira, A. K., Gonzalez-Romero, J. L., Melyan, Z., McCall, J. G., et al. (2014). Hippocampal long-term potentiation is disrupted during expression and extinction but is restored after reinstatement of morphine place preference. *Journal of Neuroscience*, *34*, 527–538.

Raybuck, J. D., & Lattal, K. M. (2014). Differential effects of dorsal hippocampal inactivation on expression of recent and remote drug and fear memory. *Neuroscience Letters*, *569*, 1–5.

Recinto, P., Samant, A. R., Chavez, G., Kim, A., Yuan, C. J., Soleiman, M., et al. (2012). Levels of neural progenitors in the hippocampus predict memory impairment and relapse to drug seeking as a function of excessive methamphetamine self-administration. *Neuropsychopharmacology*, *37*, 1275–1287.

Redila, V. A., Olson, A. K., Swann, S. E., Mohades, G., Webber, A. J., Weinberg, J., et al. (2006). Hippocampal cell proliferation is reduced following prenatal ethanol exposure but can be rescued with voluntary exercise. *Hippocampus*, *16*, 305–311.

Rivera, P. D., Raghavan, R. K., Yun, S., Latchney, S. E., McGovern, M. K., Garcia, E. F., et al. (2015). Retrieval of morphine-associated context induces cFos in dentate gyrus neurons. *Hippocampus*, *25*, 409–414.

Robbins, T. W., Cador, M., Taylor, J. R., & Everitt, B. J. (1989). Limbic-striatal interactions in reward-related processes. *Neuroscience & Biobehavioral Reviews*, *13*, 155–162.

Robinson, T. E., & Berridge, K. C. (2008). Review. The incentive sensitization theory of addiction: some current issues. *Philosophical Transactions of the Royal Society of London. Series B, Biological Sciences*, *363*, 3137–3146.

Rogers, R. D., & Robbins, T. W. (2001). Investigating the neurocognitive deficits associated with chronic drug misuse. *Current Opinion in Neurobiology*, *11*, 250–257.

Rogers, J. L., & See, R. E. (2007). Selective inactivation of the ventral hippocampus attenuates cue-induced and cocaine-primed reinstatement of drug-seeking in rats. *Neurobiology of Learning and Memory*, *87*, 688–692.

Saxe, M. D., Battaglia, F., Wang, J. W., Malleret, G., David, D. J., Monckton, J. E., et al. (2006). Ablation of hippocampal neurogenesis impairs contextual fear conditioning and synaptic plasticity in the dentate gyrus. *Proceedings of the National Academy of Sciences of the United States of America, 103*, 17501–17506.

Scheurich, A. (2005). Neuropsychological functioning and alcohol dependence. *Current Opinion in Psychiatry, 18*, 319–323.

Schwabe, L. (2013). Stress and the engagement of multiple memory systems: integration of animal and human studies. *Hippocampus, 23*, 1035–1043.

Self, D. W. (2004). Regulation of drug-taking and -seeking behaviors by neuroadaptations in the mesolimbic dopamine system. *Neuropharmacology, 47*(Suppl. 1), 242–255.

Smith, D. M., & Mizumori, S. J. (2006). Hippocampal place cells, context, and episodic memory. *Hippocampus, 16*, 716–729.

Snyder, J. S., Kee, N., & Wojtowicz, J. M. (2001). Effects of adult neurogenesis on synaptic plasticity in the rat dentate gyrus. *Journal of Neurophysiology, 85*, 2423–2431.

Taepavarapruk, P., Butts, K. A., & Phillips, A. G. (2014). Dopamine and glutamate interaction mediates reinstatement of drug-seeking behavior by stimulation of the ventral subiculum. *International Journal of Neuropsychopharmacology, 18*.

Teuchert-Noodt, G., Dawirs, R. R., & Hildebrandt, K. (2000). Adult treatment with methamphetamine transiently decreases dentate granule cell proliferation in the gerbil hippocampus. *Journal of Neural Transmission, 107*, 133–143.

Thompson, A. M., Swant, J., Gosnell, B. A., & Wagner, J. J. (2004). Modulation of long-term potentiation in the rat hippocampus following cocaine self-administration. *Neuroscience, 127*, 177–185.

Tropea, T. F., Kosofsky, B. E., & Rajadhyaksha, A. M. (2008). Enhanced CREB and DARPP-32 phosphorylation in the nucleus accumbens and CREB, ERK, and GluR1 phosphorylation in the dorsal hippocampus is associated with cocaine-conditioned place preference behavior. *Journal of Neurochemistry, 106*, 1780–1790.

Tzschentke, T. M. (1998). Measuring reward with the conditioned place preference paradigm: a comprehensive review of drug effects, recent progress and new issues. *Progress in Neurobiology, 56*, 613–672.

Vorel, S. R., Liu, X., Hayes, R. J., Spector, J. A., & Gardner, E. L. (2001). Relapse to cocaine-seeking after hippocampal theta burst stimulation. *Science, 292*, 1175–1178.

Wang, Z., Faith, M., Patterson, F., Tang, K., Kerrin, K., Wileyto, E. P., et al. (2007). Neural substrates of abstinence-induced cigarette cravings in chronic smokers. *Journal of Neuroscience, 27*, 14035–14040.

White, F. J., & Kalivas, P. W. (1998). Neuroadaptations involved in amphetamine and cocaine addiction. *Drug and Alcohol Dependence, 51*, 141–153.

Wise, R. A. (1987). The role of reward pathways in the development of drug dependence. *Pharmacology & Therapeutics, 35*, 227–263.

Wolf, M. E., Sun, X., Mangiavacchi, S., & Chao, S. Z. (2004). Psychomotor stimulants and neuronal plasticity. *Neuropharmacology, 47*(Suppl. 1), 61–79.

Zubieta, J. K., Heitzeg, M. M., Xu, Y., Koeppe, R. A., Ni, L., Guthrie, S., et al. (2005). Regional cerebral blood flow responses to smoking in tobacco smokers after overnight abstinence. *The American Journal of Psychiatry, 162*, 567–577.

CHAPTER

11

Neurological Disorders

*B.W. Man Lau**
The Hong Kong Polytechnic University, Hong Kong, China

*S.-Y. Yau**
University of Victoria, Victoria, BC, Canada

K.-T. Po
The Hong Kong Polytechnic University, Hong Kong, China

K.-F. So
The Jinan University, Guangzhou, China; The University of Hong Kong, Hong Kong, China

INTRODUCTION

The fascinating discovery of adult neurogenesis provides an important view of neural plasticity in adult brains in both physiological and neuropathological conditions. In the past decades, various studies have disclosed the involvement of adult neurogenesis in neurological disorders such as neurodegenerative diseases and brain injuries (for review, see Ruan et al., 2014). The discoveries stemming from both preclinical and clinical studies have advanced the knowledge regarding the cause and treatment of these neurological disorders. The "neurogenesis hypothesis" of neurological diseases has been suggested by emerging evidence showing the role of dysregulation in adult neurogenesis and its implications for therapeutic treatments in diseased brains. Successful translation of knowledge from laboratory to clinical practice will greatly promote the effectiveness of the current treatment regimes and reduce tremendous personal and economic burden caused by the disorders.

* Both authors contributed equally to this work.

http://dx.doi.org/10.1016/B978-0-12-801977-1.00011-8

In this chapter, we discuss an overview of dysregulation of adult neurogenesis in various neurological disorders ranging from neurodegenerative disorders and neurodevelopmental disorders to brain injuries, with current findings from both basic and clinical studies.

NEURODEGENERATIVE DISEASES

Neurodegenerative diseases like Alzheimer's disease (AD), Parkinson's disease (PD), and Huntington's disease (HD) share the common characteristic of progressive loss of structures and functions of neurons in the brain. Although neuronal degeneration predominantly starts with degeneration in specific neuronal populations like dopaminergic neurons in PD, striatal neurons in HD, and cortical and hippocampal neurons in AD, these diseases share similar cognitive declines in the later stage of the disease. However, the contribution of altered hippocampal neurogenesis to cognitive impairments in these neurodegenerative diseases is still unclear, despite that numerous animal studies have demonstrated alteration of adult neurogenesis in neurodegenerative diseases (Winner, Kohl, & Gage, 2011) (Table 11.1). Physical exercise is effective in improving and preventing age-related cognitive decline in humans, particularly in individuals with neurodegenerative diseases (Hillman, Erickson, & Kramer, 2008). Given the evidence showing that altered adult neurogenesis and interventions (eg, physical exercise) enhance neurogenesis and improves hippocampal functions in transgenic mouse models of neurodegenerative diseases (Yau, Gil-Mohapel, Christie, & So, 2014), it is possible that cognitive decline and emotional dysregulation observed in neurodegenerative diseases can be partly contributed by altered adult neurogenesis (Figs. 11.1 and 11.2).

Alzheimer's Disease

AD is characterized by progressive neuronal loss, cognitive deterioration, and behavioral changes. Accumulation of amyloid or senile plaques and formation of neurofibrillary tangles are thought to be the major cause of neuronal loss in the AD brain (Selkoe, 2001). Many of the molecules involved in development of AD could modulate neurogenesis; for example, presenilin 1 (PSEN-1) regulates neuronal differentiation (Gadadhar, Marr, & Lazarov, 2011), and soluble amyloid precursor protein α (sAPPα) regulates cell proliferation (Demars, Hollands, Zhao Kda, & Lazarov, 2013). Divergent results in alteration of hippocampal neurogenesis have been reported among different transgenic animal models of AD. This may be due to large variations in genetic mutations of AD models, for example, single gene mutation with single PSEN-1 or PSEN-2 or APP gene mutation, and knock-in or multiple mutations with combinations of the above-noted genes. Varied regimens of bromodeoxyuridine

TABLE 11.1 Summary of Representative Studies Showing Altered Hippocampal Neurogenesis in Animal Models of Neurodegenerative and Neurodevelopmental Disorders

Animal model	Proliferation	Neuronal differentiation	Maturation and survival	References
AD				
ApoE-deficient mice:	↓	↑	↑	Yang et al. (2011)
PDGF-APPsw, Ind:	↑	↑	↑	Jin, Galvan, et al. (2004), Jin, Peel, et al. (2004)
Sw,Ind:	↑		↓	Lopez-Toledano and Shelanski (2007)
Mutated APP:	↓	↓	↓	Haughey et al. (2002)
Overexpressed APP + PSEN-1:	↓			Verret et al. (2007)
3XTg-AD:	↓			Rodriguez et al. (2008)
ApoE knock-in:	↓			Li et al. (2009)
PD				
MPTP:	Transient ↑			Park and Enikolopov (2010)
6-OHDA:	↓			Hoglinger et al. (2004)
Overexpressed α-synuclein:			↓	Winner et al. (2004)
Mutant α-synuclein:	↓		↓	Crews et al. (2008)
HD				
R6/1:	↓		↓	Lazic et al. (2006)
R6/2:	↓		↓	Gil et al. (2005), Kohl et al. (2007)
YAC:	↓		↓	Gil-Mohapel et al. (2012)
Truncated HD gene:	↓			von Horsten et al. (2003)

Continued

TABLE 11.1 Summary of Representative Studies Showing Altered Hippocampal Neurogenesis in Animal Models of Neurodegenerative and Neurodevelopmental Disorders—cont'd

Animal model	Proliferation	Neuronal differentiation	Maturation and survival	References
FXS				
In vitro:	↑	↓	↓	Luo et al. (2010), Guo et al. (2011, 2012)
In vivo: (*Fmr1* Knockout)	– or ↓	↑ or ↓	↓	Eadie et al. (2012), Lazarov et al. (2012)
FASD	– or ↓	↓	↓	Kintsova et al. (2007), Redila et al. (2006); Gil-Mohapel, Boehme, et al. (2011), Gil-Mohapel, Simpson, et al. (2011), Helfer et al. (2009)

↑, Increased; ↓, Decreased; –, no change.

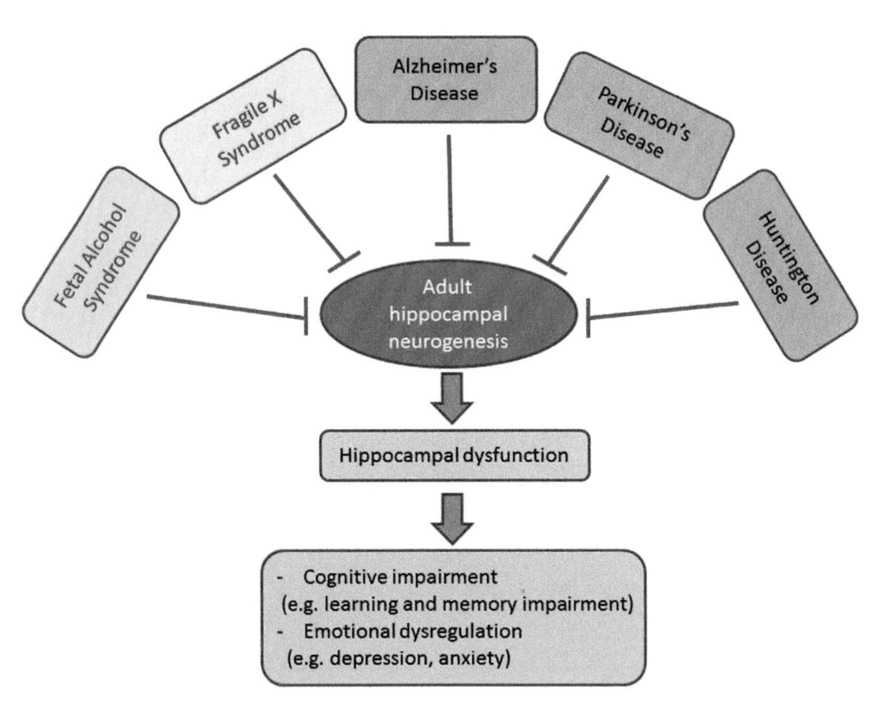

FIGURE 11.1 Alteration of hippocampal adult neurogenesis occurs in neurodevelopmental disorders and neurodegenerative diseases that may partly contribute to hippocampal dysfunction and subsequently lead to behavioral deficits observed in patients such as cognitive impairment and emotional disorders.

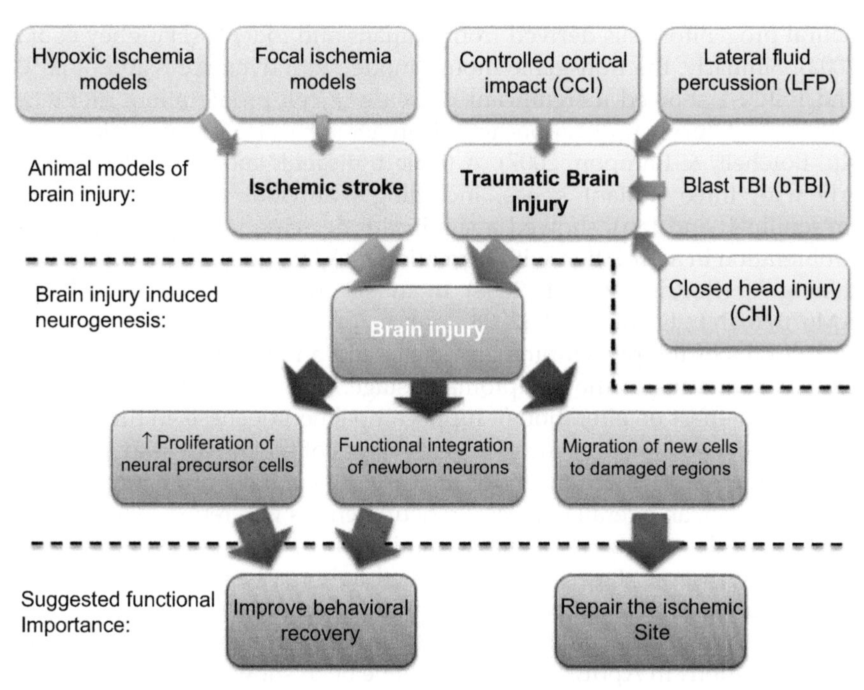

FIGURE 11.2 Alteration of hippocampal adult neurogenesis occurs in animal models of brain injury which may contribute to the recovery in physiological and behavioral aspects.

(BrdU) treatment for labeling proliferating cells, genetic background of the mice, and difference in disease stages may contribute to these divergent findings in adult neurogenesis from the literature.

In an animal model with crossing of apolipoprotein (ApoE)-deficient mice to a nestin–green fluorescence protein (GFP) reporter, ApoE deficiency increases early neural progenitor cell proliferation in the DG, but leads to depletion of neural progenitor cells at later time points (Yang, Gilley, Zhang, & Kernie, 2011). In contrast, transgenic mice (PDGF-APPSw, Ind) at 3 months old (at the stage of pre-Aβ deposition) showed an increase in hippocampal cell survival, with similar increases in cell proliferation. Furthermore, increases in survival rate have also been observed when the mice are 12 months old at the stage of post-Aβ deposition (Jin, Galvan, et al., 2004). This is echoed by a study showing a similar increase in neurogenesis in transgenic mice with overexpression of human APP (Sw, Ind) (Lopez-Toledano & Shelanski, 2007).

Contradictory results have been reported in several studies demonstrating decreased neurogenesis in the DG or/and SVZ in different transgenic mice. A significant reduction in proliferation and survival of neural progenitor cells in the hippocampal dentate region was found in 12- to 14-month-old transgenic mice with an overexpressed mutant form of amyloid precursor protein (APP). The in vitro experiment from this study demonstrated an adverse effect of APP on impairing proliferation and differentiation of neural progenitor cells derived from humans and rodents (Haughey et al., 2002). Similarly, the transgenic mouse model with overexpression of APP and PSEN-1 showed a significant decrease in cell proliferation, differentiation, and survival of hippocampal progenitor cells (Verret, Jankowsky, Xu, Borchelt, & Rampon, 2007). A triple transgenic mouse model (3XTg-AD with three mutant genes, including β-amyloid precursor protein, presenilin-1, and tau), showed a significant decrease in hippocampal cell proliferation in association with formation of Aβ plaques and neurons containing Aβ (Rodriguez et al., 2008). In an AD animal model with APPswe KM670/67INL, Ermini et al. (2008) found a decrease in immature neurons at 5 months of the presymptomatic stage, and an increase in proliferative cells at 25 months of the symptomatic stage. These data suggest an age-dependent effect on alteration of hippocampal neurogenesis in this animal model. Increase in neurogenesis at early stages of the disease may possibly be a compensatory mechanism for the early loss of newborn neuron in the DG. However, an overall decrease in neurogenesis was reported at the late stage of AD development. Detailed examination of changes in neurogenesis at different disease stages of AD will shed light on dysregulation in adult neurogenesis and its relationship to cognitive decline in AD.

Several molecules contributing to the neuropathology of AD, such as APP, mutations in ApoE, and PENS1, have been shown to regulate adult neurogenesis. The soluble APP, which is produced by APP cleavage by alpha-secretase, can rescue the age-related decline in NPCs proliferation (Demars et al., 2013). APP may promote neurogenesis by inhibiting the activity of cyclin-dependent kinase 5 and hyperphosphorylation of tau protein (Han et al., 2005). On the other hand, suppression of ApoE promotes proliferation of DG neural progenitor cells (Yang et al., 2011) while ApoE knock-in reduces hippocampal neurogenesis (Li et al., 2009). PENS-1 was reported to regulate neuronal differentiation (Gadadhar et al., 2011). PENS-1 mutation reduces hippocampal neurogenesis because PS1M146V knock-in impairs hippocampal-dependent contextual learning and reduces adult neurogenesis in the DG (Wang, Dineley, Sweatt, & Zheng, 2004). The mutant form of PENS-1 in hu-PS1P117L transgenic mice impairs the survival rate of neural progenitor cells (Wen et al., 2004). Taken together, these data suggest that mutation in ApoE or PENS1 plays an important role in regulating adult neurogenesis during the development of AD.

Adding to the above, dysregulation of γ-aminobutyric acid-ergic (GABAergic) neurons may also contribute to altered adult neurogenesis in AD animals. A decrease in GABAergic interneurons in APoE 4 transgenic mice decreases maturation of newborn neurons, whereas potentiating GABAa receptor restores neurogenesis and dendritic maturation (Li et al., 2009). Therefore, it is speculated that disrupted GABAergic transmission may partly contribute to decreased neurogenesis in these AD transgenic mice.

Similar to the results observed from different animal models of AD, varied findings on hippocampal neurogenesis in the AD brains have been reported in human studies. Increases in hippocampal protein expression of immature and mature neuronal markers including doublecortin (DCX), polysialylated–neural cell adhesion molecule (PSA–NCAM), neurogenic differentiation neuroD, and calbindin have been reported in postmortem tissues from human AD patients (Jin, Peel, et al., 2004). Another study reported no difference in cell proliferation in the DG of AD patients, but an increase in glial and vasculature-abundant area in the granule cell layer (Boekhoorn, Joels, & Lucassen, 2006). Interestingly, Crews et al. (2010) found a decrease in the number of stem cells and immature neurons in the DG.

Parkinson's Disease

PD is featured by progressive death of dopaminergic neurons in the substantia nigra (SN) and characterized by clinical symptoms including motor symptoms such as progressive impairment of movement control, and nonmotor symptoms such as cognitive decline, depression, and anxiety, and olfactory dysfunction (Tolosa & Poewe, 2009). Neurological pathology of PD includes formation of Lewy bodies and Lewy neurites (Goedert, Jakes, Anthony Crowther, & Grazia Spillantini, 2001), which contain the fibrillary and misfolded α-synuclein protein (Dauer & Przedborski, 2003). Notably, psychiatric symptoms such as depression and anxiety occur very often in PD patients (Reijnders, Ehrt, Weber, Aarsland, & Leentjens, 2008). The nonmotor symptoms, which are not directly associated with the neurodegeneration in the SN, may be contributed by alterations in neurogenesis in the hippocampus and SVZ.

The neuropathology of PD has been examined using several animal models including the lesion models induced by either the 6-hydroxydopamine (6-OHDA) or methyl-4-phenyl-1,2,3,6-tetrahydropyridine (MPTP) (Melrose, Lincoln, Tyndall, & Farrer, 2006), and the α-synuclein transgenic mouse model (Recchia et al., 2004). Ablation of dopaminergic neurons in the MPTP mouse model of PD induces a transient increase in cell proliferation in the DG (Park & Enikolopov, 2010). Using the same mouse model, Peng, Xie, Jin, Greenberg, & Andersen (2008) have reported an increase in proliferative cells and neuronal differentiation in several brain regions including the DG, SVZ, and striatum. However, contradictory results showing decreased adult neurogenesis in the SGZ and SVZ have been found in several studies. Hoglinger et al. (2004) reported that the 6-OHDA and the MPTP lesion models in rodents showed remarkable decreases in proliferation of precursor cells in the SGZ and SVZ, which can be reversed by treatment with the selective agonist of D2-like dopaminergic receptors.

Gene mutation of α-synuclein can lead to accumulation of misfolded α-synuclein that forms Lewy bodies and Lewy neurites in PD. Winner et al. (2004) have shown that survival, but not proliferation of adult-born neurons, is significantly suppressed in the DG and olfactory bulb in the transgenic mouse model of PD with overexpression of human wild-type α-synuclein. Reduction of hippocampal neurogenesis can occur with cognitive deficits in a conditional α-synuclein mouse model of PD (Nuber et al., 2008). In contrast, transgenic mice expressing mutant α-synuclein show a significant decrease in proliferation and survival of neural precursor cells (Crews et al., 2008).

In postmortem tissues of PD patients, there are decreased numbers of proliferating cell nuclear antigen (PCNA)-positive cells (a marker for proliferative cells) in SVZ, and decreased numbers of nestin and β-tubulin-positive cells in the DG of the hippocampus (Hoglinger et al., 2004), providing evidence of altered hippocampal neurogenesis in the human PD brain. Significant decreases in PCNA-positive cells and PSA–NCAM-positive cells in the SVZ were also found in PD patients (Hoglinger et al., 2004). However, there is limited evidence showing a decreased neurogenesis in PD patients to date, and future examination on the alteration of adult neurogenesis in PD patients would be necessary to show the involvement of hippocampal neurogenesis in cognitive impairment in PD.

Huntington's Disease

HD results from overexpansion of CAG trinucleotide repeats in the huntingtin gene that encodes a polyglutamine tract in the huntingtin protein (*A novel gene containing a trinucleotide repeat that is expanded and unstable on Huntington's disease chromosomes, the Huntington's Disease Collaborative Research Group*, 1993). Transgenic animal models of HD have demonstrated comparable progressive neurological phenotypes (eg, progressive dysfunction in motor ability and cognitive decline) to clinical observation in HD patients (Hodgson et al., 1999; Mangiarini et al., 1996). Cognitive deficits in HD can be detected before the onset of motor dysfunction (Duff et al., 2007; Kirkwood, Su, Conneally, & Foroud, 2001; Slaughter, Martens, & Slaughter, 2001). The significant decline in adult neurogenesis has been consistently reported in several transgenic models of HD (Gil-Mohapel, Simpson, Ghilan, & Christie, 2011), suggesting the possible linkage between cognitive decline and alteration in adult neurogenesis in HD.

Both transgenic mouse R6/1 and R6/2 lines of HD models, which express exon 1 of the human huntingtin gene with 115 and 150 CAG repeats, respectively, exhibit impairment in cognitive functions (Murphy et al., 2000; Nithianantharajah, Barkus, Murphy, & Hannan, 2008). A decrease in hippocampal neurogenesis has been found in R6/1 (Lazic et al., 2006) and R6/2 transgenic mice (Gil et al., 2005; Kohl et al., 2007). The R6/1 line mice display a normal level of hippocampal cell proliferation when they are 5 weeks old, but show a significant decrease in cell proliferation when they are 20 weeks old. In contrast, the R6/2 line HD mice, which exhibit a faster disease progression, show a reduction in hippocampal neurogenesis prior to any observable behavioral deficits when the mice are 2 weeks old (Mangiarini et al., 1996), and a robust reduction in proliferating cells in the DG when they are 12 weeks old (when they reach the late-stage HD) (Gil, Leist, Popovic, Brundin, & Petersen, 2004). Furthermore, these transgenic HD mice show a decrease in neuronal differentiation (Gil et al., 2005), suggesting that suppressed adult neurogenesis occurs in the late stage of HD in the R6/2 line.

The yeast artificial chromosome (YAC) mouse model of HD, which carries a mutation in the human huntingtin gene with 128 CAG repeats, can mimic the disease progress in clinical conditions (Gil-Mohapel, 2012; Slow et al., 2003). These mice show decreases in both adult hippocampal neurogenesis and BDNF, as well as an increase in depressive-like behavior and hippocampal-dependent cognitive deficits (Grote et al., 2005; Pang, Stam, Nithianantharajah, Howard, & Hannan, 2006; Pouladi et al., 2009; Simpson et al., 2011; Spires et al., 2004). Similarly, in a rat model of HD expressing a truncated cDNA fragment in the HD gene, there is a progressive decrease in hippocampal cell proliferation (von Horsten et al., 2003). A knock-in mice model of HD showed an alternation in neuronal maturation of newborn neurons in the DG along with increased anxiety-like behavior (Orvoen, Pla, Gardier, Saudou, & David, 2012). In contrast to the hippocampus, adult neurogenesis in the SVZ was not affected in R6/2 mice (Phillips, Morton, & Barker, 2005) and YAC128 mice (Gil et al., 2005; Lazic et al., 2006; Simpson et al., 2011).

Mechanisms underlying the suppressive effects of mutant huntingtin genes on neurogenesis remain unclear. Dephosphorylation at serines 1181 and 1201 of huntingtin protein can increase axonal transport of BDNF and hippocampal neurogenesis and reduce depression-like behavior in wild-type mice (Ben M'Barek et al., 2013), suggesting a possible mechanism of how the huntingtin gene regulates neurogenesis. The excessive huntingtin protein expression may possibly dysregulate neurogenesis-related genes (eg, NeuroD) (Marcora, Gowan, & Lee, 2003), reduce neurotrophic factor (eg, BDNF) (Ferrer, Goutan, Marin, Rey, & Ribalta, 2000; Zajac et al., 2010), and disrupt neuron transmission in dopaminergic or serotonergic systems (Hickey, Reynolds, & Morton, 2002; Petersen et al., 2002; Reynolds et al., 1999). Curtis et al. (2003, 2005) have reported a significant decrease in cell proliferation in the SVZ/OB system as indicated by a decrease in number of PCNA-positive cells and the colabeled cells with PCNA and glial cell marker, GFAP. With this limited finding on adult neurogenesis in human PD brains, the role of adult neurogenesis HD-related cognitive decline is still inconclusive.

NEURODEVELOPMENTAL DISORDERS

Neurodevelopmental disorders such as Fragile X syndrome (FXS) and fetal alcohol spectrum disorder (FASD) are associated with cognitive impairment (Kerns, Don, Mateer, & Streissguth, 1997; Lightbody & Reiss, 2009). Hippocampal deficits may play a significant role in cognitive impairment and mood disorders observed in the affected individuals. Recent studies reporting the dysregulation of adult neurogenesis in animal models of FXS and FASD have shed light on the biological mechanism underlying some forms of behavioral deficits presented in these two devastating neurodevelopmental disorders (Table 11.1).

Fragile X Syndrome

FXS is the most common form of inherited intellectual disability and the leading cause of autism spectrum disorder. It is caused by silencing of the fragile X mental retardation 1 (Fmr1) gene, which consequently leads to loss of fragile X mental retardation protein (FMRP). With its capability as an mRNA-binding protein, FMRP regulates the mRNA trafficking from the nucleus to the cytoplasm and to distal postsynaptic sites, and controls mRNA translation of synaptic proteins (Antar & Bassell, 2003; Ashley, Wilkinson, Reines, & Warren, 1993; Bassell & Warren, 2008; Zalfa et al., 2007).

The hippocampus is one of the brain regions that is greatly affected by the loss of FMRP. The transgenic mouse model of FXS, the *Fmr1* knock out (KO) mice (*Fmr1 knockout mice: a model to study fragile X mental retardation, The Dutch–Belgian Fragile X Consortium*, 1994), has demonstrated some parallel phenotypes to FXS patients, such as long thin tortuous dendritic spines (Comery et al., 1997; Nimchinsky, Oberlander, & Svoboda, 2001), increases in dendritic spine density (Comery et al., 1997), and macroorchidism (*Fmr1 knockout mice: A model to study fragile X mental retardation, The Dutch–Belgian Fragile X Consortium*, 1994; Kooy et al., 1996; Slegtenhorst-Eegdeman et al., 1998). In addition, the *Fmr1* KO mice displayed hippocampal-dependent tasks, including impairment in hippocampal-dependent learning and memory, and fear conditioning and passive avoidance tasks (Eadie et al., 2009; Kooy et al., 1996; Mineur, Sluyter, de Wit, Oostra, & Crusio, 2002; Paradee et al., 1999; Peier et al., 2000; Sabaratnam, Vroegop, & Gangadharan, 2001; Van Dam et al., 2000).

Impairment in hippocampal synaptic plasticity including impaired NMDA receptor-dependent long-term potentiation in the DG (Bostrom et al., 2013; Eadie, Cushman, Kannangara, Fanselow, & Christie, 2012; Franklin et al., 2014; Yun & Trommer, 2011) and exaggerated mGluR-dependent long-term depression in the CA1 region (Bear, Huber, & Warren, 2004; Huber, Gallagher, Warren, & Bear, 2002) have been identified and these synaptic deficits may link to cognitive impairment in FXS patients.

Alteration in adult hippocampal neurogenesis has also been found in the DG of the *Fmr1* KO mice. Hippocampal neurogenesis may play an essential role in hippocampal-dependent learning and memory (Deng, Aimone, & Gage, 2010) and mood regulation in response to antidepressant treatments (Ernst, Olson, Pinel, Lam, & Christie, 2006; Santarelli et al., 2003); therefore, deficiency in hippocampal neurogenesis may be involved in cognitive deficit and mood disorders present in the FXS patients.

An in vitro study using a neural progenitor cell culture has indicated that loss of FMRP decreases neuronal differentiation and survival, though it increases cell proliferation (Luo et al., 2010). However, ablation of FMRP in nestin-positive cells increases proliferation and astrocytic differentiation, but decreases neural differentiation (Guo et al., 2011, 2012). *Fmr1* knockout mice, Eadie et al. (2009) have revealed a decrease in cell survival and an increase in neuronal differentiation specifically in the ventral DG, but no significant differences in cell proliferation in young *Fmr1* knockout mice aged 2–4 months old. Interestingly, Lazarov et al. (2012) reported significant reduction in cell proliferation, differentiation, and survival in the DG of the *Fmr1* KO mice aged 9–12 months. Contradictory results may be due to the fact that FMRP may differentially regulate the process of neurogenesis in an age-dependent manner, or that conditional knockout of FMRP in nestin-positive cell versus complete loss of FMPR in the brain may have different influences on neurogenesis.

Impairments in neurogenesis-related signaling molecules may contribute to alteration in hippocampal neurogenesis in the *Fmr1* KO mice. The β-catenin/Wnt pathway is known to upregulate hippocampal neurogenesis. GSK3β can negatively regulate the β-catenin/Wnt pathway (Clevers, 2006; Lie et al., 2005). Disruptions in β-catenin levels and Wnt signaling, as well as elevation in glycogen synthase kinase 3β (GSK3β) levels (Min et al., 2009), have been reported in Fmr1 KO mice. Inhibition of GSK3β rescues impaired neurogenesis in the DG and improves hippocampal-dependent learning in *Fmr1* KO mice (Guo et al., 2011), suggesting that manipulating the relevant signaling molecule can rescue the deficits in adult neurogenesis observed in *Fmr1* KO mice.

Collectively, the aforementioned studies suggest that the process of neurogenesis is altered by the loss of FMRP, which may lead to negative impacts on the proper functioning of the hippocampus (Guo et al., 2011). Alteration in hippocampal neurogenesis may be partly involved in cognitive impairment in FXS.

Fetal Alcohol Spectrum Disorders

Alcohol consumption during different periods of pregnancy is the leading cause of FASD, which is characterized by neurological disorders ranging from attention deficit disorder, hyperactivity, and impaired cognitive function to mood disorder (Barr et al., 2006; Sokol, Delaney-Black, & Nordstrom, 2003; Whaley, O'Connor, & Gunderson, 2001). Accumulating evidence has shown that hippocampal neurogenesis, hippocampal-dependent spatial learning (Berman & Hannigan, 2000; Ieraci & Herrera, 2007; Wozniak et al., 2004), and synaptic plasticity (Christie et al., 2005; Izumi et al., 2005) are significantly impaired in animal models of FASD.

MRI screening reported that FASD children had smaller volumes of brain regions including the hippocampus, basal ganglia, corpus callosum, and cerebellum (Archibald et al., 2001; Autti-Ramo et al., 2002; Mattson et al., 1996; Sowell et al., 2001). The decrease in brain volume may be contributed by decreased cell proliferation or increased apoptosis of postmitotic neurons (Ikonomidou et al., 2000; Miller, 1989). Altered structure of hippocampal subfields has been reported following alcohol exposure during brain development. The binge-like alcohol exposure during the third trimester equivalent decreases cell density and cell number in hippocampal dentate gyrus, CA1, and CA3 (Bonthius & West, 1990; Livy, Miller, Maier, & West, 2003; West, Hamre, & Cassell, 1986). Neurons are particularly susceptible to alcohol exposure during the early postnatal brain developmental stage (Olney, 2004). Ieraci & Herrera (2007) have shown that a single exposure of alcohol by subcutaneous injection in postnatal day (PND) 7 decreased adult hippocampal neurogenesis at later time points and decreased the number of primary neurospheres derived from the hippocampi compared to control mice, indicating that exposure to alcohol at the early developmental stage of the brain could lead to a long-lasting defect in adult hippocampal neurogenesis. Similarly, a binge-like alcohol exposure via intragastric intubation on PNDs 4–9 (equivalent to the third trimester) robustly reduces number of mature neurons on PNDs 50 and 80 (Klintsova et al., 2007).

In contrast to the hippocampus, there is no defect in neurogenesis in the SVZ of mice exposed to alcohol (Herrera et al., 2003; Ieraci & Herrera, 2007), suggesting a region-specific effect of adult neurogenesis depletion by single alcohol exposure. In addition to region-specific effects, ethanol exposure at different trimesters differentially affects the process of hippocampal adult neurogenesis since varied results have been reported with ethanol exposure during first and second trimester equivalents (Choi, Allan, & Cunningham, 2005; Redila et al., 2006), third trimester equivalents (Helfer, Goodlett, Greenough, & Klintsova, 2009; Klintsova et al., 2007), or the entire trimester (Gil-Mohapel, Boehme, et al., 2011; Gil-Mohapel, Simpson, 2011).

Helfer et al. (2009) reported no alterations in adult-born neurons at PND 42 after ethanol exposure at PNDs 4–9 (third trimester equivalent), whereas Klintsova et al. (2007) reported no difference in cell proliferation on PNDs 50 and 80 with intragastric intubation on PNDs 4–9, but a decrease in the number of mature neurons. Conversely, Redila et al. (2006)

reported significant decreases in cell proliferation and survival rate of adult-born neurons in rats exposed to alcohol during the full gestational period via a liquid diet. In a rat model of FASD with ethanol administered via gavage throughout all three trimester equivalents, Gil-Mohapel, Boehme, et al. (2011), Gil-Mohapel, Simpson, et al. (2011) reported a significant increase in number of immature neurons, but no significant difference in hippocampal cell proliferation and survival in adult rats. The discrepancy among the different above-noted studies may be due to differential protocols such as route of ethanol administration, timing of ethanol exposures, and dosage of alcohol exposures. Gil-Mohapel, Boehme, et al. (2011), Gil-Mohapel, Simpson, et al. (2011) speculated that an increase in newborn neurons reported in their study may be due to prolonged ethanol exposure throughout three trimesters, which in turn retard neurogenic development at the early differentiation stage of newborn neurons.

Hippocampal cell proliferation, neuronal differentiation, and survival can be upregulated by interventions like physical exercise or environmental enrichment (Kempermann, Kuhn, & Gage, 1998; Nilsson, Perfilieva, Johansson, Orwar, & Eriksson, 1999; van Praag, Christie, Sejnowski, & Gage, 1999; van Praag, Kempermann, & Gage, 1999; van Praag, Shubert, Zhao, & Gage, 2005). In a rat model of FASD with voluntary consumption of 10% alcohol throughout the pregnancy, the pups displayed no defects in adult hippocampal neurogenesis, and alcohol consumption diminished enriched environment increases in progenitor cell survival and neuronal differentiation in the fetal alcohol-exposed mice (Choi et al., 2005). However, voluntary running rescued the decrease in hippocampal cytogenesis and neurogenesis in the rats fed an ethanol liquid diet throughout gestation (Redila et al., 2006). Likewise, Helfer et al. (2009) reported that adolescent wheel running increased cell proliferation and neurogenesis in rats with a binge-like alcohol exposure starting at PNDs 4–9. Since the long-lasting deficits in adult hippocampal neurogenesis can be reversed by interventions like physical exercise in FASD, these data imply that manipulation of hippocampal neurogenesis may be a promising intervention for ameliorating hippocampal-dependent learning and memory impairment in FASD.

NEUROGENESIS AND BRAIN INJURY

Stroke

Ischemic stroke causes permanent damage of the affected brain region and irreversible impairment in cognitive and motor functions (Park et al., 2009). Owing to its severe consequences and high mortality, extensive investigations have been focused on treatments that minimize the mortality and increase functional recovery in stroke patients (Chen, Venkat, & Chopp, 2014). Apart from cell-based therapy, which aims at replacing the lost neurons (Hao et al., 2014), promotion of adult neurogenesis has emerged as a source of newborn neurons for neuronal regeneration in stroke patients (Ruan et al., 2014).

Ischemic stroke is known to induce adult neurogenesis. The proliferation of neural precursor cells in both the SVZ and the hippocampus has been reported to be increased after global ischemia in animal models (Arvidsson, Collin, Kirik, Kokaia, & Lindvall, 2002; Lin et al., 2015; Liu, Solway, Messing, & Sharp, 1998). The onset of the increased proliferation starts 48h after ischemic injury, remains at a high level for 1 to 2 weeks, and returns to baseline levels 3 to 4 weeks later (Jin et al., 2001). Migration of newborn cells to the damaged brain regions suggests that these cells may be involved in repairing the ischemic site (Jin et al., 2003). The migration destination could be distant from the neurogenic zones, such as the cortex, and this process could last for a long period (more than 4 months) after the injury (Thored et al., 2006). After a transient ischemia is induced by middle cerebral artery occlusion, the neuroblasts generated from the SVZ were found to migrate to the affected striatum and expressed markers of region-specific neurons (Arvidsson, Collin, Kirik, Kokaia, & Lindvall, 2002), which suggests the potential of new cells to substitute the lost neurons. The new neuroblasts were also reported to differentiate into calretinin-positive interneurons (Doh-Ura et al., 1999). However, in the peri-infarct cortical areas, the new proliferative cells did not differentiate into mature neurons at later time points (Lichtenwalner & Parent, 2005), suggesting the failure of neuronal replacement in the cortical region. However, limited evidence from a human study has shown that adult neurogenesis occurred at the ischemic penumbra around the injured site (Jin et al., 2006).

Ischemia-induced neurogenesis can be observed in different species including laboratory rodents, gerbils, monkeys, and humans (Ruan et al., 2014). A pioneering study conducted by Liu, Solway, Messing, & Sharp (1998) described induced hippocampal cell proliferation by global ischemia in gerbils. To trace the cell fate of these proliferative cells, an enhanced green fluorescent protein (EGFP)-expressing retroviral vector was injected into the DG of gerbils to label the proliferative cells. The labeled progenitor cells migrated from the subgranular zone to the granule cell layer in the DG of the hippocampus. These cells expressed early neuronal markers like doublecortin, and differentiated into the mature granule cells with dendritic growth (Tanaka et al., 2004). While most of the new cells followed the neuronal lineage, about 10–20% of the new cells differentiated into astrocytes (Komitova, Perfilieva, Mattsson, Eriksson, & Johansson, 2002). Interestingly, the migrating neuroblasts were also shown to migrate to the CA1 region of the hippocampus (Nakatomi et al., 2002), which was hypothesized to improve spatial learning and memory after ischemic injury. These studies collectively suggest that mature dentate gyrus neurons could be generated from proliferative neural progenitor cells after ischemic injury, and these new neurons may function in the nonneurogenic regions under healthy conditions.

Functional integration of newborn neurons was shown after stroke. Following ischemia, newborn neurons show neuronal polarity and establish synapses with existing neurons (Hou et al., 2008). These newborn cells are also able to integrate into the injured striatum and display functional electrophysiological properties (Yamashita et al., 2006), suggesting that these new neurons are functional. From the functional perspective, treatment with erythropoietin after stroke enhanced neurogenesis in association with an improved neurological functioning (Wang, Zhang, Wang, Zhang, & Chopp, 2004). Other neurotrophic factors or medications that promote neurogenesis also improved behavioral recovery when administrated after stroke (Sun et al., 2003; Tang et al., 2013; Zhang et al., 2006). Notably, behavioral outcome of the stroke animals can be improved not only with an increase in neurotrophic factors but also with antidepressant or mobilization treatments (McCann et al., 2014; Qu et al., 2014). In contrast, inhibition of neurogenesis caused a long-term impairment in functional recovery (Wang et al., 2012). These findings indicate the necessity of neurogenesis in the recovery after ischemic stroke, and provide the grounds for further study on the functional significance of neurogenesis.

Traumatic Brain Injury

After a traumatic brain injury (TBI), the patients suffer from neuronal loss brought about by apoptosis or necrosis, which may be localized or diffused (Smith et al., 2000). The hippocampus is a commonly affected site of diffuse injury in humans, and damage in this area usually causes deficits in learning and memory (Kotapka, Graham, Adams, & Gennarelli, 1992). Being similar to a stroke, adult neurogenesis in the hippocampus raises the hope for replacing the neuronal loss in TBI, and hence the functional recovery in cognition related to the hippocampus.

There are several commonly used experimental models of TBI including controlled cortical impact, lateral fluid percussion, blast TBI, and closed head injury (Osier, Carlson, DeSana, & Dixon, 2014). From the result obtained from different models, various histological disruptions were found after TBI, like ventricular enlargement, shrinkage of brain regions, and inflammation and changes of neurogenesis (Osier et al., 2014). Being similar to a stroke, various experimental models of TBI showed an increase of cell proliferation in both neurogenic zones in an early stage after injury (Rice et al., 2003), which could be up to three- to fourfold change. The newly proliferating cells can differentiate into both glial cells and neurons (Sun et al., 2005), indicating that neurogenesis and gliogenesis could be found in the TBI model. Although it is hypothesized that the new cells attempt to repair damage and lead to recovery, there is still lack of concrete evidence to support the functional significance of new neurons (Leuner, Gould, & Shors, 2006).

As early as 3 days after the injury, DCX-positive neuroblasts could be found in various sites including the striatum, corpus callosum, cortex, and hippocampus (Ramaswamy, Goings, Soderstrom, Szele, & Kozlowski, 2005), while the appearance of the neuroblasts could last for at least 42 days postinjury (Lu, Mahmood, Zhang, & Copp, 2003). Proneurogenic treatment augments the increase of the new neurons in SVZ and other nonneurogenic regions (Lu et al., 2003). Further evidence showed that neural precursor cells in the SVZ may serve as the source of new neurons. Labeling of proliferative cells in the SVZ with fluorescent microspheres showed the migration of the progenitor cells from the SVZ to the affected cortex via the corpus callosum (Ramaswamy et al., 2005).

Neuroblasts generated in the subgranular zone did not migrate to the hippocampal CA regions, and thus the new neurons found in the injured site may be derived from the SVZ (Ruan et al., 2014). There are studies reporting a two- to fourfold increase of neuroblasts after TBI (Lu et al., 2003; Ramaswamy et al., 2005), which is comparable to the increase in the SVZ. The cortex is another site where the activated neural precursor cells could be found after TBI. Magavi, Leavitt, & Macklis (2000) found that after neuronal death, the dormant NPCs in the cortex activate and differentiate into mature neurons and establish synaptic connections in the adult rodent. The nestin, a marker of neural stem cells, is found in the peri-injured cortex 7 days post injury, suggesting the possibility of NPC activation (Kernie, Erwin, & Parada, 2001). In animals of the controlled cortical impact model, when the cortical tissue was isolated, nestin expression peaks in vitro 3 days after injury, and is associated with the formation of neurospheres (Itoh, Satou, Hashimoto, & Ito, 2005). The capacity of human cortical NPCs to generate new neurons was also found after TBI (Arsenijevic et al., 2001; Richardson, Holloway, Bullock, Broaddus, & Fillmore, 2006), which suggests that the cortical dormant NPCs are a possible source of new neurons after injury.

Animal studies showing improvement in functional outcome by augmenting neurogenesis following TBI have suggested the functional role of newborn neurons in this injury. In a physiological condition, neural progenitors in the SVZ migrate to the olfactory bulb and differentiate into inhibitory interneurons or dopaminergic periglomerular neurons (Ruan et al., 2014). These progenitors of the SVZ can differentiate into calbindin-expressing neurons when they are engrafted into the DG (Richardson et al., 2005). Exogenously transplanted neural precursor cells in the TBI animal brains can differentiate into mature neurons and promote functional and cognitive outcomes (Blaya, Tsoulfas, Bramlett, & Dietrich, 2015). Furthermore, treatment of TBI animals with proneurogenic medications such as statins, erythropoietin, and fibroblast growth factor 2 enhances neurogenesis in the SVZ and promotes functional outcomes (Lu et al., 2005, 2007; Sun et al., 2009; Xie et al., 2014). In contrast, inhibition of the injury-induced neurogenesis suppresses cognitive recovery in TBI models (Sun, Daniels, Rofle, Waters, & Hamm, 2014). From the above-noted findings, it is known that new neurons could potentially develop into mature and functional neurons in the proper niche. The neurons may integrate into the existing neural circuit directly or benefit by secreting neurotrophic factors to promote survival and functioning of the affected neurons. Since both TBI and stroke are associated with promoted endogenous neurogenesis that may benefit rehabilitation, understanding the underlying mechanisms can be beneficial to the treatment regimen of both diseases.

CONCLUSIONS

Despite that the significance of adult neurogenesis in pathological conditions has been investigated for decades, its role in the etiology and therapeutic treatments are still under investigation. Evidence from both preclinical and clinical studies supports the hypothesis that neurogenesis participates in disease conditions in the CNS. Different aspects of neurogenesis have been intensively investigated, for instance, the behavioral/physiological significance of neurogenesis, the regulatory mechanisms of the neurogenesis process, modalities that regulate neurogenesis, roles of neurogenesis in neurological diseases and recovery, and potential application of proneurogenic treatment in human studies. In general, alternations in adult neurogenesis have been suggested to be related to neurological disorders, while treatments which could modulate neurogenesis are hoped to be the future treatments. Preclinical studies would provide valuable information for understanding the process of adult neurogenesis, efficacy of proneurogenic treatment on the aforementioned diseases, and the mechanisms underlying the pathology. For potential translation of the knowledge obtained from animal studies, clinical studies would need to reveal more about changes and functions of adult neurogenesis in neurological disorders.

While both preclinical and clinical studies are crucial for the final application of treatment and understanding of CNS diseases, the translation of knowledge from basic science to clinical studies will be a great challenge. Promoting collaboration between both parties would greatly facilitate the development of effective treatment for neurogenesis-related brain disorders.

LIST OF ABBREVIATIONS

6-OHDA 6-hydroxydopamine
AD Alzheimer's disease
ApoE Apolipoprotein
APP Amyloid precursor protein
BDNF Brain-derived neurotrophic factor
BrdU Bromodeoxyuridine
CA Cornus ammonis
CDK5 Cyclin-dependent kinase 5
DA Dopamine
DCX Doublecortin
DG Dentate gyrus
FASD Fetal alcohol spectrum disorders
FMRP Fragile X mental retardation protein
FXS Fragile X syndrome
GABA γ-aminobutyric acid
GFAP Glial fibrillary acidic protein
GSK3β Glycogen synthase kinase 3β
HD Huntington disease
MPTP Methyl-4-phenyl-1,2,3,6-tetrahydropyridine
PCNA Proliferating cell nuclear antigen
PD Parkinson's disease
PSA–NCAM Polysialylated–neural cell adhesion molecule
PSEN-1 Presenilin 1
PSEN-2 Presenilin 2
SGZ Subgranular zone
SN Substantia nigra
SVZ Subventricular zone
YAC Yeast artificial chromosome

References

Antar, L. N., & Bassell, G. J. (2003). Sunrise at the synapse: the FMRP mRNP shaping the synaptic interface. *Neuron, 37*(4), 555–558.

Archibald, S. L., Fennema-Notestine, C., Gamst, A., Riley, E. P., Mattson, S. N., & Jernigan, T. L. (2001). Brain dysmorphology in individuals with severe prenatal alcohol exposure. *Developmental Medicine and Child Neurology, 43*(3), 148–154.

Arsenijevic, Y., Villemure, J.-G., Brunet, J.-F., Bloch, J. J., Déglon, N., Kostic, C., et al. (2001). Isolation of multipotent neural precursors residing in the cortex of the adult human brain. *Experimental Neurology, 170*(1), 48–62.

Arvidsson, A., Collin, T., Kirik, D., Kokaia, Z., & Lindvall, O. (2002). Neuronal replacement from endogenous precursors in the adult brain after stroke. *Nature Medicine, 8*(9), 963–970.

Ashley, C. T., Jr., Wilkinson, K. D., Reines, D., & Warren, S. T. (1993). FMR1 protein: conserved RNP family domains and selective RNA binding. *Science, 262*(5133), 563–566.

Autti-Ramo, I., Autti, T., Korkman, M., Kettunen, S., Salonen, O., & Valanne, L. (2002). MRI findings in children with school problems who had been exposed prenatally to alcohol. *Developmental Medicine and Child Neurology, 44*(2), 98–106.

Barr, H. M., Bookstein, F. L., O'Malley, K. D., Connor, P. D., Huggins, J. E., & Streissguth, A. P. (2006). Binge drinking during pregnancy as a predictor of psychiatric disorders on the structured clinical interview for DSM-IV in young adult offspring. *The American Journal of Psychiatry, 163*(6), 1061–1065.

Bassell, G. J., & Warren, S. T. (2008). Fragile X syndrome: loss of local mRNA regulation alters synaptic development and function. *Neuron, 60*(2), 201–214.

Bear, M. F., Huber, K. M., & Warren, S. T. (2004). The mGluR theory of fragile X mental retardation. *Trends in Neurosciences, 27*(7), 370–377.

Ben M'Barek, K., Pla, P., Orvoen, S., Benstaali, C., Godin, J. D., Gardier, A. M., et al. (2013). Huntingtin mediates anxiety/depression-related behaviors and hippocampal neurogenesis. *The Journal of Neuroscience, 33*(20), 8608–8620.

Berman, R. F., & Hannigan, J. H. (2000). Effects of prenatal alcohol exposure on the hippocampus: spatial behavior, electrophysiology, and neuroanatomy. *Hippocampus, 10*(1), 94–110.

Blaya, M. O., Tsoulfas, P., Bramlett, H. M., & Dietrich, W. D. (2015). Neural progenitor cell transplantation promotes neuroprotection, enhances hippocampal neurogenesis, and improves cognitive outcomes after traumatic brain injury. *Experimental Neurology, 264*, 67–81.

Boekhoorn, K., Joels, M., & Lucassen, P. J. (2006). Increased proliferation reflects glial and vascular-associated changes, but not neurogenesis in the presenile Alzheimer hippocampus. *Neurobiology of Disease, 24*(1), 1–14.

Bonthius, D. J., & West, J. R. (1990). Alcohol-induced neuronal loss in developing rats: increased brain damage with binge exposure. *Alcoholism, Clinical and Experimental Research, 14*(1), 107–118.

Bostrom, C. A., Majaess, N. M., Morch, K., White, E., Eadie, B. D., & Christie, B. R. (2013). Rescue of NMDAR-dependent synaptic plasticity in Fmr1 knock-out mice. *Cerebral Cortex, 25*(1), 271–279.

Chen, J., Venkat, P., & Chopp, M. (2014). Neurorestorative therapy for stroke. *Frontiers in Human Neuroscience, 8*, 382.

Choi, I. Y., Allan, A. M., & Cunningham, L. A. (2005). Moderate fetal alcohol exposure impairs the neurogenic response to an enriched environment in adult mice. *Alcoholism, Clinical and Experimental Research, 29*(11), 2053–2062.

Christie, B. R., Swann, S. E., Fox, C. J., Froc, D., Lieblich, S. E., Redila, V., et al. (2005). Voluntary exercise rescues deficits in spatial memory and long-term potentiation in prenatal ethanol-exposed male rats. *The European Journal of Neuroscience, 21*(6), 1719–1726.

Clevers, H. (2006). Wnt/beta-catenin signaling in development and disease. *Cell, 127*(3), 469–480.

Comery, T. A., Harris, J. B., Willems, P. J., Oostra, B. A., Irwin, S. A., Weiler, I. J., et al. (1997). Abnormal dendritic spines in fragile X knockout mice: maturation and pruning deficits. *Proceedings of the National Academy of Sciences of the United States of America, 94*(10), 5401–5404.

Crews, L., Adame, A., Patrick, C., Delaney, A., Pham, E., Rockenstein, E., et al. (2010). Increased BMP6 levels in the brains of Alzheimer's disease patients and APP transgenic mice are accompanied by impaired neurogenesis. *The Journal of Neuroscience, 30*(37), 12252–12262.

Crews, L., Mizuno, H., Desplats, P., Rockenstein, E., Adame, A., Patrick, C., et al. (2008). Alpha-synuclein alters notch-1 expression and neurogenesis in mouse embryonic stem cells and in the hippocampus of transgenic mice. *The Journal of Neuroscience, 28*(16), 4250–4260.

Curtis, M. A., Penney, E. B., Pearson, J., Dragunow, M., Connor, B., & Faull, R. L. (2005). The distribution of progenitor cells in the subependymal layer of the lateral ventricle in the normal and huntington's disease human brain. *Neuroscience, 132*(3), 777–788.

Curtis, M. A., Penney, E. B., Pearson, A. G., van Roon-Mom, W. M., Butterworth, N. J., Dragunow, M., et al. (2003). Increased cell proliferation and neurogenesis in the adult human huntington's disease brain. *Proceedings of the National Academy of Sciences of the United States of America, 100*(15), 9023–9027.

Dauer, W., & Przedborski, S. (2003). Parkinson's disease: mechanisms and models. *Neuron, 39*(6), 889–909.

Demars, M. P., Hollands, C., Zhao Kda, T., & Lazarov, O. (2013). Soluble amyloid precursor protein-alpha rescues age-linked decline in neural progenitor cell proliferation. *Neurobiology of Aging*, *34*(10), 2431–2440.

Deng, W., Aimone, J. B., & Gage, F. H. (2010). New neurons and new memories: how does adult hippocampal neurogenesis affect learning and memory? *Nature Reviews. Neuroscience*, *11*(5), 339–350.

Doh-Ura, K., Mohri, S., Tashiro, H., Kawashima, T., Kikuchi, H., & Iwaki, T. (1999). Brain injury does not modify transmissible spongiform encephalopathy caused by intraperitoneal inoculation with fukuoka-1 strain. *The Journal of General Virology*, *80*(6), 1551–1556.

Duff, K., Paulsen, J. S., Beglinger, L. J., Langbehn, D. R., Stout, J. C., Predict, H. D., & Investigators of the Huntington Study Group. (2007). Psychiatric symptoms in huntington's disease before diagnosis: the predict-HD study. *Biological Psychiatry*, *62*(12), 1341–1346.

Eadie, B. D., Cushman, J., Kannangara, T. S., Fanselow, M. S., & Christie, B. R. (2012). NMDA receptor hypofunction in the dentate gyrus and impaired context discrimination in adult Fmr1 knockout mice. *Hippocampus*, *22*(2), 241–254.

Eadie, B. D., Zhang, W. N., Boehme, F., Gil-Mohapel, J., Kainer, L., Simpson, J. M., et al. (2009). Fmr1 knockout mice show reduced anxiety and alterations in neurogenesis that are specific to the ventral dentate gyrus. *Neurobiology of Disease*, *36*(2), 361–373.

Ermini, F. V., Grathwohl, S., Radde, R., Yamaguchi, M., Staufenbiel, M., Palmer, T. D., et al. (2008). Neurogenesis and alterations of neural stem cells in mouse models of cerebral amyloidosis. *The American Journal of Pathology*, *172*(6), 1520–1528.

Ernst, C., Olson, A. K., Pinel, J. P., Lam, R. W., & Christie, B. R. (2006). Antidepressant effects of exercise: evidence for an adult-neurogenesis hypothesis? *Journal of Psychiatry and Neuroscience*, *31*(2), 84–92.

Ferrer, I., Goutan, E., Marin, C., Rey, M. J., & Ribalta, T. (2000). Brain-derived neurotrophic factor in huntington disease. *Brain Research*, *866*(1–2), 257–261.

Fmr1 knockout mice: a model to study fragile X mental retardation. the Dutch-Belgian Fragile X Consortium. *Cell*, *78*(1), (1994), 23–33.

Franklin, A. V., King, M. K., Palomo, V., Martinez, A., McMahon, L. L., & Jope, R. S. (2014). Glycogen synthase kinase-3 inhibitors reverse deficits in long-term potentiation and cognition in fragile X mice. *Biological Psychiatry*, *75*(3), 198–206.

Gadadhar, A., Marr, R., & Lazarov, O. (2011). Presenilin-1 regulates neural progenitor cell differentiation in the adult brain. *The Journal of Neuroscience*, *31*(7), 2615–2623.

Gil-Mohapel, J. M. (2012). Screening of therapeutic strategies for huntington's disease in YAC128 transgenic mice. *CNS Neuroscience and Therapeutics*, *18*(1), 77–86.

Gil-Mohapel, J., Boehme, F., Patten, A., Cox, A., Kainer, L., Giles, E., et al. (2011). Altered adult hippocampal neuronal maturation in a rat model of fetal alcohol syndrome. *Brain Research*, *1384*, 29–41.

Gil-Mohapel, J., Simpson, J. M., Ghilan, M., & Christie, B. R. (2011). Neurogenesis in huntington's disease: can studying adult neurogenesis lead to the development of new therapeutic strategies? *Brain Research*, *1406*, 84–105.

Gil, J. M., Leist, M., Popovic, N., Brundin, P., & Petersen, A. (2004). Asialoerythropoietin is not effective in the R6/2 line of huntington's disease mice. *BMC Neuroscience*, *5*, 17.

Gil, J. M., Mohapel, P., Araujo, I. M., Popovic, N., Li, J. Y., Brundin, P., et al. (2005). Reduced hippocampal neurogenesis in R6/2 transgenic huntington's disease mice. *Neurobiology of Disease*, *20*(3), 744–751.

Goedert, M., Jakes, R., Anthony Crowther, R., & Grazia Spillantini, M. (2001). Parkinson's disease, dementia with lewy bodies, and multiple system atrophy as alpha-synucleinopathies. *Methods in Molecular Medicine*, *62*, 33–59.

Grote, H. E., Bull, N. D., Howard, M. L., van Dellen, A., Blakemore, C., Bartlett, P. F., et al. (2005). Cognitive disorders and neurogenesis deficits in huntington's disease mice are rescued by fluoxetine. *The European Journal of Neuroscience*, *22*(8), 2081–2088.

Guo, W., Allan, A. M., Zong, R., Zhang, L., Johnson, E. B., Schaller, E. G., et al. (2011). Ablation of fmrp in adult neural stem cells disrupts hippocampus-dependent learning. *Nature Medicine, 17*(5), 559–565.

Guo, W., Murthy, A. C., Zhang, L., Johnson, E. B., Schaller, E. G., Allan, A. M., et al. (2012). Inhibition of GSK3beta improves hippocampus-dependent learning and rescues neurogenesis in a mouse model of fragile X syndrome. *Human Molecular Genetics, 21*(3), 681–691.

Han, P., Dou, F., Li, F., Zhang, X., Zhang, Y. W., Zheng, H., et al. (2005). Suppression of cyclin-dependent kinase 5 activation by amyloid precursor protein: a novel excitoprotective mechanism involving modulation of tau phosphorylation. *The Journal of Neuroscience, 25*(50), 11542–11552.

Hao, L., Zou, Z., Tian, H., Zhang, Y., Zhou, H., & Liu, L. (2014). Stem cell-based therapies for ischemic stroke. *BioMed Research International, 2014*.

Haughey, N. J., Nath, A., Chan, S. L., Borchard, A. C., Rao, M. S., & Mattson, M. P. (2002). Disruption of neurogenesis by amyloid ß-peptide, and perturbed neural progenitor cell homeostasis, in models of Alzheimer's disease. *Journal of Neurochemistry, 83*(6), 1509–1524.

Helfer, J. L., Goodlett, C. R., Greenough, W. T., & Klintsova, A. Y. (2009). The effects of exercise on adolescent hippocampal neurogenesis in a rat model of binge alcohol exposure during the brain growth spurt. *Brain Research, 1294*, 1–11.

Herrera, D. G., Yague, A. G., Johnsen-Soriano, S., Bosch-Morell, F., Collado-Morente, L., Muriach, M., et al. (2003). Selective impairment of hippocampal neurogenesis by chronic alcoholism: protective effects of an antioxidant. *Proceedings of the National Academy of Sciences of the United States of America, 100*(13), 7919–7924.

Hickey, M. A., Reynolds, G. P., & Morton, A. J. (2002). The role of dopamine in motor symptoms in the R6/2 transgenic mouse model of huntington's disease. *Journal of Neurochemistry, 81*(1), 46–59.

Hillman, C. H., Erickson, K. I., & Kramer, A. F. (2008). Be smart, exercise your heart: exercise effects on brain and cognition. *Nature Reviews. Neuroscience, 9*(1), 58–65.

Hodgson, J. G., Agopyan, N., Gutekunst, C. A., Leavitt, B. R., LePiane, F., Singaraja, R., et al. (1999). A YAC mouse model for huntington's disease with full-length mutant huntingtin, cytoplasmic toxicity, and selective striatal neurodegeneration. *Neuron, 23*(1), 181–192.

Hoglinger, G. U., Rizk, P., Muriel, M. P., Duyckaerts, C., Oertel, W. H., Caille, I., et al. (2004). Dopamine depletion impairs precursor cell proliferation in parkinson disease. *Nature Neuroscience, 7*(7), 726–735.

von Horsten, S., Schmitt, I., Nguyen, H. P., Holzmann, C., Schmidt, T., Walther, T., et al. (2003). Transgenic rat model of huntington's disease. *Human Molecular Genetics, 12*(6), 617–624.

Hou, S. W., Wang, Y. Q., Xu, M., Shen, D. H., Wang, J. J., Huang, F., et al. (2008). Functional integration of newly generated neurons into striatum after cerebral ischemia in the adult rat brain. *Stroke, 39*(10), 2837–2844.

Huber, K. M., Gallagher, S. M., Warren, S. T., & Bear, M. F. (2002). Altered synaptic plasticity in a mouse model of fragile X mental retardation. *Proceedings of the National Academy of Sciences of the United States of America, 99*(11), 7746–7750.

Ieraci, A., & Herrera, D. G. (2007). Single alcohol exposure in early life damages hippocampal stem/progenitor cells and reduces adult neurogenesis. *Neurobiology of Disease, 26*(3), 597–605.

Ikonomidou, C., Bittigau, P., Ishimaru, M. J., Wozniak, D. F., Koch, C., Genz, K., et al. (2000). Ethanol-induced apoptotic neurodegeneration and fetal alcohol syndrome. *Science, 287*(5455), 1056–1060.

Itoh, T., Satou, T., Hashimoto, S., & Ito, H. (2005). Isolation of neural stem cells from damaged rat cerebral cortex after traumatic brain injury. *Neuroreport, 16*(15), 1687–1691.

Izumi, Y., Kitabayashi, R., Funatsu, M., Izumi, M., Yuede, C., Hartman, R. E., et al. (2005). A single day of ethanol exposure during development has persistent effects on bi-directional plasticity, N-methyl-D-aspartate receptor function and ethanol sensitivity. *Neuroscience, 136*(1), 269–279.

Jin, K., Galvan, V., Xie, L., Mao, X. O., Gorostiza, O. F., Bredesen, D. E., et al. (2004). Enhanced neurogenesis in Alzheimer's disease transgenic (PDGF-APPSw,ind) mice. *Proceedings of the National Academy of Sciences of the United States of America*, *101*(36), 13363–13367.

Jin, K., Minami, M., Lan, J. Q., Mao, X. O., Batteur, S., Simon, R. P., et al. (2001). Neurogenesis in dentate subgranular zone and rostral subventricular zone after focal cerebral ischemia in the rat. *Proceedings of the National Academy of Sciences of the United States of America*, *98*(8), 4710–4715.

Jin, K., Peel, A. L., Mao, X. O., Xie, L., Cottrell, B. A., Henshall, D. C., et al. (2004). Increased hippocampal neurogenesis in Alzheimer's disease. *Proceedings of the National Academy of Sciences of the United States of America*, *101*(1), 343–347.

Jin, K., Sun, Y., Xie, L., Peel, A., Mao, X. O., Batteur, S., et al. (2003). Directed migration of neuronal precursors into the ischemic cerebral cortex and striatum. *Molecular and Cellular Neurosciences*, *24*(1), 171–189.

Jin, K., Wang, X., Xie, L., Mao, X. O., Zhu, W., Wang, Y., et al. (2006). Evidence for stroke-induced neurogenesis in the human brain. *Proceedings of the National Academy of Sciences of the United States of America*, *103*(35), 13198–13202.

Kempermann, G., Kuhn, H. G., & Gage, F. H. (1998). Experience-induced neurogenesis in the senescent dentate gyrus. *The Journal of Neuroscience*, *18*(9), 3206–3212.

Kernie, S. G., Erwin, T. M., & Parada, L. F. (2001). Brain remodeling due to neuronal and astrocytic proliferation after controlled cortical injury in mice. *Journal of Neuroscience Research*, *66*(3), 317–326.

Kerns, K. A., Don, A., Mateer, C. A., & Streissguth, A. P. (1997). Cognitive deficits in nonretarded adults with fetal alcohol syndrome. *Journal of Learning Disabilities*, *30*(6), 685–693.

Kirkwood, S. C., Su, J. L., Conneally, P., & Foroud, T. (2001). Progression of symptoms in the early and middle stages of huntington disease. *Archives of Neurology*, *58*(2), 273–278.

Klintsova, A. Y., Helfer, J. L., Calizo, L. H., Dong, W. K., Goodlett, C. R., & Greenough, W. T. (2007). Persistent impairment of hippocampal neurogenesis in young adult rats following early postnatal alcohol exposure. *Alcoholism, Clincal and Experimental Research*, *31*(12), 2073–2082.

Kohl, Z., Kandasamy, M., Winner, B., Aigner, R., Gross, C., Couillard-Despres, S., et al. (2007). Physical activity fails to rescue hippocampal neurogenesis deficits in the R6/2 mouse model of huntington's disease. *Brain Research*, *1155*, 24–33.

Komitova, M., Perfilieva, E., Mattsson, B., Eriksson, P. S., & Johansson, B. B. (2002). Effects of cortical ischemia and postischemic environmental enrichment on hippocampal cell genesis and differentiation in the adult rat. *Journal of Cerebral Blood Flow and Metabolism*, *22*(7), 852–860.

Kooy, R. F., D'Hooge, R., Reyniers, E., Bakker, C. E., Nagels, G., De Boulle, K., et al. (1996). Transgenic mouse model for the fragile X syndrome. *American Journal of Medical Genetics*, *64*(2), 241–245.

Kotapka, M. J., Graham, D. I., Adams, J. H., & Gennarelli, T. A. (1992). Hippocampal pathology in fatal non-missile human head injury. *Acta Neuropathologica*, *83*(5), 530–534.

Lazarov, O., Demars, M. P., Zhao Kda, T., Ali, H. M., Grauzas, V., Kney, A., et al. (2012). Impaired survival of neural progenitor cells in dentate gyrus of adult mice lacking fMRP. *Hippocampus*, *22*(6), 1220–1224.

Lazic, S. E., Grote, H. E., Blakemore, C., Hannan, A. J., van Dellen, A., Phillips, W., et al. (2006). Neurogenesis in the R6/1 transgenic mouse model of huntington's disease: effects of environmental enrichment. *The European Journal of Neuroscience*, *23*(7), 1829–1838.

Leuner, B., Gould, E., & Shors, T. J. (2006). Is there a link between adult neurogenesis and learning? *Hippocampus*, *16*(3), 216–224.

Li, G., Bien-Ly, N., Andrews-Zwilling, Y., Xu, Q., Bernardo, A., Ring, K., et al. (2009). GABAergic interneuron dysfunction impairs hippocampal neurogenesis in adult apolipoprotein E4 knockin mice. *Cell Stem Cell*, *5*(6), 634–645.

Lichtenwalner, R. J., & Parent, J. M. (2005). Adult neurogenesis and the ischemic forebrain. *Journal of Cerebral Blood Flow & Metabolism, 26*(1), 1–20.

Lie, D., Colamarino, S. A., Song, H., Désiré, L., Mira, H., Consiglio, A., et al. (2005). Wnt signalling regulates adult hippocampal neurogenesis. *Nature, 437*(7063), 1370–1375.

Lightbody, A. A., & Reiss, A. L. (2009). Gene, brain, and behavior relationships in fragile X syndrome: evidence from neuroimaging studies. *Developmental Disabilities Research Reviews, 15*(4), 343–352.

Lin, R., Cai, J., Nathan, C., Wei, X., Schleidt, S., Rosenwasser, R., et al. (2015). Neurogenesis is enhanced by stroke in multiple new stem cell niches along the ventricular system at sites of high BBB permeability. *Neurobiology of Disease, 74*, 229–239.

Liu, J., Solway, K., Messing, R. O., & Sharp, F. R. (1998). Increased neurogenesis in the dentate gyrus after transient global ischemia in gerbils. *Journal of Neuroscience, 18*(19), 7768–7778.

Livy, D. J., Miller, E. K., Maier, S. E., & West, J. R. (2003). Fetal alcohol exposure and temporal vulnerability: effects of binge-like alcohol exposure on the developing rat hippocampus. *Neurotoxicology and Teratology, 25*(4), 447–458.

Lopez-Toledano, M. A., & Shelanski, M. L. (2007). Increased neurogenesis in young transgenic mice overexpressing human APP(sw, ind). *Journal of Alzheimer's Disease, 12*(3), 229–240.

Lu, D., Mahmood, A., Qu, C., Goussev, A., Schallert, T., & Chopp, M. (2005). Erythropoietin enhances neurogenesis and restores spatial memory in rats after traumatic brain injury. *Journal of Neurotrauma, 22*(9), 1011–1017.

Lu, D., Mahmood, A., Zhang, R., & Copp, M. (2003). Upregulation of neurogenesis and reduction in functional deficits following administration of DEtA/NONOate, a nitric oxide donor, after traumatic brain injury in rats. *Journal of Neurosurgery, 99*(2), 351–361.

Lu, D., Qu, C., Goussev, A., Jiang, H., Lu, C., Schallert, T., et al. (2007). Statins increase neurogenesis in the dentate gyrus, reduce delayed neuronal death in the hippocampal CA3 region, and improve spatial learning in rat after traumatic brain injury. *Journal of Neurotrauma, 24*(7), 1132–1146.

Luo, Y., Shan, G., Guo, W., Smrt, R. D., Johnson, E. B., Li, X., et al. (2010). Fragile x mental retardation protein regulates proliferation and differentiation of adult neural stem/progenitor cells. *PLoS Genetics, 6*(4), e1000898.

Magavi, S. S., Leavitt, B. R., & Macklis, J. D. (2000). Induction of neurogenesis in the neocortex of adult mice. *Nature, 405*(6789), 951–955.

Mangiarini, L., Sathasivam, K., Seller, M., Cozens, B., Harper, A., Hetherington, C., et al. (1996). Exon 1 of the HD gene with an expanded CAG repeat is sufficient to cause a progressive neurological phenotype in transgenic mice. *Cell, 87*(3), 493–506.

Marcora, E., Gowan, K., & Lee, J. E. (2003). Stimulation of NeuroD activity by huntingtin and huntingtin-associated proteins HAP1 and MLK2. *Proceedings of the National Academy of Sciences of the United States America, 100*(16), 9578–9583.

Mattson, S. N., Riley, E. P., Sowell, E. R., Jernigan, T. L., Sobel, D. F., & Jones, K. L. (1996). A decrease in the size of the basal ganglia in children with fetal alcohol syndrome. *Alcoholism, Clinical and Experimental Research, 20*(6), 1088–1093.

McCann, S. K., Irvine, C., Mead, G. E., Sena, E. S., Currie, G. L., Egan, K. E., et al. (2014). Efficacy of antidepressants in animal models of ischemic stroke: a systematic review and meta-analysis. *Stroke, 45*(10), 3055–3063.

Melrose, H. L., Lincoln, S. J., Tyndall, G. M., & Farrer, M. J. (2006). Parkinson's disease: a rethink of rodent models. *Experimental Brain Research, 173*(2), 196–204.

Miller, M. W. (1989). Effects of prenatal exposure to ethanol on neocortical development: II. cell proliferation in the ventricular and subventricular zones of the rat. *The Journal of Comparative Neurology, 287*(3), 326–338.

Mineur, Y. S., Sluyter, F., de Wit, S., Oostra, B. A., & Crusio, W. E. (2002). Behavioral and neuroanatomical characterization of the Fmr1 knockout mouse. *Hippocampus, 12*(1), 39–46.

Min, W. W., Yuskaitis, C. J., Yan, Q., Sikorski, C., Chen, S., Jope, R. S., et al. (2009). Elevated glycogen synthase kinase-3 activity in fragile X mice: key metabolic regulator with evidence for treatment potential. *Neuropharmacology*, *56*(2), 463–472.

Murphy, K. P., Carter, R. J., Lione, L. A., Mangiarini, L., Mahal, A., Bates, G. P., et al. (2000). Abnormal synaptic plasticity and impaired spatial cognition in mice transgenic for exon 1 of the human huntington's disease mutation. *The Journal of Neuroscience*, *20*(13), 5115–5123.

Nakatomi, H., Kuriu, T., Okabe, S., Yamamoto, S. -, Hatano, O., Kawahara, N., et al. (2002). Regeneration of hippocampal pyramidal neurons after ischemic brain injury by recruitment of endogenous neural progenitors. *Cell*, *110*(4), 429–441.

Nilsson, M., Perfilieva, E., Johansson, U., Orwar, O., & Eriksson, P. S. (1999). Enriched environment increases neurogenesis in the adult rat dentate gyrus and improves spatial memory. *Journal of Neurobiology*, *39*(4), 569–578.

Nimchinsky, E. A., Oberlander, A. M., & Svoboda, K. (2001). Abnormal development of dendritic spines in FMR1 knock-out mice. *The Journal of Neuroscience*, *21*(14), 5139–5146.

Nithianantharajah, J., Barkus, C., Murphy, M., & Hannan, A. J. (2008). Gene-environment interactions modulating cognitive function and molecular correlates of synaptic plasticity in huntington's disease transgenic mice. *Neurobiology of Disease*, *29*(3), 490–504.

A novel gene containing a trinucleotide repeat that is expanded and unstable on huntington's disease chromosomes. the Huntington's Disease Collaborative Research Group. *Cell*, *72*(6), (1993), 971–983.

Nuber, S., Petrasch-Parwez, E., Winner, B., Winkler, J., von Horsten, S., Schmidt, T., et al. (2008). Neurodegeneration and motor dysfunction in a conditional model of parkinson's disease. *The Journal of Neuroscience*, *28*(10), 2471–2484.

Olney, J. W. (2004). Fetal alcohol syndrome at the cellular level. *Addiction Biology*, *9*(2), 137–149 discussion 151.

Orvoen, S., Pla, P., Gardier, A. M., Saudou, F., & David, D. J. (2012). Huntington's disease knock-in male mice show specific anxiety-like behaviour and altered neuronal maturation. *Neuroscience Letters*, *507*(2), 127–132.

Osier, N., Carlson, S. W., DeSana, A. J., & Dixon, C. E. (2014). Chronic histopathological and behavioral outcomes of experimental traumatic brain injury in adult male animals. *Journal of Neurotrauma*, *32*(23), 1861–1882.

Pang, T. Y., Stam, N. C., Nithianantharajah, J., Howard, M. L., & Hannan, A. J. (2006). Differential effects of voluntary physical exercise on behavioral and brain-derived neurotrophic factor expression deficits in huntington's disease transgenic mice. *Neuroscience*, *141*(2), 569–584.

Paradee, W., Melikian, H. E., Rasmussen, D. L., Kenneson, A., Conn, P. J., & Warren, S. T. (1999). Fragile X mouse: strain effects of knockout phenotype and evidence suggesting deficient amygdala function. *Neuroscience*, *94*(1), 185–192.

Park, D., Borlongan, C. V., Willing, A. E., Eve, D. J., Cruz, L. E., Sanberg, C. D., et al. (2009). Human umbilical cord blood cell grafts for brain ischemia. *Cell Transplantation*, *18*(9), 985–998.

Park, J. H., & Enikolopov, G. (2010). Transient elevation of adult hippocampal neurogenesis after dopamine depletion. *Experimental Neurology*, *222*(2), 267–276.

Peier, A. M., McIlwain, K. L., Kenneson, A., Warren, S. T., Paylor, R., & Nelson, D. L. (2000). (Over)correction of FMR1 deficiency with YAC transgenics: behavioral and physical features. *Human Molecular Genetics*, *9*(8), 1145–1159.

Peng, J., Xie, L., Jin, K., Greenberg, D. A., & Andersen, J. K. (2008). Fibroblast growth factor 2 enhances striatal and nigral neurogenesis in the acute 1-methyl-4-phenyl-1,2,3,6-tetrahydropyridine model of parkinson's disease. *Neuroscience*, *153*(3), 664–670.

Petersen, A., Puschban, Z., Lotharius, J., NicNiocaill, B., Wiekop, P., O'Connor, W. T., et al. (2002). Evidence for dysfunction of the nigrostriatal pathway in the R6/1 line of transgenic huntington's disease mice. *Neurobiology of Disease*, *11*(1), 134–146.

Phillips, W., Morton, A. J., & Barker, R. A. (2005). Abnormalities of neurogenesis in the R6/2 mouse model of huntington's disease are attributable to the in vivo microenvironment. *The Journal of Neuroscience*, *25*(50), 11564–11576.

Pouladi, M. A., Graham, R. K., Karasinska, J. M., Xie, Y., Santos, R. D., Petersen, A., et al. (2009). Prevention of depressive behaviour in the YAC128 mouse model of huntington disease by mutation at residue 586 of huntingtin. *Brain*, *132*(Pt 4), 919–932.

van Praag, H., Christie, B. R., Sejnowski, T. J., & Gage, F. H. (1999). Running enhances neurogenesis, learning, and long-term potentiation in mice. *Proceedings of the National Academy of Sciences of the United States of America*, *96*(23), 13427–13431.

van Praag, H., Kempermann, G., & Gage, F. H. (1999). Running increases cell proliferation and neurogenesis in the adult mouse dentate gyrus. *Nature Neuroscience*, *2*(3), 266–270.

van Praag, H., Shubert, T., Zhao, C., & Gage, F. H. (2005). Exercise enhances learning and hippocampal neurogenesis in aged mice. *The Journal of Neuroscience*, *25*(38), 8680–8685.

Qu, H., Zhao, M., Zhao, S., Xiao, T., Tang, X., Zhao, D., et al. (2014). Forced limb-use enhances brain plasticity through the cAMP/PKA/CREB signal transduction pathway after stroke in adult rats. *Restorative Neurology and Neuroscience*, *32*(5), 597–609.

Ramaswamy, S., Goings, G. E., Soderstrom, K. E., Szele, F. G., & Kozlowski, D. A. (2005). Cellular proliferation and migration following a controlled cortical impact in the mouse. *Brain Research*, *1053*(1–2), 38–53.

Recchia, A., Debetto, P., Negro, A., Guidolin, D., Skaper, S. D., & Giusti, P. (2004). Alpha-synuclein and parkinson's disease. *FASEB Journal*, *18*(6), 617–626.

Redila, V. A., Olson, A. K., Swann, S. E., Mohades, G., Webber, A. J., Weinberg, J., et al. (2006). Hippocampal cell proliferation is reduced following prenatal ethanol exposure but can be rescued with voluntary exercise. *Hippocampus*, *16*(3), 305–311.

Reijnders, J. S., Ehrt, U., Weber, W. E., Aarsland, D., & Leentjens, A. F. (2008). A systematic review of prevalence studies of depression in parkinson's disease. *Movement Disorders*, *23*(2), 183–189 quiz 313.

Reynolds, G. P., Dalton, C. F., Tillery, C. L., Mangiarini, L., Davies, S. W., & Bates, G. P. (1999). Brain neurotransmitter deficits in mice transgenic for the huntington's disease mutation. *Journal of Neurochemistry*, *72*(4), 1773–1776.

Rice, A. C., Khaldi, A., Harvey, H. B., Salman, N. J., White, F., Fillmore, H., et al. (2003). Proliferation and neuronal differentiation of mitotically active cells following traumatic brain injury. *Experimental Neurology*, *183*(2), 406–417.

Richardson, R. M., Broaddus, W. C., Holloway, K. L., Sun, D., Bullock, M. R., & Fillmore, H. L. (2005). Heterotypic neuronal differentiation of adult subependymal zone neuronal progenitor cells transplanted to the adult hippocampus. *Molecular and Cellular Neuroscience*, *28*(4), 674–682.

Richardson, R. M., Holloway, K. L., Bullock, M. R., Broaddus, W. C., & Fillmore, H. L. (2006). Isolation of neuronal progenitor cells from the adult human neocortex. *Acta Neurochirurgica*, *148*(7), 773–777.

Rodriguez, J. J., Jones, V. C., Tabuchi, M., Allan, S. M., Knight, E. M., LaFerla, F. M., et al. (2008). Impaired adult neurogenesis in the dentate gyrus of a triple transgenic mouse model of Alzheimer's disease. *PLoS One*, *3*(8), e2935.

Ruan, L., Lau, B. W., Wang, J., Huang, L., ZhuGe, Q., Wang, B., et al. (2014). Neurogenesis in neurological and psychiatric diseases and brain injury: from bench to bedside. *Progress in Neurobiology; 2013 Pangu Meeting on Neurobiology of Stroke and CNS Injury: Progresses and Perspectives of Future*, *115*, 116–137.

Sabaratnam, M., Vroegop, P. G., & Gangadharan, S. K. (2001). Epilepsy and EEG findings in 18 males with fragile X syndrome. *Seizure*, *10*(1), 60–63.

Santarelli, L., Saxe, M., Gross, C., Surget, A., Battaglia, F., Dulawa, S., et al. (2003). Requirement of hippocampal neurogenesis for the behavioral effects of antidepressants. *Science*, *301*(5634), 805–809.

Selkoe, D. J. (2001). Alzheimer's disease: genes, proteins, and therapy. *Physiological Review, 81*(2), 741–766.
Simpson, J. M., Gil-Mohapel, J., Pouladi, M. A., Ghilan, M., Xie, Y., Hayden, M. R., et al. (2011). Altered adult hippocampal neurogenesis in the YAC128 transgenic mouse model of huntington disease. *Neurobiology of Disease, 41*(2), 249–260.
Slaughter, J. R., Martens, M. P., & Slaughter, K. A. (2001). Depression and huntington's disease: prevalence, clinical manifestations, etiology, and treatment. *CNS Spectrums, 6*(4), 306–326.
Slegtenhorst-Eegdeman, K. E., de Rooij, D. G., Verhoef-Post, M., van de Kant, H. J., Bakker, C. E., Oostra, B. A., et al. (1998). Macroorchidism in FMR1 knockout mice is caused by increased sertoli cell proliferation during testicular development. *Endocrinology, 139*(1), 156–162.
Slow, E. J., van Raamsdonk, J., Rogers, D., Coleman, S. H., Graham, R. K., Deng, Y., et al. (2003). Selective striatal neuronal loss in a YAC128 mouse model of huntington disease. *Human Molecular Genetics, 12*(13), 1555–1567.
Smith, F. M., Raghupathi, R., MacKinnon, M., McIntosh, T. K., Saatman, K. E., Meaney, D. F., et al. (2000). TUNEL-positive staining of surface contusions after fatal head injury in man. *Acta Neuropathologica, 100*(5), 537–545.
Sokol, R. J., Delaney-Black, V., & Nordstrom, B. (2003). Fetal alcohol spectrum disorder. *JAMA, 290*(22), 2996–2999.
Sowell, E. R., Thompson, P. M., Mattson, S. N., Tessner, K. D., Jernigan, T. L., Riley, E. P., et al. (2001). Voxel-based morphometric analyses of the brain in children and adolescents prenatally exposed to alcohol. *Neuroreport, 12*(3), 515–523.
Spires, T. L., Grote, H. E., Varshney, N. K., Cordery, P. M., van Dellen, A., Blakemore, C., et al. (2004). Environmental enrichment rescues protein deficits in a mouse model of huntington's disease, indicating a possible disease mechanism. *The Journal of Neuroscience, 24*(9), 2270–2276.
Sun, D., Bullock, M. R., McGinn, M. J., Zhou, Z., Altememi, N., Hagood, S., et al. (2009). Basic fibroblast growth factor-enhanced neurogenesis contributes to cognitive recovery in rats following traumatic brain injury. *Experimental Neurology, 216*(1), 56–65.
Sun, D., Colello, R. J., Daugherty, W. P., Kwon, T. H., McGinn, M. J., Harvey, H. B., et al. (2005). Cell proliferation and neuronal differentiation in the dentate gyrus in juvenile and adult rats following traumatic brain injury. *Journal of Neurotrauma, 22*(1), 95–105.
Sun, D., Daniels, T. E., Rofle, A., Waters, M., & Hamm, R. J. (2014). Inhibition of injury-induced cell proliferation in the dentate gyrus of the hippocampus impairs spontaneous cognitive recovery following traumatic brain injury. *Journal of Neurotrauma, 2*(7), 495–505.
Sun, Y., Jin, K., Xie, L., Childs, J., Mao, X. O., Logvinova, A., et al. (2003). VEGF-induced neuroprotection, neurogenesis, and angiogenesis after focal cerebral ischemia. *The Journal of Clinical Investigation, 111*(12), 1843–1851.
Tanaka, R., Yamashiro, K., Mochizuki, H., Cho, N., Onodera, M., Mizuno, Y., et al. (2004). Neurogenesis after transient global ischemia in the adult hippocampus visualized by improved retroviral vector. *Stroke, 35*(6), 1454–1459.
Tang, Y., Cai, B., Yuan, F., He, X., Lin, X., Wang, J., et al. (2013). Melatonin pretreatment improves the survival and function of transplanted mesenchymal stem cells after focal cerebral ischemia. *Cell Transplantation, 23*(10), 1279–1291.
Thored, P., Arvidsson, A., Cacci, E., Ahlenius, H., Kallur, T., Darsalia, V., et al. (2006). Persistent production of neurons from adult brain stem cells during recovery after stroke. *Stem Cells, 24*(3), 739–747.
Tolosa, E., & Poewe, W. (2009). Premotor parkinson disease. *Neurology, 72*(Suppl. 7), S1.
Van Dam, D., D'Hooge, R., Hauben, E., Reyniers, E., Gantois, I., Bakker, C. E., et al. (2000). Spatial learning, contextual fear conditioning and conditioned emotional response in Fmr1 knockout mice. *Behavioural Brain Research, 117*(1–2), 127–136.
Verret, L., Jankowsky, J. L., Xu, G. M., Borchelt, D. R., & Rampon, C. (2007). Alzheimer's-type amyloidosis in transgenic mice impairs survival of newborn neurons derived from adult hippocampal neurogenesis. *The Journal of Neuroscience, 27*(25), 6771–6780.

Wang, R., Dineley, K. T., Sweatt, J. D., & Zheng, H. (2004). Presenilin 1 familial Alzheimer's disease mutation leads to defective associative learning and impaired adult neurogenesis. *Neuroscience, 126*(2), 305–312.

Wang, X., Yang, F., Liu, C., Zhou, H., Wu, G., Qiao, S., et al. (2012). Dietary supplementation with the probiotic *Lactobacillus fermentum* I5007 and the antibiotic aureomycin differentially affects the small intestinal proteomes of weanling piglets. *The Journal of Nutrition, 142*(1), 7–13.

Wang, L., Zhang, Z., Wang, Y., Zhang, R., & Chopp, M. (2004). Treatment of stroke with erythropoietin enhances neurogenesis and angiogenesis and improves neurological function in rats. *Stroke, 35*(7), 1732–1737.

Wen, P. H., Hof, P. R., Chen, X., Gluck, K., Austin, G., Younkin, S. G., et al. (2004). The presenilin-1 familial Alzheimer disease mutant P117L impairs neurogenesis in the hippocampus of adult mice. *Experimental Neurology, 188*(2), 224–237.

West, J. R., Hamre, K. M., & Cassell, M. D. (1986). Effects of ethanol exposure during the third trimester equivalent on neuron number in rat hippocampus and dentate gyrus. *Alcoholism, Clinical and Experimental Research, 10*(2), 190–197.

Whaley, S. E., O'Connor, M. J., & Gunderson, B. (2001). Comparison of the adaptive functioning of children prenatally exposed to alcohol to a nonexposed clinical sample. *Alcoholism, Clinical and Experimental Research, 25*(7), 1018–1024.

Winner, B., Kohl, Z., & Gage, F. H. (2011). Neurodegenerative disease and adult neurogenesis. *The European Journal of Neuroscience, 33*(6), 1139–1151.

Winner, B., Lie, D. C., Rockenstein, E., Aigner, R., Aigner, L., Masliah, E., et al. (2004). Human wild-type alpha-synuclein impairs neurogenesis. *Journal of Neuropathology and Experimental Neurology, 63*(11), 1155–1166.

Wozniak, D. F., Hartman, R. E., Boyle, M. P., Vogt, S. K., Brooks, A. R., Tenkova, T., et al. (2004). Apoptotic neurodegeneration induced by ethanol in neonatal mice is associated with profound learning/memory deficits in juveniles followed by progressive functional recovery in adults. *Neurobiology of Disease, 17*(3), 403–414.

Xie, C., Cong, D., Wang, X., Wang, Y., Liang, H., Zhang, X., et al. (2014). The effect of simvastatin treatment on proliferation and differentiation of neural stem cells after traumatic brain injury. *Brain Research, 1602*, 1–8.

Yamashita, T., Ninomiya, M., Hernandez Acosta, P., Garcia-Verdugo, J. M., Sunabori, T., Sakaguchi, M., et al. (2006). Subventricular zone-derived neuroblasts migrate and differentiate into mature neurons in the post-stroke adult striatum. *Journal of Neuroscience, 26*(24), 6627–6636.

Yang, C. P., Gilley, J. A., Zhang, G., & Kernie, S. G. (2011). ApoE is required for maintenance of the dentate gyrus neural progenitor pool. *Development, 138*(20), 4351–4362.

Yau, S. Y., Gil-Mohapel, J., Christie, B. R., & So, K. F. (2014). Physical exercise-induced adult neurogenesis: a good strategy to prevent cognitive decline in neurodegenerative diseases? *Biomed Research International, 2014*, 403120.

Yun, S. H., & Trommer, B. L. (2011). Fragile X mice: reduced long-term potentiation and N-methyl-D-aspartate receptor-mediated neurotransmission in dentate gyrus. *Journal of Neuroscience Research, 89*(2), 176–182.

Zajac, M. S., Pang, T. Y., Wong, N., Weinrich, B., Leang, L. S., Craig, J. M., et al. (2010). Wheel running and environmental enrichment differentially modify exon-specific BDNF expression in the hippocampus of wild-type and pre-motor symptomatic male and female huntington's disease mice. *Hippocampus, 20*(5), 621–636.

Zalfa, F., Eleuteri, B., Dickson, K. S., Mercaldo, V., De Rubeis, S., di Penta, A., et al. (2007). A new function for the fragile X mental retardation protein in regulation of PSD-95 mRNA stability. *Nature Neuroscience, 10*(5), 578–587.

Zhang, R. L., Zhang, Z., Zhang, L., Wang, Y., Zhang, C., & Chopp, M. (2006). Delayed treatment with sildenafil enhances neurogenesis and improves functional recovery in aged rats after focal cerebral ischemia. *Journal of Neuroscience Research, 83*(7), 1213–1219.

Index

'*Note*: Page numbers followed by "f" indicate figures and "t" indicate tables.'

B

Printed and bound by CPI Group (UK) Ltd, Croydon, CR0 4YY

08/05/2025

01865022-0002